Structural Glycobiology

Structural Glycobiology

EDITED BY

Elizabeth Yuriev
Paul A. Ramsland

CRC Press
Taylor & Francis Group
Boca Raton London New York

CRC Press is an imprint of the
Taylor & Francis Group, an **informa** business

CRC Press
Taylor & Francis Group
6000 Broken Sound Parkway NW, Suite 300
Boca Raton, FL 33487-2742

© 2013 by Taylor & Francis Group, LLC
CRC Press is an imprint of Taylor & Francis Group, an Informa business

No claim to original U.S. Government works

Printed in the United States of America on acid-free paper
Version Date: 20120801

International Standard Book Number: 978-1-4398-5460-0 (Hardback)

Library of Congress Cataloging-in-Publication Data

Structural glycobiology / editors, Elizabeth Yuriev, Paul A. Ramsland.
    p. ; cm.
  Includes bibliographical references and index.
  ISBN 978-1-4398-5460-0 (hardcover : alk. paper)
  I. Yuriev, Elizabeth. II. Ramsland, Paul A.
  [DNLM: 1. Glycomics--methods. 2. Carbohydrates. 3. Computational Biology--methods. 4. Glycoproteins. 5. Protein Binding. QU 75]

572'.56--dc23   2012025728

Visit the Taylor & Francis Web site at
http://www.taylorandfrancis.com

and the CRC Press Web site at
http://www.crcpress.com

Contents

SECTION I   Experimental Techniques to Determine Three-Dimensional Structures

SECTION II   Theortical/Modeling Techniques to Predict Three-Dimensional Structures

SECTION III   Alternative Approaches for Yielding Detailed Structural Information

SECTION IV   Carbohydrates in Medicine

Preface

Structural glycobiology is a rapidly progressing field of research, where the diverse structural and functional roles of carbohydrates (oligo- and polysaccharides, glycolipids, and glycoproteins) are examined using a wide variety of experimental as well as theoretical (predictive) approaches. Carbohydrates are key molecules in diverse biological processes that include, but are not limited to, metabolic pathways, cell–cell interactions, carbohydrate–protein interactions, host–pathogen interactions, and immunity.

Although there are several well-written and comprehensive textbooks on glycobiology and the chemistry, biochemistry, and microbiology of carbohydrates, no current book focuses on the specific topic of structural glycobiology. We believe that this book fills the gap by bringing together world-recognized authors to contribute chapters that cover their own specialties in the experimental, theoretical, and emerging technologies employed in this field.

In this book, individual chapters are written by expert authors who are active research scientists in the field and are specialists in key techniques that are relevant to modern structural glycobiology. The book provides concise overviews of the application of specialized technologies to the study of carbohydrates in biology, reviews of relevant and current research in the field, and is illustrated throughout by specific examples of how research investigations have yielded key structural and associated biological information on carbohydrates, glycolipids, and glycoproteins.

The topics covered in this book are broadly divided into four sections. Section I covers well-established, but often challenging, experimental approaches for structure determinations of carbohydrate–protein complexes and large glycoprotein assemblies and explores the techniques of x-ray crystallography and small-angle scattering (Chapter 1), nuclear magnetic resonance (NMR) (Chapter 2), and cryoelectron microscopy (Chapter 3). Jeffries, Farrugia, and Ramsland (Chapter 1) discuss two complementary approaches examining the high-resolution three-dimensional structures (x-ray crystallography) and the solution shapes and conformations (small-angle x-ray and neutron scattering) of carbohydrate binding proteins and glycoproteins. They outline the general features of carbohydrate–protein interactions and discuss the importance of multivalent carbohydrate binding and the role of oligomerization in carbohydrate recognition by proteins. Koharudin and Gronenborn (Chapter 2) provide an easily accessible and educational outline of how NMR can be applied to the study of protein–glycan interactions. The power of the different NMR methodologies for investigating carbohydrate binding, bound carbohydrate conformations, and detailed structures of carbohydrate–protein complexes is beautifully illustrated with carbohydrate–lectin systems. The state-of-the-art in cryoelectron microscopy (cryoEM) for studying very large assemblies of glycoproteins is presented by Zeev-Ben-Mordehai and Grünewald (Chapter 3). Recent advances in sample preparation, image collection, and processing are greatly accelerating the use of cryoEM for structural studies of glycoproteins under near native settings (e.g., in viruses, cells, and tissues). They provide stunning examples of where cryoEM, particularly tomographic methods, are providing unprecedented structural information on biological assemblies, such as membrane channels, intracellular junctions, and viral glycoproteins.

Section II covers theoretical, or modeling-based, approaches, such as molecular mechanics, molecular dynamics, and free energy calculations (Chapter 4) and carbohydrate docking (Chapter 5). Sarkar and Pérez (Chapter 4) give an excellent overview of the computational approaches used to study protein–carbohydrate interactions. They complement the description of traditional methods with a brief foray into alternative methods used for the enhancement of conformational sampling, such as molecular robotics. Agostino, Ramsland, and Yuriev (Chapter 5) demonstrate the usefulness of molecular docking in structural glycobiology by considering recent docking validation studies on a range of protein targets. They also describe very recent developments in the modeling of water-mediated carbohydrate–protein interactions.

Section III covers alternative techniques for yielding structural information on carbohydrates from complex biological samples (fluids/secretions, cells, and tissues). Here, the rapid advances in mass spectrometry (Chapter 6) are being complemented with glycan-based arrays (Chapter 7) for the study of carbohydrate specificity and recognition. Kolarich and Packer (Chapter 6) provide a detailed overview of the latest mass spectrometric methods for characterization of complex N- and O-linked oligosaccharides and how the exciting subfield of glycoproteomics is emerging for studying diverse protein glycoforms that are relevant to health and disease. A brief discussion of the need for uniform collection and reporting standards for glycomics-based mass spectrometric investigations is followed by a useful overview of the bioinformatics resources available to researchers using mass spectrometry in glycobiology. The contribution by Song, Smith, and Cummings (Chapter 7) illustrates glycan array technology, focusing on the Consortium for Functional Glycomics methodologies and also introduces the cutting-edge shotgun glycomics approach, which has enormous potential to accelerate research into carbohydrate-mediated interactions from diverse and complex biological samples.

Section IV deals with carbohydrates in medicine. Although carbohydrates are centrally involved in many physiological, biochemical, and cellular processes, three areas of modern medicine (organ transplantation, cancer immunotherapy, and infection treatment) have been directly impacted by our understanding of the structural role of carbohydrates in immune recognition. Brockhausen and Gao (Chapter 8) focus on a range of cancer-related structural and enzymatic glycoaberrations. Christiansen et al. (Chapter 9) deal with carbohydrate antigens implicated in organ rejection. Specifically, they discuss the biochemical, genetic, and immunological characteristics of these carbohydrates, their origins, and interactions with antibodies. Xu and Wilson (Chapter 10) highlight the role of protein–carbohydrate binding for viral adhesion and invasion. They focus on three paradigm systems of viral proteins that recognize sialic acid in cell surface glycans as receptors for viral attachment. They demonstrate, using high-resolution x-ray structures, the subtle structural variations governing the recognition process; for example, for avian versus human influenza A hemagglutinin. Gandhi and Mancera (Chapter 11) describe molecules designed to mimic the biological activity of glycosaminoglycans (GAGs) through modifications of structure, composition, and sulfation patterns. These new generation GAG-mimetics offer rich potential as therapeutics for treating cancer, inflammation, and infection.

Although each chapter could be a useful stand-alone introduction to a specific technique or area of structural glycobiology, several themes are consistent throughout the book, namely, the role of specific proteins in carbohydrate recognition and function: lectins (Chapters 1, 2, 4, and 5), antibodies (Chapters 1, 4, 5, 9, and 10), and glycosyltransferases (Chapters 4, 5, 8, and 9). From the ligand point of view, the structure and biological roles of two particular types of carbohydrates are of interest in several areas of study: sialic acid and its derivatives (Chapters 1, 8, and 10) and glycosaminoglycans (Chapters 5, 7, 8, 10, and 11).

In summary, this book covers the experimental, theoretical, and alternative technologies that are being applied to the study of the structural basis for the diverse biological roles of carbohydrates. This should be a valuable reference for researchers, graduate students, postdoctoral scientists, and academics with an interest in glycobiology. Researchers from other fields, such as medicinal chemists, biochemists, immunologists, and microbiologists, should also find this a relevant and up-to-date reference and a suitable introduction to the field.

Editors

Elizabeth Yuriev, BSc (Hons) (Kharkov State University, Ukraine), PhD (Victoria University, Melbourne), is a senior lecturer of medicinal chemistry at the Monash Institute of Pharmaceutical Sciences, Monash University, Melbourne, Australia. She is a foundation member and the secretary of the Association of Molecular Modellers of Australasia. She is a member of the editorial board of the *Journal of Molecular Recognition* (John Wiley & Sons). Her current research is focused on protein recognition of carbohydrate and peptide ligands, GPCR modeling, and molecular docking. She has 63 publications in peer-reviewed journals/books, including 47 primary research papers, 8 invited reviews/book chapters, and 8 editorial notes/commentaries.

Paul A. Ramsland, BSc (Hons), PhD (University of Technology, Sydney), is a senior research fellow and a group leader in the Centre for Immunology at the Burnet Institute, Melbourne, Australia. In 2011, he was awarded the Sir Zelman Cowen Fellowship for excellence in medical research relating to cancer. He is an honorary principal fellow (associate professor) in the Department of Surgery Austin Health at the University of Melbourne and is an adjunct senior lecturer in the Department of Immunology at Monash University, Melbourne, Australia. He is an associate editor for the *Journal of Molecular Recognition* and *Frontiers in Immunology* (*Molecular Innate Immunity Specialty*), and a section editor for *BMC Structural Biology* as well as serving on the editorial boards of *Molecular Immunology* and *Molecular Biotechnology*. His current research is focused on examining the three-dimensional structural roles of carbohydrates and glycoproteins in immunity and infection. He has 87 publications in peer-reviewed journals/books, including 67 primary research papers, 16 invited reviews/book chapters, and 4 editorial notes/commentaries.

.

Contributors

Mark Agostino
Medicinal Chemistry and Drug Action
Monash Institute of Pharmaceutical
 Sciences
Monash University
Melbourne, Victoria, Australia

Inka Brockhausen
Division of Rheumatology
Department of Biomedical and Molecular
 Sciences and Department of Medicine
Queen's University
Kingston, Ontario, Canada

Dale Christiansen
Department of Surgery
Austin Health/Northern Health
University of Melbourne
Heidelberg, Victoria, Australia

Richard D. Cummings
Department of Biochemistry
School of Medicine
Emory University
Atlanta, Georgia

William Farrugia
Centre for Immunology
Burnet Institute
Melbourne, Victoria, Australia

Neha S. Gandhi
Curtin Health Innovation Research
 Institute
and
Western Australian Biomedical
 Research Institute
and
School of Biomedical Sciences
Curtin University
Perth, Western Australia, Australia

Yin Gao
Division of Rheumatology
Department of Biomedical and
 Molecular Sciences and Department
 of Medicine
Queen's University
Kingston, Ontario, Canada

Angela M. Gronenborn
Department of Structural Biology
School of Medicine
University of Pittsburgh
Pittsburgh, Pennsylvania

Kay Grünewald
Oxford Particle Imaging Centre
Division of Structural Biology
Wellcome Trust Centre for Human
 Genetics
University of Oxford
Oxford, United Kingdom

Cy M. Jeffries
Bragg Institute
Australian Nuclear Science and
 Technology Organisation
Kirrawee DC, New South Wales,
 Australia

Leonardus M. I. Koharudin
Department of Structural Biology
School of Medicine
University of Pittsburgh
Pittsburgh, Pennsylvania

Daniel Kolarich
Department of Biomolecular
 Systems
Max-Planck-Institute of Colloids and
 Interfaces
Potsdam, Germany

Ricardo L. Mancera
Curtin Health Innovation Research
    Institute
and
Western Australian Biomedical
    Research Institute
and
School of Biomedical Sciences
and
School of Pharmacy
Curtin University
Perth, Western Australia, Australia

Nicolle H. Packer
Biomolecular Frontiers Research Centre
Department of Chemistry and
    Biomolecular Sciences
Macquarie University
Sydney, New South Wales, Australia

Serge Pérez
Centre de Recherches sur les
    Macromolécules Végétales
Centre National de la Recherche
    Scientifique
and
European Synchrotron Research
    Facility
Grenoble, France

Paul A. Ramsland
Centre for Immunology
Burnet Institute
and
Department of Surgery Austin Health
University of Melbourne
and
Department of Immunology
Monash University
Melbourne, Victoria, Australia

Mauro S. Sandrin
Department of Surgery
Austin Health/Northern Health
University of Melbourne
Heidelberg, Victoria, Australia

Anita Sarkar
Centre de Recherches sur les
    Macromolécules Végétales
Centre National de la Recherche
    Scientifique
Grenoble, France

David F. Smith
Department of Biochemistry
School of Medicine
Emory University
Atlanta, Georgia

Xuezheng Song
Department of Biochemistry
School of Medicine
Emory University
Atlanta, Georgia

Ian A. Wilson
Department of Molecular Biology
and
Skaggs Institute for Chemical Biology
The Scripps Research Institute

Rui Xu
Department of Molecular Biology
The Scripps Research Institute
La Jolla, California

Elizabeth Yuriev
Medicinal Chemistry
Monash Institute of Pharmaceutical
    Sciences
Monash University
Melbourne, Victoria, Australia

Tzviya Zeev-Ben-Mordehai
Oxford Particle Imaging Centre
Division of Structural Biology
Wellcome Trust Centre for Human
    Genetics
University of Oxford
Oxford, United Kingdom

Section I

Experimental Techniques to Determine Three-Dimensional Structures

1 Crystallography and Small-Angle Scattering of Carbohydrate– Protein Complexes and Glycoproteins

Cy M. Jeffries, William Farrugia, and Paul A. Ramsland

CONTENTS

1.1   INTRODUCTION

Carbohydrate-binding proteins and glycoproteins are the focus of intense scientific investigation due to their central role in diverse biological processes that include, but are not limited to, immunity and infection, cellular adhesion, and cellular communication and signaling. Yet, our understanding of the fundamental molecular mechanisms through which carbohydrate-binding proteins and glycoproteins realize their functions still remains underdeveloped. Of note, while it is estimated that over 50% of all eukaryotic proteins are glycosylated (Apweiler et al. 1999), only around 5% of the three-dimensional (3D) structures deposited in structural databases such as the Protein Data Bank (PDB) (Berman et al. 2000) include proteins with *N*- or *O*-linked carbohydrates (often called glycans). Similarly, only around 7% of all PDB entries contain information on protein/carbohydrate systems (covalently or noncovalently bound to proteins), and there are even fewer examples of high-resolution structures where the associated carbohydrate components have been fully resolved (Lutteke 2009).

In this chapter, we discuss two highly complementary experimental approaches for probing the 3D structures of carbohydrate-binding proteins and glycoproteins: (1) x-ray crystallography that can provide high-resolution details of macromolecular 3D structures and (2) small-angle scattering that provides global structural parameters and shape information from proteins in solution. Using select examples, we summarize the structural basis for carbohydrate recognition and the role of multivalency through oligomerization as revealed by x-ray crystallography, while small-angle scattering is highlighted as a powerful strategy to probe the states and shapes of the intact glycoproteins without the conformational constraints imposed by the crystal matrix.

1.2   STRUCTURE DETERMINATION BY X-RAY CRYSTALLOGRAPHY

X-ray crystallography is a powerful means and currently the most commonly used experimental methodology for determining 3D structures of biological macromolecules. The basic approach and methodology for determining 3D structures by crystallography is similar for all biological macromolecules and has been described in detail elsewhere. For a comprehensive and easily accessible reference on macromolecular crystallography, the reader is referred to the excellent textbook on *Biomolecular Crystallography* by Bernhard Rupp (2010). Herein, we provide a brief overview of the steps involved in 3D structure determination by x-ray crystallography and some of the specific considerations required when working with carbohydrate–protein complexes and glycoproteins.

The first and most crucial step in any crystallography project is growing a single crystal that diffracts x-rays to suitably high resolution (in practice, normally between 3.0 and 1.0 Å) for 3D structure determination. This largely empirical process is achieved via screening highly purified material against numerous crystallization conditions that typically contain dehydrating or precipitating agents (e.g., polyethylene glycol and ammonium sulfate) and a variety of additives (e.g., buffers and metal ions).

The most common crystallization method is vapor diffusion where a small droplet containing the sample and a crystallization solution is equilibrated against a larger reservoir of the same crystallization media. Crystallization screening against hundreds of individual conditions is performed in parallel in multi-well plastic plates and can be highly automated using robotics. Once a crystal is obtained—often from one or a handful of specific conditions—x-ray diffraction data can be collected. However, to determine a structure, the crystallization process may require a number of rounds of optimization to produce crystals with improved diffraction intensity and resolution (for further reading, see the comprehensive textbook on *Protein Crystallization* by McPherson [1999]). Since simple carbohydrates are typically highly soluble and are relatively small, it is possible to either co-crystallize these ligands with the target protein or soak into the hydrated crystals of the carbohydrate-binding protein. The affinity of carbohydrates for protein-binding sites is generally quite low (K_d values in the millimolar to micromolar range); thus, it is often beneficial to use a molar excess (e.g., 10- to 100-fold) of the carbohydrate over the protein to ensure that high occupancy is achieved by the ligand in the carbohydrate-binding site during the crystallization or crystal-soaking process.

The natural heterogeneity of *N*- and *O*-linked glycans results in glycoproteins being mixtures of glycosylated variants or glycoforms (Marino et al. 2010). Consequently, generating crystals of glycoproteins with a well-ordered (uniform) crystal matrix and which diffract x-rays to high resolution can often be a very frustrating enterprise. Approaches to crystallize glycoproteins have included the production of recombinant proteins in bacterial systems such as *Eschericha coli* that essentially lack glycosylation machinery or in eukaryotic systems such as insect cell lines (or engineered mammalian cell lines) that add carbohydrates of reduced complexity and increased homogeneity compared to unmodified mammalian cells (Nettleship et al. 2010). Alternatively, site-directed mutagenesis of the glycoprotein can be used to remove some or all of the glycosylation motifs from the protein to obtain crystals for structure determination. Another method of increasing the quality of crystals is to truncate the carbohydrates with the most accessible technique being the removal of terminal sialic acid residues with neuraminidase (Lustbader et al. 1989). However, the removal of sialic acids seems to have been often overlooked as a simple method for generating high-quality crystals of glycoproteins. We suggest that neuraminidase treatment should be routinely trialed for crystallizing mammalian glycoproteins (particularly with proteins purified directly from primary sources) such as we found useful for generating crystals of a glycosylated antigen-binding fragment (Fab) from an IgM cryoglobulin, which was purified from the plasma of a Waldenström's macroglobulinemia patient (Ramsland et al. 2006). For a detailed example of the wide variety and potential of glycosylation modification strategies, see the excellent study by Lee and colleagues (Lee et al. 2009) who successfully determined the crystal structure of the Ebola virus trimeric spike glycoprotein (Lee et al. 2008).

Most macromolecular diffraction data is currently collected using laboratory or high-intensity synchrotron x-ray sources from crystals that have been cryoprotected at low temperature (around 100 K using liquid N_2 cooling systems) to reduce the effects of ionizing radiation and thermal damage. Diffraction data is often obtained from a single crystal that is precisely rotated (around at least one axis) to collect a series of diffraction images that result from passing an intense x-ray beam through the crystal. The positions and intensities of numerous diffraction "spots" obtained from the crystal at each angle are integrated into a unique dataset using readily available computer algorithms. The electron density is reconstructed by combining the Fourier transformation of these diffraction data with the derived phases, which may need to be determined experimentally or calculated using molecular replacement (MR) methods that are based on fitting previously determined homologous protein structures to the experimental data (Rupp 2010).

One possible approach to solving the "phase problem" with carbohydrate-binding proteins has been proposed that uses selenium derivatives of the native carbohydrate ligands for multi-wavelength anomalous dispersion (MAD) phasing experiments. This strategy was successfully used for the three-wavelength MAD phasing of a bacterial adhesin F17-G in complex with an *N*-acetyl-d-glucosamine derivative, where the anomeric oxygen was replaced by a selenium atom (Buts et al. 2003). However, 3D structures of most carbohydrate-binding proteins have been determined from phases obtained using heavy-atom crystal derivatives, MAD phasing from selenomethionine-substituted recombinant proteins, or MR.

The final step in crystal structure determination is the iterative process of crystallographic refinement where the 3D electron density map is progressively fitted (automatically and manually) with a molecular model and a variety of parameters are optimized that describe the correlation between that 3D model and the observed experimental data (e.g., atom positions, temperature B-values, and structure factor intensities or amplitudes). Well-established computational approaches are available for crystallographic model building and refinement (Rupp 2010) and these are not described here.

1.3   CRYSTAL STRUCTURES OF CARBOHYDRATE–PROTEIN COMPLEXES

A wide range of carbohydrate-binding proteins have been now characterized by x-ray crystallography such as antibodies, lectins (from plants, fungi, and animals), carbohydrate-binding proteins of pathogens, transport proteins, and enzymes. This section illustrates the basic principles of carbohydrate recognition using select examples of crystal structures of carbohydrate–protein complexes, including anti-carbohydrate antibodies, mammalian lectins involved in innate immunity, and proteins from pathogens.

1.3.1   GENERAL FEATURES OF CARBOHYDRATE–PROTEIN INTERACTIONS

Carbohydrate-binding sites are generally located on the surface of proteins and form cavities or grooves. Most amino acids can participate in binding carbohydrates, although there is a frequent over-representation of amino acids with polar, charged, and aromatic side-chains lining carbohydrate-binding sites. Hydrophobic interactions primarily between aromatic residue side-chains (e.g., Tyr and Trp) and the more hydrophobic regions (faces) of carbohydrate rings are known to be important contributors to the affinity of carbohydrate–protein interactions. In addition to amino acids, the relatively solvent-exposed binding sites contain numerous water molecules, which play a pivotal role in carbohydrate–protein interactions. Bound metal ions also provide further carbohydrate coordination centers for molecular recognition events and can act in parallel as critical structural components that help maintain the binding site shape. The multivalent binding of carbohydrates is also a frequent feature of carbohydrate-binding proteins and is typified by the repetition of carbohydrate recognition domains within a polypeptide chain and/or the oligomerization of protein subunits to generate multiple binding sites for carbohydrate recognition. Thus, a relatively low-affinity carbohydrate–protein binding site interaction is converted into a high-strength (avidity) interaction through multivalent carbohydrate binding.

1.3.2 Common Carbohydrate-Binding Modes

Two common binding modes that have been repeatedly observed in crystal structures of carbohydrate–protein complexes are end-on insertion and groove-type binding (Figure 1.1). End-on insertion involves the terminal groups of the carbohydrate ligand, normally a terminal monosaccharide unit, entering first and most deeply into the carbohydrate binding site. End-on insertion has been observed for antibodies in binding small molecules such as haptens and carbohydrates and appears to be the predominant manner in which carbohydrate epitopes are recognized by antibodies (Ramsland et al. 2003). Such binding allows the antibody to specifically interact with unique determinants (epitopes) that are presented near the terminal ends of longer carbohydrate chains conjugated to proteins or lipids. Frequently, the epitopes targeted by antibodies consist of minimal determinants, often ranging in size from disaccharides to tetrasaccharides, which are easily accommodated by the combining site that is formed by the association of the heavy and light chain variable domains (Agostino et al. 2012).

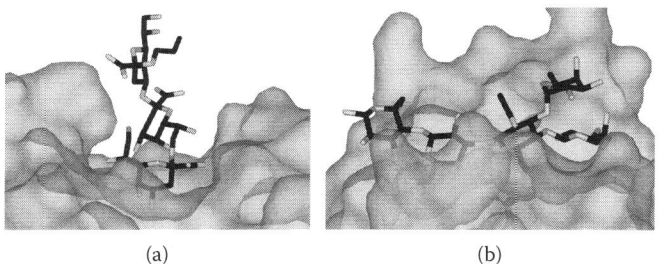

(a)   (b)

FIGURE 1.1 Two common binding modes involved in carbohydrate recognition by proteins. (a) End-on insertion of a trisaccharide, Kdoα(2-8)Kdoα(2-4)Kdo, in the binding site of an antibody, Se25-2 Fab, determined at 1.49 Å resolution, PDB ID: 3SY0 (Nguyen et al. 2003). (b) Groove-type interaction of a pentasaccharide, Galβ(1-4)[Fucα(1-3)]GlcNAcβ(1-3) Galβ(1-4)Glc (lacto-N-fucopentaose III, a Le^x tumor-associated antigen), in the binding site of the amino-terminal domain of human Galectin-8 at 1.33 Å resolution, PDB ID: 3AP9 (Ideo et al. 2011). Molecular surfaces are shown for the proteins and the bound carbohydrate ligands are in stick representations with carbon atoms in black and polar atoms in gray.

An example of where end-on insertion is used to recognize carbohydrates is the Se25-2 antibody that uses a germline-encoded binding site that interacts with the terminal 3-Deoxy-D-*manno*-oct-2-ulosonic acid (Kdo) residues used by certain bacteria to form lipopolysaccharides (LPS). The 1.49 Å resolution structure of Kdoα(2-8) Kdoα(2-4)Kdo in complex with Se25-2 Fab (Nguyen et al. 2003) shows how the terminal Kdo residue penetrates a cavity while antibody combining site residues could potentially participate in further interactions with the second and third carbohydrate residues in the chain (Figure 1.1a).

Lectins have also been shown to utilize end-on insertion binding for recognition, but in addition often employ groove-type binding, where an extended carbohydrate chain is bound in a solvent-filled groove (Yuriev et al. 2009). Binding of carbohydrate chains in a groove allows proteins to interact with internal carbohydrate moieties, which can partially explain the cross-reactivity with different types of complex carbohydrates by many lectins. It should be noted that antibodies can also participate in groove-type binding, but that this appears to be less frequent than for lectins and other classes of carbohydrate-binding proteins.

Galectins are a class of lectins found in mammals that form part of the innate immune system and recognize β-galactoside (βGal)-containing carbohydrates. Human Galectin-8 has been crystallized with a pentasaccharide ligand, Lacto-*N*-fucopentaose III (LNFIII, a carbohydrate containing the Lewis x, Le^x, trisaccharide epitope). The crystal structure of the Galectin-8 complex with LNFIII was determined at a resolution of 1.33 Å as well as complexes with other related carbohydrate ligands (Ideo et al. 2011). The LNFIII pentasaccharide, Galβ(1-4)[Fucα(1-3)] GlcNAcβ(1-3)Galβ(1-4)Glc, is bound in an extended conformation by an elongated groove that is open at both ends and located in the amino-terminal domain of Galectin-8 (Figure 1.1b). Conserved binding interactions (seen in the other Galectin-8 complexes) occur with the lactose Galβ(1-4)Glc disaccharide portion of the ligand (Ideo et al. 2011), while additional interactions occur with the central GlcNAc residue and the terminal βGal stacks against a Tyr side-chain in the binding site. The α1,3-linked fucose residue that is part of the terminal Le^x epitope does not contact the protein, but participates in a further stacking interaction with the terminal βGal (opposite face to the binding site Tyr residue) as expected for Lewis-type carbohydrate antigens (Yuriev et al. 2005).

1.3.3 ANCHORED BINDING OF CARBOHYDRATE LIGANDS

Carbohydrates are often anchored in the binding site by tight interaction with a monosaccharide subunit of the carbohydrate chain. Two major types of anchored binding are metal ion mediated (e.g., calcium) and charge neutralization or compensation of terminal sialic acid residues (Figure 1.2).

Metal ion–mediated anchoring of carbohydrate ligands is exemplified by a family of innate effector molecules called collectins, which are members of the larger group of C-type (Ca^{2+}-dependent) lectins (Seaton et al. 2010; Veldhuizen et al. 2011). Up to four Ca^{2+} ions can be bound to C-type lectin domains and not all these metal ions directly interact with carbohydrates and have been proposed to have roles in stabilizing the domains. In particular, a single Ca^{2+} is held in place in the

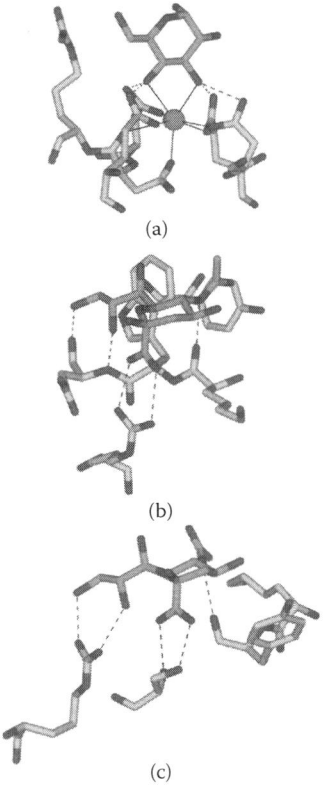

(a)

(b)

(c)

FIGURE 1.2 **(See color insert.)** Anchored binding of terminal carbohydrate residues. (a) Calcium-mediated coordination of a glucose residue of maltose in the binding site of human lung surfactant protein D at 1.40 Å resolution, PDB ID: 1PWB (Shrive et al. 2003). (b) Sialic acid (Neu5Ac) in the binding site of Siglec-7 at 1.90 Å resolution, PDB ID: 2DF3 (Attrill et al. 2006). (c) Sialic acid (Neu5Ac) in the binding site of rhesus rotavirus protein VP4 at 1.40 Å resolution, PDB ID: 1KQR (Dormitzer et al. 2002). Atoms are colored by type: C, yellow (protein) and cyan (carbohydrate); N, steel blue; O, red; Ca, green. Hydrogen bonds are shown as dashed black lines and metallic ion coordination bonds shown as solid black lines. Only the terminal carbohydrate residues are displayed for clarity.

carbohydrate-binding site by six coordination bonds with conserved residues (normally Asp, Glu, Asn, or Gln) and is involved in the coordination of two waters or the hydroxyl groups of bound monosaccharides. Thus, carbohydrates are anchored in the site by the strong pairing with Ca^{2+} and specificity for different carbohydrates (e.g., mannose or galactose) is determined by subtle differences in the residues in the Ca^{2+} binding pocket (Weis and Drickamer 1996). The crystal structure of human lung surfactant protein D (SP-D) has been determined in complex with maltose, Glcα(1-4)Glc, at 1.40 Å resolution (Shrive et al. 2003). The coordination of the terminal αGlc (the C2 epimer of mannose that also binds SP-D) occurs between the 3- and 4-hydroxyls and the bound Ca^{2+} ion, which provides a clear example of metal ion–mediated anchoring of carbohydrate antigens (Figure 1.2a).

Sialic acids are widely distributed in animal tissues (mostly Neu5Ac and Neu5Gc), as glycolipids (e.g., gangliosides) or at the ends of complex *N*-glycans, and are important biological ligands for many physiological recognition events and host–pathogen interactions. Their location at the termini of carbohydrate chains and the negatively charged carboxylate group make them ideal candidates for anchored binding to proteins. The most obvious anchoring mechanism is charge neutralization by formation of ion-pairs between the sialic acid carboxylate and basic residues (Arg and Lys) of the carbohydrate-binding protein. Ion-pairing between the sialic acid carboxylate anion and the guanidinium cation of an Arg residue is a key interaction of Siglec (sialic acid immunoglobulin-like lectin) receptors and is illustrated with the Siglec-7 interaction with Neu5Ac (Figure 1.2b). Siglec-7 was co-crystallized with a larger tetrasaccharide ligand, but only the terminal Neu5Ac is depicted, as this is involved in anchored binding to Siglec-7 (Attrill et al. 2006). An alternate mode of sialic acid recognition involves charge compensation of the carboxylate anion by the formation of strong hydrogen bonds with this portion of the ligand. The rhesus rotavirus VP4 carbohydrate recognition domain binds sialic acid by anchoring through multiple hydrogen bonding interactions and the sugar and the carboxylate form two hydrogen bonds with the side-chain hydroxyl and main chain amide of a serine residue in the binding site (Dormitzer et al. 2002). Thus, VP4 is an example where a protein uses hydrogen bonding for charge compensation to anchor sialic acid residues (Figure 1.2c).

1.3.4   ROLE OF WATER IN CARBOHYDRATE–PROTEIN INTERACTIONS

Water is a critical component that both drives carbohydrate binding and contributes to the specificity of carbohydrate–protein interactions within carbohydrate binding pockets. The strength of x-ray crystallography is it has allowed investigators to show that, while most of the bulk solvent is displaced from a carbohydrate-binding site when a target carbohydrate binds, certain ordered water molecules remain and are integral to maintaining architecture and specificity of a carbohydrate-binding site. In particular, ordered water molecules frequently participate in extensive hydrogen bonding networks that form the carbohydrate–protein interaction. For example, we previously observed the involvement of seven water molecules in forming a hydrogen bonding network linking the Le[y] tetrasaccharide to the binding site of a humanized antibody (hu3S193), for which the hu3S193 Fab complex with Le[y] was determined at 1.90 Å resolution (Ramsland et al. 2004). The role of water in this and other Lewis carbohydrate systems was further examined by independent molecular dynamics studies (Reynolds et al. 2008), which was in general agreement that the water is directly involved both in determining specificity and maintaining the conformation of carbohydrate antigens.

Recently, Saraboji and colleagues have determined a series of ultra-high-resolution crystal structures of Galectin-3 both in its unliganded (apo) form and in complexes with lactose and glycerol (Saraboji et al. 2012). The 0.86 Å resolution crystal structure of the Galectin-3 complex with lactose contains at least 10 ordered water molecules directly engaging the carbohydrate ligand or acting as hydrogen-bonded bridges between carbohydrate- and protein-binding site residues (Figure 1.3a). When the 1.08 Å apo-structure of Galectin-8 was compared, five of the same ordered water molecules involved in bridging carbohydrate–protein were

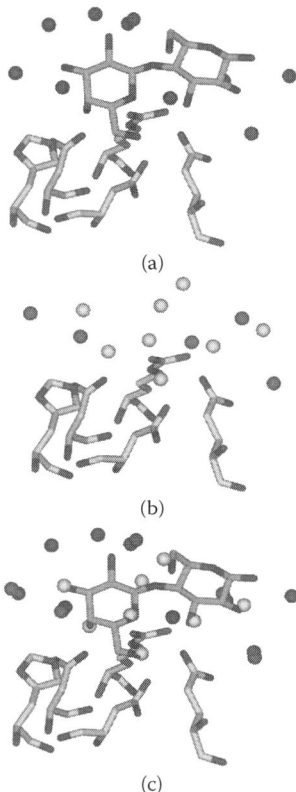

(a)

(b)

(c)

FIGURE 1.3 **(See color insert.)** Central role of water in protein recognition of carbohydrates. (a) Close-up of water (red spheres) network around the lactose interaction with Galectin-3 determined at 0.86 Å resolution, PDB ID: 3ZSJ. (b) Key binding-site waters (steel blue and light gray spheres) in the apo-structure (unliganded) of human Galectin-3 determined at 1.08 Å resolution, PDB ID: 3ZSL. (c) Structural overlay of the lactose-bound and apo crystal forms of human Galectin-3 depicted in panels A and B. Note that waters in the apo form that correspond to key binding-site waters from the lactose complex with Galectin-3 are shown in steel blue and those that superimpose with carbohydrate atoms are in light gray. The crystal structures depicted in this figure and related PDB entries are described in detail elsewhere (Saraboji et al. 2012).

maintained (Figure 1.3b, steel blue spheres). Interestingly, a further eight water molecules were observed that closely matched the positions of oxygen atoms from the bound lactose molecule (Figure 1.3b, light gray spheres). Thus, key water molecules, important for determining carbohydrate specificity, are maintained in what appears to be a pre-configured binding site ready to engage carbohydrate ligands. In addition, waters clearly occupy the same location as oxygen atoms in the bound carbohydrate (so may mimic the carbohydrate ligand), and are displaced upon entry of the carbohydrate into the binding site (see overlays in Figure 1.3c). Many other examples for the involvement of water in carbohydrate–protein interactions can be found in the literature, but are beyond the scope of this chapter.

1.3.5 MULTIVALENCY OF CARBOHYDRATE-BINDING PROTEINS

A quick survey of entries in the PDB that contain the keyword "carbohydrate" reveals that for antibodies and lectins there appears to be a modular type of assembly and a capacity for the protein subunits to associate as multimers/oligomers (Figure 1.4). For the antibody crystal structures, there are two major populations at around 50 kDa (Fab and Fc regions) and at 100 kDa (dimers of Fab and Fc), which indicate that in

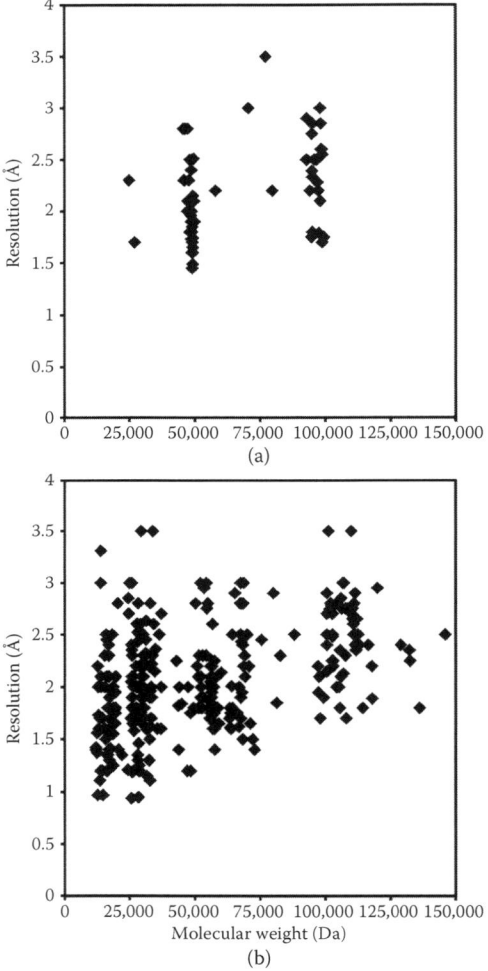

FIGURE 1.4 Distributions of resolutions and molecular weights for representative crystal structures of antibodies and lectins within the Protein Data Bank (PDB). (a) PDB entries containing the keywords "antibody" and "carbohydrate." The modular nature of antibody structures shows that most structures fall into monomers (Fab or Fc) or dimers within the crystals. (b) PDB entries containing the keywords "lectin" and "carbohydrate." Lectins also often contain repeating carbohydrate-binding domain architectures or form oligomers that appear to fall within 4–5 predominant size populations.

crystals Fab–Fab or Fc–Fc pairing is common (Figure 1.4a). For the lectins, there are at least four or five size groupings that correspond to the known tandem arrangement of carbohydrate-binding modules and stable oligomers that occur in this class of carbohydrate-binding protein (Figure 1.4b). While this survey is not comprehensive and contains some non-carbohydrate-bound structures, it is clear that there is a propensity for carbohydrate-binding proteins to form discrete oligomers in crystals. This observation is in agreement with the concept that multivalent binding to carbohydrates is often required for biological function (Dam and Brewer 2010; Weis and Drickamer 1996). The multivalency of carbohydrate–protein interactions results in high avidity and can overcome the low to modest affinities of most carbohydrate interactions with individual binding sites.

A variety of protein oligomers observed in crystals of carbohydrate–protein complexes are illustrated in Figure 1.5. The anti-carbohydrate antibody 2G12 binds to the high density of branched N-glycans on the human immunodeficiency virus (HIV) envelope protein, gp120. The crystal structures of the 2G12 Fab dimer, with mannose-containing saccharides, reveal how this antibody is geared toward binding the high densities of gp120 N-glycans (Calarese et al. 2003). The swapping of the variable domains of the heavy chains in 2G12 Fab forces a stable dimer, in crystals and the intact IgG antibody, with the carbohydrate-binding sites closely packed together to form a useful surface for multivalent carbohydrate recognition (Figure 1.5a). While this is an elegant immunological solution to carbohydrate cluster recognition, it is not widespread among antibodies. Other mechanisms such as through metal-ion-mediated Fab–Fab pairing (Farrugia et al. 2009) may be involved in carbohydrate cluster recognition by antibodies. In addition, antibodies form larger oligomers due to their covalent structure where IgG has two binding sites and polymeric IgM has at least 10 binding sites with the potential to interact with carbohydrates.

Similar to other collectins, the carbohydrate recognition domain "heads" of lung SP-D associate as a stable trimer through the presence of a coiled-coil "neck" region (Shrive et al. 2003). The carbohydrate ligands recognized by SP-D are located in binding pockets at one end of the molecule separated by around 45 to 50 Å (Figure 1.5b). The relatively flat carbohydrate-binding surface is suitable for multivalent recognition of repeating carbohydrate ligands on the surfaces of invading pathogens. Collectins can also form larger oligomers where several of the trimeric units associate to form oligomers or "fuzzy balls," which are highly multivalent macromolecular assemblies (Veldhuizen et al. 2011).

Carbohydrate-binding proteins from pathogens also tend to form as multivalent oligomers suitable for avid binding to host carbohydrate determinants. The AB_5 toxins are virulence factors of many bacterial pathogens and consist of a catalytic domain (A-subunit) and a multimeric host receptor-binding domain (pentamers of B-subunits), which display high avidity binding to glycans on target cells (Beddoe et al. 2010). The crystal structure of the pentameric B subunit of shiga-like toxin I (SLT I) of *Escherichia coli* in complex with an analog of the glycolipid Gb3 (trisaccharide epitope) shows how the AB_5 toxins can bind a large number of glycan ligands (Ling et al. 1998). Each B-subunit of SLT-1 interacts with three Gb3 molecules so that 15 glycans are bound on one face of the B5 oligomer (Figure 1.5c). A second example of a pathogen carbohydrate-binding protein is the sigma 1 (σ1) trimeric attachment

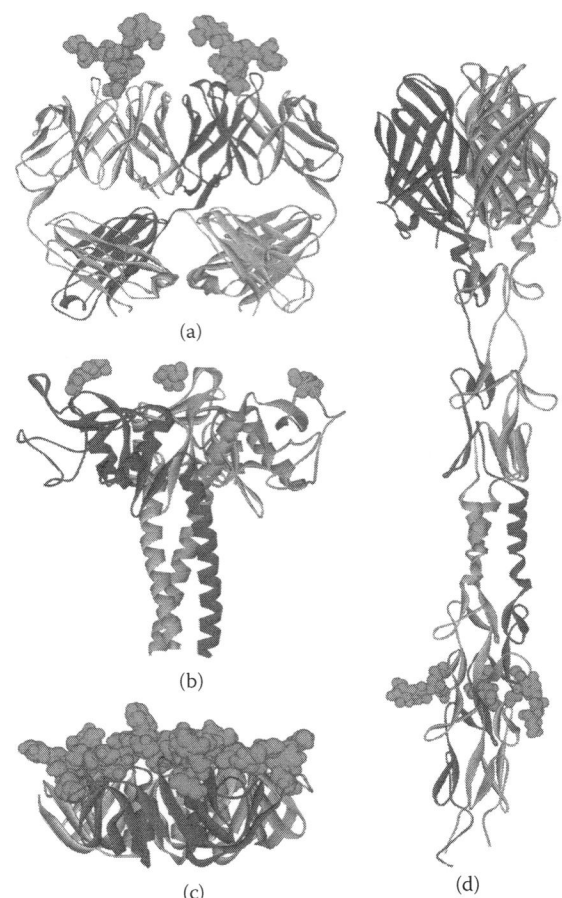

(a)

(b)

(c)   (d)

FIGURE 1.5 (See color insert.) Examples of crystal structures of protein oligomers involved in binding to carbohydrates. (a) The domain-swapped 2G12 Fab dimer with two bound high-mannose-branched complex oligosaccharides (Man$_9$GlcNAc$_3$) determined at 3.0 Å resolution, PDB ID: 1OP5 (Calarese et al. 2003). (b) Trimeric head and neck regions of human SP-D in complex with maltose determined at 1.40 Å resolution, PDB ID: 1PWB (Shrive et al. 2003). (c) Pentameric B subunit of shiga-like toxin I (SLT-I) of *Escherichia coli* in complex with an analog of the glycolipid Gb3 (globotriaosyl ceramide) determined at 2.80 Å resolution, PDB ID: 1BOS (Ling et al. 1998). (d) Trimeric reovirus attachment protein, sigma 1 (σ1), in complex with 3′-sialyllactose, Neu5Acα(2-3)Galβ(1-4)Glc, determined at 2.25 Å resolution, PDB ID: 3S6X (Reiter et al. 2011). Proteins are shown as ribbon-style representations and separate polypeptide chains in the oligomers are colored differently. Carbohydrate ligands are displayed as space-filling spheres in green.

protein of reovirus, which binds to sialylated glycans on host cells. The crystal structure of σ1 in complex with 3′-sialyllactose (Reiter et al. 2011) shows an unusual arrangement where the carbohydrates are bound to the extended trimeric "stalk" regions rather than the ends of the globular head domains of the attachment protein (Figure 1.5d). In the intact σ1 protein, the stalk is further extended by a long trimeric coiled-coil region placing the carbohydrate-binding sites around the middle of this approximately 400 Å long pathogen cell attachment protein (Reiter et al. 2011).

Clearly, the diversity of oligomeric assemblies within host and pathogen carbohydrate-binding proteins (illustrated here by only a few examples) highlights the need to study the full-length proteins in their near physiological states. Such structural problems often require techniques complementary to x-ray crystallography such as small-angle scattering (of x-rays and neutrons) in aqueous solutions, which is discussed in the second part of this chapter (Sections 1.4 and 1.5).

1.4 PROBING MACROMOLECULAR STRUCTURES BY SMALL-ANGLE X-RAY AND NEUTRON SCATTERING

Small-angle x-ray and neutron scattering (SAXS and SANS) are powerful techniques that can complement x-ray crystallography to obtain global size and shape information from biological macromolecular systems in solution (Jeffries and Trewhella 2012; Mertens and Svergun 2010; Neylon 2008; Svergun 2010). Importantly, SAXS and SANS experiments require minimal amounts of purified material, can be performed in dilute solutions, and the conditions can be readily adjusted to simulate physiological environments (e.g., ionic strength and pH). Thus, conformations of macromolecules can be probed and cross-checked against the restricted and often single conformation observed in crystals. Structural parameters such as the radius of gyration (R_g), maximum dimension (D_{max}), and the probable distribution of atom-pair distances ($P(r)$ vs. r) of, or within, a macromolecule can be evaluated from the small-angle scattering data. Furthermore, it is possible to monitor (1) the oligomeric state of a protein, (2) how changes in sample environment or ligand binding can influence global protein conformation (He et al. 2009, 2003), and (3) how the formation of higher-order macromolecular complexes effects the overall structure of a macromolecule in solution (Wall et al. 2000).

Both SAXS and SANS are based on a difference existing either between the average electron density of a macromolecule relative to a supporting solvent (SAXS), or between the "isotopic composition/density" per unit volume of a macromolecule relative to a supporting solvent (SANS; in particular, hydrogen, ^1H, content per unit volume). This difference is known as contrast ($\Delta\rho$, where $\Delta\rho_{macromolecule} = \rho_{macromolecule} - \Delta\rho_{solvent}$). Assuming that contrast is present, and given that the total population of macromolecules within a sample is monodisperse, the rate at which scattering intensities decrease with increasing angle (after having subtracted solvent scattering contributions) will reflect the distribution of distances between scattering centers within a single particle. Models can be generated and fitted against the solvent-subtracted data to yield information on the overall volume and shape of a macromolecule in solution as well as the spatial dispositions of subunits within oligomers or even macromolecular complexes comprised of different subunits.

One of the advantages of SANS over SAXS (Lakey 2009) is that the isotopic composition/density, specifically the ratio of ^1H to deuterium (^2H) per unit volume in the solvent, can be easily altered so that $\Delta\rho$ can either be maximized to increase scattering signals from a macromolecule, or be reduced to minimize scattering signal contributions (Jeffries and Trewhella 2012). Therefore, SANS is particularly useful for investigating higher-order macromolecular complexes. Relative to each other, different classes of macromolecules—proteins, lipids, DNA, and carbohydrates—have different ^1H content per unit volume and hence different neutron scattering "power." Therefore, it is possible to selectively "match out" neutron scattering contributions of one component of a complex relative to another by altering the contrast through adjustments of $^1H_2O{:}^2H_2O$ ratios in the solvent. Consequently, using SANS with contrast variation, shape information can be obtained from a whole complex as well as the shapes of the individual components within a complex: each different class of macromolecule will have a different match point at a particular $^1H_2O{:}^2H_2O$ ratio where their scattering contributions are minimized from the overall scattering profile. However, depending on the mass ratios of the components within a complex (scattering intensity is proportionate to the volume of a particle squared) the separation of each component's match point maybe narrow. The quality of the SANS scattering signal can be drastically improved by deuterating one component of a complex so as to radically alter the ratio of ^1H per unit volume. The selective deuteration of one component of a complex is very powerful for probing the shape of protein–protein complexes in solution as the incorporation of nonexchangeable ^2H into one protein enables a contrast difference to be set up between the ^1H and deuterated components of a complex and subsequent separation of their match points in solution (Jeffries and Trewhella 2012).

Aside from obtaining R_g, D_{max}, and $P(r)$ versus r, that in themselves can provide valuable insights into the solution states of macromolecules, what is of most interest to structural biologists is obtaining a sensible consensus model, or series of models, that best fit their scattering data. Advances in computational methods (Petoukhov and Svergun 2007) have seen the routine application of restoring molecular volumes and shapes of a macromolecule from SAXS and SANS data using easy-to-use *ab initio* methods (Franke and Svergun 2009) that do not require any prior knowledge regarding the structure of a macromolecule of interest. When combined with models derived from x-ray crystallography, NMR, or homology modeling (Petoukhov and Svergun 2005, 2006; Svergun et al. 1995) SAXS and SANS begin to open new frontiers with respect to building representative protein structures that cannot otherwise be accessed using x-ray crystallography alone. SAXS and SANS are especially powerful for the analysis of modular proteins with inherent structural flexibility, large macromolecular multi-component complexes, assemblies, and glycoproteins.

1.5 EXAMPLES OF GLYCOPROTEINS STUDIED BY CRYSTALLOGRAPHY AND SMALL-ANGLE SCATTERING TECHNIQUES

An increasing number of structural biologists are beginning to integrate SAXS and SANS as complementary techniques into their research programs that focus on the molecular foundations of glycoprotein structure and function. As SAXS and SANS can be performed on natively glycosylated proteins and usually require similar or smaller amounts of sample when compared to crystallography (using synchrotron-SAXS it is possible to obtain quality x-ray scattering data from as little as 10 µL of protein at 1–5 mg·mL^{-1} in 1–10 seconds), then, at the very least, scattering techniques provide an invaluable complementary tool to probe the native global states and the shapes of glycoproteins in solution.

1.5.1 CRYSTALLOGRAPHY OF AMIGO-I COMBINED WITH SAXS OF AMIGO-II AND -III

Kajander et al. (2011a) provided an elegant example of where SAXS has been employed in combination with crystallography to probe the structures of a group of related glycoproteins. For example, the transmembrane AMIGO proteins that are required for regulating neuronal growth, mobility, and adhesion (Kuja-Panula et al. 2003), which share a similar leucine-rich repeat (LRR) domain that is responsible for modulating protein–protein interactions (Chen et al. 2006; Kajava and Kobe 2002). The crystal structure was determined for the glycosylated neuronal protein AMIGO-I and this was used to develop SAXS-based models for the related and more heavily glycosylated protein AMIGO-II as well as AMIGO-III (Kajander et al. 2011a).

AMIGO-I crystallized as a horseshoe-shaped dimer with a twofold rotational axis between each AMIGO-I monomer. The LRR domains of opposing sub-units pack together to form the base of the horseshoe while separate LRRCT capping and immunoglobulin (Ig) domains extend from the base to form the individual arms of the U-shaped horseshoe (Figure 1.6a). Interestingly, SAXS data revealed that AMIGO-II and AMIGO-III have significantly increased radii of gyration when compared to the AMIGO-I crystal structure (AMIGO-I, R_g crystal = 30 Å; AMIGO-II, R_g = 46 Å; AMIGO-III, R_g = 40 Å). These results indicate that all three related proteins have very different mass distributions and hence different domain orientations with respect to each other. Indeed, the resulting estimates of D_{max} derived from the $P(r)$ versus r profiles of AMIGO-II and AMIGO-III show them to be significantly more extended than the AMIGO-I crystal form (AMIGO-I, D_{max} crystal = 100 Å; AMIGO-II, D_{max} = 170 Å; AMIGO-III, D_{max} = 135 Å). Unfortunately, the crystallizable AMIGO-I isoform could not be studied by SAXS (for a direct comparison to SAXS of AMIGO-II and AMIGO-III) due to nonspecific protein aggregation issues. However, rigid-body modeling of AMIGO-II and AMIGO-III against their respective SAXS datasets, using the crystal structure of AMIGO-I as a rigid-body template, shows that the best-fit models are those where the position of the LRR domains at the dimer interface are reasonably preserved across all three proteins and that the structural extension involves the arms of the U-shaped molecule opening up through large movements of the C-terminal Ig domains (Figure 1.6a). Since the Ig domain of the AMIGO proteins is a primary site of glycosylation, one of the interesting hypotheses from the crystallographic/SAXS study is that the flexibility between the Ig domains and LRRs of AMIGO-I, AMIGO-II, and AMIGO-III is modulated by glycosylation, which may affect the orientation of the LRRs within the intracellular space to facilitate the formation of AMIGO LRR-LRR mediated intercellular *trans*-dimers and, consequently, promote neuronal cell adhesion (Kajander et al. 2011a).

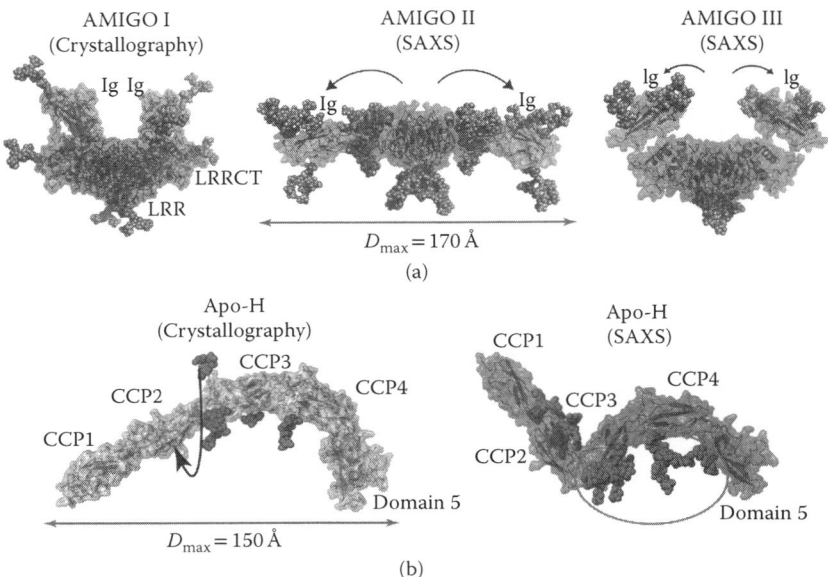

FIGURE 1.6 (**See color insert.**) Crystallography in combination with solution SAXS of glycoproteins: AMIGO proteins and Apo-H. (a) x-ray crystal structure of AMIGO-I and a comparison with the AMIGO-II and AMIGO-III models determined from solution SAXS data. The LRR domains (blue) self-associate to drive dimerization between AMIGO monomers and are capped off with the LRRCT domains (red) from which extends an Ig domain (teal). The spatial orientation of the Ig domains differs between each of the three AMIGO isoforms. The N-glycans of the AMIGO proteins were modeled and are represented by gray spheres. The images were generated from PDB coordinates kindly provided by Dr. Tommi Kajander (Kajander et al. 2011a). (b) The crystal structure of Apo-H, PDB ID: 1C1Z (Schwarzenbacher et al. 1999) shows that the CCP modules and domain 5 of the protein adopt a J-shaped conformation (left), while SAXS reveals that Apo-H undergoes a conformational shift in solution to form an S-shape and that the N-glycans (red spheres) appear to form a carbohydrate patch (marked with an ellipse) along one side of the proteins. The SAXS data is represented as a model derived using the crystallographic coordinates (PDB ID: 1C1Z) as well as information and a figure presented in the original Apo-H paper (Hammel et al. 2002).

1.5.2 Different Conformations of Apolipoprotein H (β₂-Glycoprotein I) Observed by Crystallography and SAXS

Apolipoprotein H (Apo-H or β_2-glycoprotein I) is a highly glycosylated protein (~19% w/w) that comprises five domains: four complement control protein (CCP) domains and a unique fifth domain, domain V (Bouma et al. 1999; Schwarzenbacher et al. 1999). Apo-H is involved in triggering the blood coagulation cascade (Brighton et al. 1996) and has apparent high affinity for heparin, cell membranes, macrophages, and phospholipids (Balasubramanian and Schroit 1998; Del Papa et al. 1998; Schousboe and Rasmussen, 1988). Complexes of Apo-H and phospholipids have been suggested to act as antigens for autoimmune phospholipid antibodies that are associated with a number of pathological conditions (e.g., lupus) (Guerin et al. 1997; Kandiah et al. 1998; McNeil et al. 1990). A key feature proposed to promote epitope exposure on ligand binding is the capacity for Apo-H to undergo conformational changes (Wang et al. 2000).

The crystal structure of Apo-H indicates that the five domains adopt a modular and extended "J-shaped" conformation (Figure 1.6b) with a calculated R_g of 46 Å and an approximate maximum dimension of ~150 Å (Bouma et al. 1999; Schwarzenbacher et al. 1999). Hammel et al. (2002) used SAXS to analyze the solution state of Apo-H and showed that the protein adopts a very different conformation in solution compared to the crystal form as evidenced by a 3 Å reduction in R_g for the solution state (43 Å), which represents a significant mass redistribution. Results derived from modeling the shape of the protein in solution via *ab initio* methods, as well as rigid-body refinement against the SAXS data, revealed that the protein adopts a more compact S-shaped conformation in solution (D_{max} = 140 Å) that is due to a significant rotation in the linker connecting CCP domains 2 and 3 (Figure 1.6b). Furthermore, the N-glycan extensions from CCP domains 3 and 4 "spatially-coalesce" to form a carbohydrate-rich patch along one side of the protein so that the overall structure is more bulky than that observed in the crystal form (cross section of crystal, R_g = 14.2 Å; SAXS = 17.8 Å). Importantly, as Apo-H comprises five modules connected by linkers, the authors stress that the interpretation of the SAXS model should be viewed as a representation of well-definable, but one of several domain dispositions within the protein as opposed to a singular static/rigid structure (Hammel et al. 2002). Indeed, differences between the crystal and solution states automatically indicate that the protein can adopt multiple conformations. The authors suggest that discrepancies between the solution state and crystal form of Apo-H are, in part, influenced by crystal packing forces and crystallization conditions that were well away from physiological conditions. The high ionic strengths used to crystallize the protein may have destabilized interdomain hydrogen bonds and strengthened interdomain hydrophobic interactions to cause the observed extension of the Apo-H structure in the crystalline state. What the combined SAXS/crystallographic analyses have captured (that crystallography in isolation could not readily access) is that large conformational transitions can occur in Apo-H. The bending and stretching of the protein are thought to be key mechanisms by which epitope presentation is realized for the recognition and interaction of the protein with autoimmune phospholipid antibodies. The SAXS analyses have provided conclusive data to support the hypothesis that such global structural transitions of Apo-H are indeed possible.

1.5.3 SAXS AND SANS OF COMPLEMENT SYSTEM COMPONENTS

The structure of a number of immune system glycoproteins has been probed using SAXS or SANS, which includes the analysis of components of the complement system (Esser et al. 1993; Morgan et al. 2011; Perkins et al. 2002; 2011; Smith et al. 1992) as well as some of the proteins through which pathogens evade complement-mediated immune responses (e.g., as observed in *Staphylococcus aureus* infection) (Burman et al. 2008; Chen et al. 2011). Currently, around 30 proteins have been identified to comprise the complement system that is divided into three main pathways (classical, alternate, and lectin pathways), which in combination are responsible for the rapid targeting and elimination of pathogens as well as regulating the removal of immune complexes and cell debris. The difficulty in analyzing the structure of the protein components of complement, in particular using x-ray crystallography, is that a significant number of proteins comprising the complement system (including many that are heavily glycosylated) undergo diverse combinatorial interactions and cleavage events to realize their biological function. Formation of higher-order complexes and oligomers is further complicated by the structural adaptability of many of the protein constituents, both in terms of structural flexibility and conversions from pro-protein to active-protein constituents via regulated and targeted proteolytic cascades, the order and timing of which is pivotal to the function of the complement system (Daha 2010; Ricklin et al. 2010). Of particular interest to structural biologists is the major task of decoding the molecular interactions that are fundamental to complement in order to understand the molecular basis of innate immune responses, homeostasis, and pathogen defense systems (Burman et al. 2008; Chen et al. 2011). Small-angle scattering, as complementary to crystallography, has provided unique insights into the molecular foundations of the complement system, with select examples outlined below.

Factor H is a large (155 kDa) modular glycoprotein that consists of 20 CCP domains (Ripoche et al. 1986). Note that the glycoprotein Apo-H has only four of these CCP domains (see Section 1.5.2). Factor H is at high concentration in blood plasma (0.5 mg·mL^{-1}) and is central to regulating the rapid response cascade of the alternate complement pathway to pathogens, allows for host/foreign cell delineation and the maintenance of tissues (Clark et al. 2010; Ferreira et al. 2010). Among a number of Factor H-binding interactions (Perkins et al. 2010), it binds to another key complement protein 3Cb, which is produced from C3 via interaction with Factor B. Interestingly, the bimolecular C3bBb complex converts even more C3 into 3Cb that exposes an activated thioester in the thioester containing domain (TED) region of 3Cb that rapidly attaches to nearby particles, including cell surfaces, to instigate a potent effector response (e.g., the formation of membrane attack complexes or macrophage targeting) (Morgan et al. 2011). Factor H is also responsible for accelerating the decay of the C3bBb complex. Of note, the interaction between C3b or C3bBb and Factor H is amplified on host-cell self-surfaces carrying glycosaminoglycan or sialic acid markers that are otherwise absent on the surface of most pathogenic organisms (Pangburn 2000; Varki 2007). Consequently, this allows the immune system to discriminate between host and pathogen, although some pathogens have developed sialic acid markers that mimic self-surfaces (Vimr et al. 2004) as well as Factor H capturing mechanisms that result in immune system evasion such as glycoproteins gp41 and gp120 of HIV (Pinter et al. 1995a, 1995b; Stoiber et al. 1996).

Initial SAXS and SANS investigations showed that Factor H is a flexible protein and that the CCP modules are organized so that the protein folds back on itself as opposed to adopting a highly extended conformation in solution, suggesting that the C- and N-terminal domains of the protein could come into close spatial proximity (Aslam and Perkins 2001). More recently, Morgan et al. (2011) have shed light on the structural basis underpinning the interaction between

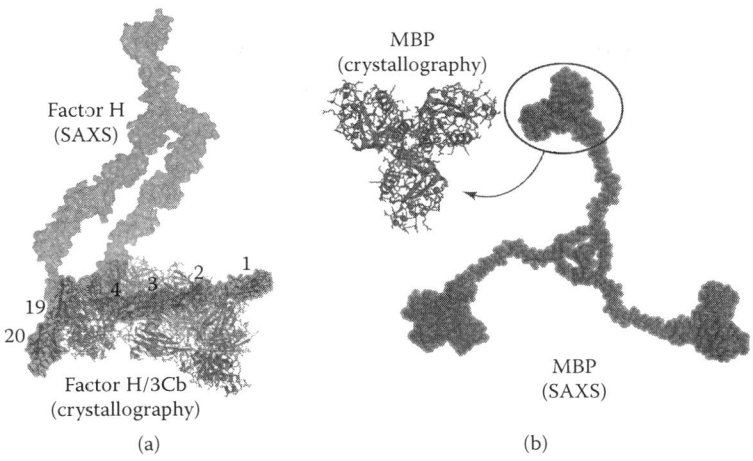

FIGURE 1.7 **(See color insert.)** Building complex models from crystallographic and small-angle scattering data: Factor H and mannan-binding protein (MBP). (a) Using a combination of crystal structures and solution SAXS data, Morgan et al. (2011) have built a structural model representing the engagement between Factor H (red) and 3Cb (blue). SAXS data indicated that Factor H folds back on itself (red surface) with the conformation of the protruding portions (CCP modules 5-18) representing one of multiple possible folded-back conformations, within the ensembles of Factor H solution conformations previously interpreted from SAXS and SANS data (Aslam and Perkins 2001). (b) Solution SAXS reveals that MBP forms a large-molecular-weight oligomer in solution containing 9 subunits arranged like the spokes of a wheel. Crystallography has shown that the carbohydrate-binding domains of MBP form a trimer (inset) with four Ca^{2+} ions (blue spheres) bound per protein monomer, PDB ID: 1HUP (Sheriff et al. 1994). The trimeric MBP carbohydrate recognition domains are located at the end of each spoke in the SAXS-derived structure of intact MBP. Models of the complexes shown here were derived from information and figures provided in Dong et al. (2007) and Morgan et al. (2011) and the reader should refer to these works for specific details. The crystal structures of Factor H CCP1-4/3Cb and Factor H CCP19-20/3Cd are derived from PDB ID: 2WII (Wu et al. 2009) and 2XQW (Kajander et al. 2011b) respectively.

Factor H and 3Cb on the cell surface by synthesizing data obtained from x-ray crystallography, NMR, and SAXS to produce a structural model of the Factor H/3Cb complex (Morgan et al. 2011). The folded-back conformation of Factor H appears to be integral for the N- and C-terminal domains to both bind 3Cb so that the N-terminus can perform cofactor and decay accelerating activities of the C3bBb complex, while the C-terminal domain can interact with the TED of 3Cb in combination with glycosaminoglycans to modulate self-surface attachment (Figure 1.7a).

1.5.4 SAXS OF AN OLIGOMERIC COLLECTIN: MANNAN-BINDING PROTEIN OR LECTIN

A branch of the innate complement system is the lectin pathway (Fujita et al. 2004) that results in the initiation of 3Cb-based complement cascades and, ultimately, the lysis or phagocytosis of the pathogen. The collectin family member mannan-binding protein (MBP) or lectin (MBL) is the archetypal initiator of the lectin pathway of complement activation and displays binding activity for mannose and fucose residues on the surface of pathogens like bacteria (Takahashi et al. 2006). MBP undergoes a number of posttranslational modifications (including glycosylation), but more significantly it exists in an oligomeric state in solution capable of multivalent carbohydrate recognition (Jensen et al. 2005; Teillet et al. 2005). Dong et al. (2007) have performed an extensive SAXS analysis of MBP in solution. Molecular weight determinations based on SAXS data show that the protein consists of a trimer of trimers (i.e., nine MBP monomers) with a molecular weight of 266 kDa (Dong et al. 2007). A reconstruction of MBL from the SAXS data shows that the oligomer adopts a threefold symmetrical shape in solution, which is relatively flat in one dimension but is splayed out over 320 Å in the other so that the protein adopts a conformation like wheel spokes connected to a central hub (Figure 1.7b). MBP carbohydrate recognition domains form a trimer (as revealed by crystallography, inset of Figure 1.7b [Sheriff et al. 1994]) and are located at the ends of each spoke of intact MBP. Note the structural similarity of the carbohydrate-binding heads of MBP with the related collectin family member SP-D shown in Figure 1.5b. Interestingly, on comparison to the solution state, subsequent atomic force microscopy showed that MBP undergoes further "spoke extensions" in response to binding surface-conjugated ligands (e.g., mannosamine) suggesting that surface-induced conformational changes in MBP maybe important for protein function and regulation (Dong et al. 2007). This study, as well as other examples discussed previously, highlights the enormous potential of small-angle scattering to probe the global states of carbohydrate-binding proteins and glycoproteins in solution so as to greatly enhance high-resolution 3D structural information derived from x-ray crystallographic investigations of domains or smaller fragments of these proteins.

1.6   CONCLUDING REMARKS

A structural and functional understanding of multivalent recognition of carbohydrates is essential for appreciating the biological activity of carbohydrate-binding proteins in a wide variety of biological systems. While x-ray crystallography continues to provide many fundamental detailed 3D structural insights into carbohydrate–protein interactions and glycoprotein structure, the technique is often limited by the availability of crystals and the essentially conformationally restrained crystalline environment. Thus, it is often important to consider other complementary structural biology methods that allow intact glycoproteins to be examined in solution or in a less conformationally restrained environment. Both SAXS and SANS have been successfully employed as complementary techniques to x-ray crystallography and together have and will continue to expand the frontiers of our knowledge regarding the 3D structure of glycoproteins and carbohydrate-binding proteins and their diverse multimeric assemblies.

ACKNOWLEDGMENTS

We kindly thank Dr. Tommi Kajander (Structural Biology and Biophysics, Institute of Biotechnology, University of Helsinki, Finland) for supplying coordinates of the three AMIGO glycoproteins derived by crystallography and SAXS. P.A.R. is the Sir Zelman Cowen Senior Research Fellow (Sir Zelman Cowen Fellowship Fund, Burnet Institute). The authors gratefully acknowledge the contribution to this work of the Victorian Operational Infrastructure Support Program received by the Burnet Institute.

REFERENCES

Agostino, M., Farrugia, W., Sandrin, M. S. et al. 2012. Structural glycobiology of antibody recognition in xenotransplantation and cancer immunotherapy. In *Anticarbohydrate Antibodies*, Kosma, P. and Muller-Loennies, S. (eds), pp. 203–228. Wien: Spinger-Verlag.

Apweiler, R., Hermjakob, H., and Sharon, N. 1999. On the frequency of protein glycosylation, as deduced from analysis of the SWISS-PROT database. *Biochim Biophys Acta*, 1473, 4–8.

Aslam, M. and Perkins, S. J. 2001. Folded-back solution structure of monomeric factor H of human complement by synchrotron x-ray and neutron scattering, analytical ultracentrifugation and constrained molecular modelling. *J Mol Biol*, 309, 1117–1138.

Attrill, H., Takazawa, H., Witt, S. et al. 2006. The structure of siglec-7 in complex with sialosides: Leads for rational structure-based inhibitor design. *Biochem J*, 397, 271–278.

Balasubramanian, K. and Schroit, A. J. 1998. Characterization of phosphatidylserine-dependent beta2-glycoprotein I macrophage interactions. Implications for apoptotic cell clearance by phagocytes. *J Biol Chem*, 273, 29272–29277.

Beddoe, T., Paton, A. W., Le Nours, J., Rossjohn, J., and Paton, J. C. 2010. Structure, biological functions and applications of the AB5 toxins. *Trends Biochem Sci*, 35, 411–418.

Berman, H. M., Westbrook, J., Feng, Z. et al. 2000. The protein data bank. *Nucleic Acids Res*, 28, 235–242.

Bouma, B., de Groot, P. G., van den Elsen, J. M. et al. 1999. Adhesion mechanism of human beta(2)-glycoprotein I to phospholipids based on its crystal structure. *EMBO J*, 18, 5166–5174.

Brighton, T. A., Hogg, P. J., Dai, Y. P. et al. 1996. Beta 2-glycoprotein I in thrombosis: evidence for a role as a natural anticoagulant. *Br J Haematol*, 93, 185–194.

Burman, J. D., Leung, E., Atkins, K. L. et al. 2008. Interaction of human complement with Sbi, a staphylococcal immunoglobulin-binding protein: Indications of a novel mechanism of complement evasion by Staphylococcus aureus. *J Biol Chem*, 283, 17579–17593.

Buts, L., Loris, R., De Genst, E. et al. 2003. Solving the phase problem for carbohydrate-binding proteins using selenium derivatives of their ligands: A case study involving the bacterial F17-G adhesin. *Acta Crystallogr D Biol Crystallogr*, 59, 1012–1015.

Calarese, D. A., Scanlan, C. N., Zwick, M. B. et al. 2003. Antibody domain exchange is an immunological solution to carbohydrate cluster recognition. *Science*, 300, 2065–2071.

Chen, H., Ricklin, D., Hammel, M. et al. 2011. Allosteric inhibition of complement function by a staphylococcal immune evasion protein. *Proc Natl Acad Sci USA*, 107, 17621–17626.

Chen, Y., Aulia, S., Li, L., and Tang, B. L. 2006. AMIGO and friends: An emerging family of brain-enriched, neuronal growth modulating, type I transmembrane proteins with leucine-rich repeats (LRR) and cell adhesion molecule motifs. *Brain Res Rev*, 51, 265–274.

Clark, S. J., Bishop, P. N., and Day, A. J. 2010. Complement factor H and age-related macular degeneration: The role of glycosaminoglycan recognition in disease pathology. *Biochem Soc Trans*, 38, 1342–1348.

Daha, M. R. 2010. Role of complement in innate immunity and infections. *Crit Rev Immunol*, 30, 47–52.

Dam, T. K. and Brewer, C. F. 2010. Lectins as pattern recognition molecules: The effects of epitope density in innate immunity. *Glycobiology*, 20, 270–279.

Del Papa, N., Sheng, Y. H., Raschi, E. et al. 1998. Human beta 2-glycoprotein I binds to endothelial cells through a cluster of lysine residues that are critical for anionic phospholipid binding and offers epitopes for anti-beta 2-glycoprotein I antibodies. *J Immunol*, 160, 5572–5578.

Dong, M., Xu, S., Oliveira, C. L. et al. 2007. Conformational changes in mannan-binding lectin bound to ligand surfaces. *J Immunol*, 178, 3016–3022.

Dormitzer, P. R., Sun, Z. Y., Wagner, G., and Harrison, S. C. 2002. The rhesus rotavirus VP4 sialic acid binding domain has a galectin fold with a novel carbohydrate binding site. *EMBO J*, 21, 885–897.

Esser, A. F., Thielens, N. M., and Zaccai, G. 1993. Small angle neutron scattering studies of C8 and C9 and their interactions in solution. *Biophys J*, 64, 743–748.

Farrugia, W., Scott, A. M., and Ramsland, P. A. 2009. A possible role for metallic ions in the carbohydrate cluster recognition displayed by a Lewis Y specific antibody. *PLoS One*, 4, e7777.

Ferreira, V. P., Pangburn, M. K., and Cortes, C. 2010. Complement control protein factor H: The good, the bad, and the inadequate. *Mol Immunol*, 47, 2187–2197.

Franke, D. and Svergun, D. 2009. DAMMIF, a program for rapid ab-initio shape determination in small-angle scattering. *J Appl Cryst*, 42, 342–346.

Fujita, T., Matsushita, M., and Endo, Y. 2004. The lectin-complement pathway: Its role in innate immunity and evolution. *Immunol Rev*, 198, 185–202.

Guerin, J., Feighery, C., Sim, R. B., and Jackson, J. 1997. Antibodies to beta2-glycoprotein I: A specific marker for the antiphospholipid syndrome. *Clin Exp Immunol*, 109, 304–309.

Hammel, M., Kriechbaum, M., Gries, A. et al. 2002. Solution structure of human and bovine beta(2)-glycoprotein I revealed by small-angle x-ray scattering. *J Mol Biol*, 321, 85–97.

He, L., Andre, S., Garamus, V. M. et al. 2009. Small angle neutron scattering as sensitive tool to detect ligand-dependent shape changes in a plant lectin with beta-trefoil folding and their dependence on the nature of the solvent. *Glycoconj J*, 26, 111–116.

He, L., Andre, S., Siebert, H. C. et al. 2003. Detection of ligand- and solvent-induced shape alterations of cell-growth-regulatory human lectin galectin-1 in solution by small angle neutron and x-ray scattering. *Biophys J*, 85, 511–524.

Ideo, H., Matsuzaka, T., Nonaka, T., Seko, A., and Yamashita, K. 2011. Galectin-8-N-domain recognition mechanism for sialylated and sulfated glycans. *J Biol Chem*, 286, 11346–11355.

Jeffries, C. M. and Trewhella, J. 2012. Small-angle scattering: For structural molecular biologists. In *Quantitative Biology: From Molecular to Cellular Systems* (Chapman & Hall/ CRC Mathematical & Computational Biology), Wall, M. E. (ed). London: Taylor & Francis, LLC. Chapter 6, pp. 111–149.

Jensen, P. H., Weilguny, D., Matthiesen, F. et al. 2005. Characterization of the oligomer structure of recombinant human mannan-binding lectin. *J Biol Chem*, 280, 11043–11051.

Kajander, T., Kuja-Panula, J., Rauvala, H., and Goldman, A. 2011a. Crystal structure and role of glycans and dimerization in folding of neuronal leucine-rich repeat protein AMIGO-1. *J Mol Biol*, 413, 1001–1015.

Kajander, T., Lehtinen, M. J., Hyvarinen, S. et al. 2011b. Dual interaction of factor H with C3d and glycosaminoglycans in host-nonhost discrimination by complement. *Proc Natl Acad Sci USA*, 108, 2897–2902.

Kajava, A. V. and Kobe, B. 2002. Assessment of the ability to model proteins with leucine-rich repeats in light of the latest structural information. *Protein Sci*, 11, 1082–1090.

Kandiah, D. A., Sali, A., Sheng, Y. et al. 1998. Current insights into the "antiphospholipid" syndrome: Clinical, immunological, and molecular aspects. *Adv Immunol*, 70, 507–563.

Kuja-Panula, J., Kiiltomaki, M., Yamashiro, T., Rouhiainen, A., and Rauvala, H. 2003. AMIGO, a transmembrane protein implicated in axon tract development, defines a novel protein family with leucine-rich repeats. *J Cell Biol*, 160, 963–973.

Lakey, J. H. 2009. Neutrons for biologists: A beginner's guide, or why you should consider using neutrons. *J R Soc Interface*, 6 Suppl 5, S567–573.

Lee, J. E., Fusco, M. L., Abelson, D. M. et al. 2009. Techniques and tactics used in determining the structure of the trimeric ebolavirus glycoprotein. *Acta Crystallogr D Biol Crystallogr*, 65, 1162–1180.

Lee, J. E., Fusco, M. L., Hessell, A. J. et al. 2008. Structure of the Ebola virus glycoprotein bound to an antibody from a human survivor. *Nature*, 454, 177–182.

Ling, H., Boodhoo, A., Hazes, B. et al. 1998. Structure of the shiga-like toxin I B-pentamer complexed with an analogue of its receptor Gb3. *Biochemistry*, 37, 1777–1788.

Lustbader, J. W., Birken, S., Pileggi, N. F. et al. 1989. Crystallization and characterization of human chorionic gonadotropin in chemically deglycosylated and enzymatically desialylated states. *Biochemistry*, 28, 9239–9243.

Lutteke, T. 2009. Analysis and validation of carbohydrate three-dimensional structures. *Acta Crystallogr D Biol Crystallogr*, 65, 156–168.

Marino, K., Bones, J., Kattla, J. J., and Rudd, P. M. 2010. A systematic approach to protein glycosylation analysis: A path through the maze. *Nat Chem Biol*, 6, 713–723.

McNeil, H. P., Simpson, R. J., Chesterman, C. N., and Krilis, S. A. 1990. Anti-phospholipid antibodies are directed against a complex antigen that includes a lipid-binding inhibitor of coagulation: Beta 2-glycoprotein I (apolipoprotein H). *Proc Natl Acad Sci USA*, 87, 4120–4124.

McPherson, A. 1999. *Crystallization of Biological Macromolecules*. Cold Spring Harbor, NY: Cold Spring Harbor Laboratory Press.

Mertens, H. D. and Svergun, D. I. 2010. Structural characterization of proteins and complexes using small-angle x-ray solution scattering. *J Struct Biol*, 172, 128–141.

Morgan, H. P., Schmidt, C. Q., Guariento, M. et al. 2011. Structural basis for engagement by complement factor H of C3b on a self surface. *Nat Struct Mol Biol*, 18, 463–470.

Nettleship, J. E., Assenberg, R., Diprose, J. M., Rahman-Huq, N., and Owens, R. J. 2010. Recent advances in the production of proteins in insect and mammalian cells for structural biology. *J Struct Biol*, 172, 55–65.

Neylon, C. 2008. Small angle neutron and x-ray scattering in structural biology: Recent examples from the literature. *Eur Biophys J*, 37, 531–541.

Nguyen, H. P., Seto, N. O., MacKenzie, C. R. et al. 2003. Germline antibody recognition of distinct carbohydrate epitopes. *Nat Struct Biol*, 10, 1019–1025.

Pangburn, M. K. 2000. Host recognition and target differentiation by factor H, a regulator of the alternative pathway of complement. *Immunopharmacology*, 49, 149–157.

Perkins, S. J., Gilbert, H. E., Aslam, M. et al. 2002. Solution structures of complement components by x-ray and neutron scattering and analytical ultracentrifugation. *Biochem Soc Trans*, 30, 996–1001.

Perkins, S. J., Nan, R., Li, K., Khan, S., and Abe, Y. 2011. Analytical ultracentrifugation combined with x-ray and neutron scattering: Experiment and modelling. *Methods*, 54, 181–199.

Perkins, S. J., Nan, R., Okemefuna, A. I. et al. 2010. Multiple interactions of complement Factor H with its ligands in solution: A progress report. *Adv Exp Med Biol*, 703, 25–47.

Petoukhov, M. V. and Svergun, D. I. 2005. Global rigid body modeling of macromolecular complexes against small-angle scattering data. *Biophys J*, 89, 1237–1250.

Petoukhov, M. V. and Svergun, D. I. 2006. Joint use of small-angle x-ray and neutron scattering to study biological macromolecules in solution. *Eur Biophys J*, 35, 567–576.

Petoukhov, M. V. and Svergun, D. I. 2007. Analysis of x-ray and neutron scattering from biomacromolecular solutions. *Curr Opin Struct Biol*, 17, 562–571.

Pinter, C., Siccardi, A. G., Longhi, R., and Clivio, A. 1995a. Direct interaction of complement factor H with the C1 domain of HIV type 1 glycoprotein 120. *AIDS Res Hum Retroviruses*, 11, 577–588.

Pinter, C., Siccardi, A. G., Lopalco, L., Longhi, R., and Clivio, A. 1995b. HIV glycoprotein 41 and complement factor H interact with each other and share functional as well as antigenic homology. *AIDS Res Hum Retroviruses*, 11, 971–980.

Ramsland, P. A., Farrugia, W., Bradford, T., Hogarth, P. M., and Scott, A. M. 2004. Structural convergence of antibody binding of carbohydrate determinants in Lewis Y tumor antigens. *J Mol Biol*, 340, 809–818.

Ramsland, P. A., Farrugia, W., Yuriev, E., Edmundson, A. B., and Sandrin, M. S. 2003. Evidence for structurally conserved recognition of the major carbohydrate xenoantigen by natural antibodies. *Cell Mol Biol*, 49, 307–317.

Ramsland, P. A., Terzyan, S. S., Cloud, G. et al. 2006. Crystal structure of a glycosylated Fab from an IgM cryoglobulin with properties of a natural proteolytic antibody. *Biochem J*, 395, 473–481.

Reiter, D. M., Frierson, J. M., Halvorson, E. E. et al. 2011. Crystal structure of reovirus attachment protein sigma1 in complex with sialylated oligosaccharides. *PLoS Pathog*, 7, e1002166.

Reynolds, M., Fuchs, A., Lindhorst, T. K., and Perez, S. 2008. The hydration features of carbohydrate determinants of Lewis antigens. *Mol Simul*, 34, 447–460.

Ricklin, D., Hajishengallis, G., Yang, K., and Lambris, J. D. 2010. Complement: A key system for immune surveillance and homeostasis. *Nat Immunol*, 11, 785–797.

Ripoche, J., Day, A. J., Willis, A. C. et al. 1986. Partial characterization of human complement factor H by protein and cDNA sequencing: Homology with other complement and non-complement proteins. *Biosci Rep*, 6, 65–72.

Rupp, B. 2010. *Biomolecular Crystallography: Principles, Practice, and Application to Structural Biology*. New York: Garland Science, Taylor & Francis Group, LLC.

Saraboji, K., Hakansson, M., Genheden, S. et al. 2012. The carbohydrate-binding site in galectin-3 is preorganized to recognize a sugarlike framework of oxygens: Ultra-high-resolution structures and water dynamics. *Biochemistry*, 51, 296–306.

Schousboe, I. and Rasmussen, M. S. 1988. The effect of beta 2-glycoprotein I on the dextran sulfate and sulfatide activation of the contact system (Hageman factor system) in the blood coagulation. *Int J Biochem*, 20, 787–792.

Schwarzenbacher, R., Zeth, K., Diederichs, K. et al. 1999. Crystal structure of human beta2-glycoprotein I: Implications for phospholipid binding and the antiphospholipid syndrome. *EMBO J*, 18, 6228–6239.

Seaton, B. A., Crouch, E. C., McCormack, F. X. et al. 2010. Review: Structural determinants of pattern recognition by lung collectins. *Innate Immun*, 16, 143–150.

Sheriff, S., Chang, C. Y., and Ezekowitz, R. A. 1994. Human mannose-binding protein carbohydrate recognition domain trimerizes through a triple alpha-helical coiled-coil. *Nat Struct Biol*, 1, 789–794.

Shrive, A. K., Tharia, H. A., Strong, P. et al. 2003. High-resolution structural insights into ligand binding and immune cell recognition by human lung surfactant protein D. *J Mol Biol*, 331, 509–523.

Smith, K. F., Harrison, R. A., and Perkins, S. J. 1992. Molecular modeling of the domain structure of C9 of human complement by neutron and x-ray solution scattering. *Biochemistry*, 31, 754–764.

Stoiber, H., Pinter, C., Siccardi, A. G., Clivio, A., and Dierich, M. P. 1996. Efficient destruction of human immunodeficiency virus in human serum by inhibiting the protective action of complement factor H and decay accelerating factor (DAF, CD55). *J Exp Med*, 183, 307–310.

Svergun, D., Barberato, C., and Koch, M. H. 1995. CRYSOL: A program to evaluate x-ray solution scattering of biological macromolecules from atomic coordinates. *J Appl Cryst*, 28, 768–773.

Svergun, D. I. 2010. Small-angle x-ray and neutron scattering as a tool for structural systems biology. *Biol Chem*, 391, 737–743.

Takahashi, K., Ip, W. E., Michelow, I. C., and Ezekowitz, R. A. 2006. The mannose-binding lectin: A prototypic pattern recognition molecule. *Curr Opin Immunol*, 18, 16–23.

Teillet, F., Dublet, B., Andrieu, J. P. et al. 2005. The two major oligomeric forms of human mannan-binding lectin: Chemical characterization, carbohydrate-binding properties, and interaction with MBL-associated serine proteases. *J Immunol*, 174, 2870–2877.

Varki, A. 2007. Glycan-based interactions involving vertebrate sialic-acid-recognizing proteins. *Nature*, 446, 1023–1029.

Veldhuizen, E. J., van Eijk, M., and Haagsman, H. P. 2011. The carbohydrate recognition domain of collectins. *FEBS J*, 278, 3930–3941.

Vimr, E. R., Kalivoda, K. A., Deszo, E. L., and Steenbergen, S. M. 2004. Diversity of microbial sialic acid metabolism. *Microbiol Mol Biol Rev*, 68, 132–153.

Wall, M. E., Gallagher, S. C., and Trewhella, J. 2000. Large-scale shape changes in proteins and macromolecular complexes. *Annu Rev Phys Chem*, 51, 355–380.

Wang, S. X., Sun, Y. T., and Sui, S. F. 2000. Membrane-induced conformational change in human apolipoprotein H. *Biochem J*, 348 Pt 1, 103–106.

Weis, W. I. and Drickamer, K. 1996. Structural basis of lectin-carbohydrate recognition. *Annu Rev Biochem*, 65, 441–473.

Wu, J., Wu, Y. Q., Ricklin, D. et al. 2009. Structure of complement fragment C3b-factor H and implications for host protection by complement regulators. *Nat Immunol*, 10, 728–733.

Yuriev, E., Agostino, M., Farrugia, W. et al. 2009. Structural biology of carbohydrate xenoantigens. *Expert Opin Biol Ther*, 9, 1017–1029.

Yuriev, E., Farrugia, W., Scott, A. M., and Ramsland, P. A. 2005. Three-dimensional structures of carbohydrate determinants of Lewis system antigens: Implications for effective antibody targeting of cancer. *Immunol Cell Biol*, 83, 709–717.

2 Nuclear Magnetic Resonance Studies of Carbohydrate– Protein Interactions

Leonardus M. I. Koharudin and
Angela M. Gronenborn

CONTENTS

2.1 INTRODUCTION

Lectins, carbohydrate-binding proteins, are involved in numerous molecular recognition events associated with the immune response, tumorigenesis, and apoptosis that include cell–cell interaction and adhesion, trafficking of glycoproteins, and glycosylation (Sharon 2008; Sharon and Lis 1989, 2004). Unfortunately, for a large number of these important biological activities detailed mechanisms of the carbohydrate-mediated functions are not fully understood. Considering solely glycosylation of proteins, a highly sophisticated enzymatic machinery exists in eukaryotic systems (e.g., human mucin-type O-glycosylation alone involves 20 different isoenzymes) (Hassan et al. 2000; Ten Hagen et al. 2003), and the addition of carbohydrates is ideally suited to generate diversity of biological molecules. Indeed, the coding capacity of oligosaccharides is several orders of magnitude larger than that of oligonucleotides or peptides. This arises from their inherent configurational variability, created by two anomeric configurations (α/β), different linkages (1–2, 1–3, 1–4, 1–6), different ring sizes (pyranose/furanose), as well as branching and additional modification through acetylation, phosphorylation, and sulfation (Herget et al. 2008; Marino et al. 2010; Richards and Lowary 2009; Varki 2009). Although such complexity seems daunting, there is a growing appreciation for the need to decipher "sugar-codes" and elucidate glycan shape and dynamics, as well as specific features pivotal to their specific recognition. To that end, investigating protein–carbohydrate interactions at the atomic level is becoming a necessity.

Nuclear magnetic resonance (NMR) spectroscopy is a powerful technique to study protein–glycan recognition. Different from the detailed atomic picture of molecules arranged in a crystalline array that is provided by x-ray crystallography (Chapter 1), NMR spectroscopy yields structural and dynamic information about molecules in solution. Both methods, naturally, have advantages and disadvantages. The first and most unpredictable requirement for any crystallographic study is the necessity to grow well-diffracting single crystals of the complexes. This is a difficult process and for protein–glycan complexes, weak binding and, sometimes, the transient nature of the encounter further confounds this task. In contrast, NMR spectroscopy is uniquely suited for such studies, since it is possible to characterize both strong and weak interactions, even those involving transient complexes. On the other hand, solution NMR spectroscopy is restricted with regard to molecular size of a complex: large systems tumble slowly in solution, resulting in fast relaxation and hence large linewidths of resonances. In addition, the extensive number of resonances in macromolecular complexes causes severe overlap of signals, making spectral analysis difficult at best and impossible at worse. However, solution NMR spectroscopic methods are still advancing and innovative isotopic labeling strategies (Gardner and Kay 1998), sophisticated spin manipulations (Pervushin et al. 1997, 2000; Riek et al. 2000; Tjandra and Bax 1997; Tjandra et al. 1997), and cryogenic probe technology all contribute to recent progress.

This chapter describes the application of NMR spectroscopy to the study of intermolecular protein–glycan interactions with a focus on lectin–carbohydrate systems. Since such systems involve noncovalent binding between two molecules it is opportune to introduce and define the three different NMR regimes that can be encountered: (1) fast exchange, (2) intermediate exchange, and (3) slow exchange on the chemical shift scale. The parameters that define these regimes are the exchange rate between bound and free molecules and the difference in resonance frequencies for atoms in the bound and free states. Appreciating these different scenarios is critical when considering a particular NMR method, given that certain methodologies are more suited to particular exchange regimes.

In the following, we describe four major approaches that are frequently used to characterize intermolecular interactions by NMR (Zuiderweg 2004; Clarkson and

Campbell 2003; Haselhorst et al. 2009; Mayer and Meyer 1999, 2001; Jimenez-Barbero and Peters 2003; Siebert et al. 2003; Burgering et al. 1993; Folkers et al. 1993; Zwahlen et al. 1997). These include chemical shift perturbation studies, saturation transfer difference NMR (STD-NMR), transferred nuclear Overhauser effect (TrNOE), and standard nuclear Overhauser effect spectroscopy (NOESY) experiments. The description of the methods will be illustrated with examples from our own and other laboratories. The different approaches can involve detection via the glycan resonances (ligand-based method) or via protein resonances (protein-based method). TrNOE and STD-NMR are ligand-based methods, while chemical shift perturbation mapping and NOESY involve mainly protein resonances, although the latter can also monitor glycan resonances. Indeed, the three-dimensional (3D) structure determination of a protein–carbohydrate complex will encompass extensive NOESY analysis of edited/filtered experiments, involving protein and ligand resonances. In addition to the above four approaches, we will also briefly discuss WaterLOGSY or diffusion filtering experiments.

2.2 EXCHANGE RATE AND NMR CHEMICAL SHIFT DIFFERENCE

For any binding reaction, a molecule exchanges between free and bound states with a rate k (expressed in per second). This implies that any nucleus in the molecule will exist in two magnetically distinct environments, corresponding to the free and bound states. The resonance frequencies for the bound and free molecules are different, with $\Delta\nu$ representing the chemical shift difference (expressed in Hz) (Figure 2.1). In the fast exchange regime, $k \gg \Delta\nu$, the resonance will be a population-weighted average signal between free and bound resonances. This allows one to follow the binding event using the chemical shift perturbation during titration experiments. The binding affinity K_d for the protein–glycan interaction can be obtained by plotting the average chemical shift change, $\Delta\delta$, as a function of the carbohydrate/protein molar

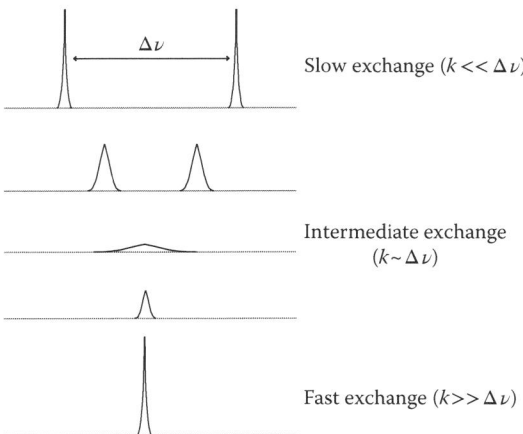

FIGURE 2.1 NMR chemical shift exchange. The three possible regimes are slow, intermediate, and fast for a given molecule between free and bound states in a binding reaction.

ratio. In the slow exchange regime, $k \ll \Delta\nu$, two distinct resonances are observed, corresponding to the free and bound states. For increasing concentrations of the carbohydrate, the intensities of the protein's free signals decrease while those of the protein's bound form increase. The most unfavorable case occurs when the exchange rate is comparable to the frequency difference between the two states, $k \sim \Delta\nu$. In this intermediate exchange regime, extensive line broadening, sometimes beyond detection, is observed. Occasionally, it may be possible to escape from this undesirable intermediate exchange regime by changing temperature (although NMR can only cover a limited temperature range), by recording spectra at different field strengths (since the frequency difference between the two signals depends on the magnetic field strength), using different buffer conditions, such as pH, or salt concentration, all with the aim of pushing the exchange into either the fast or the slow exchange regime.

2.3   CHEMICAL SHIFT PERTURBATION STUDIES

Chemical shift perturbation mapping obtained via NMR titration experiments is a sensitive and relatively straightforward NMR approach to identify those residues in the protein that are involved in carbohydrate binding (Zuiderweg 2004; Clarkson and Campbell 2003). This protein-based detection method can be employed in all three NMR exchange regimes. It relies on the fact that the resonance frequency, or the chemical shift of an individual nucleus, is very sensitive to its conformational, chemical and electronic environment. Upon complex formation, the carbohydrate modifies the environment of the protein nuclei that reside at the interface in the complex. As a result, atoms involved in binding will exhibit a change in their chemical shifts.

Chemical shift changes in titration experiments can be monitored with 1D ^{1}H or two-dimensional (2D) spectra, with the 2D ^{1}H–^{15}N HSQC experiment probably most widely used for this purpose. Reasons for this are manifold: first, only ^{15}N-labeling of the protein is required, and this is easily and inexpensively achieved using *Escherichia coli* expression systems. Second, only backbone assignments are necessary, avoiding time-consuming side-chain-focused experiments. Finally, ^{1}H and ^{15}N amide chemical shifts are exquisitely sensitive to any environmental changes, permitting the detection of even small differences.

For chemical shift perturbation mapping, a series of NMR spectra of the lectin, first without and then with increasing amounts of the carbohydrate ligand, are recorded, keeping the concentration of the protein constant. Five to ten different sugar concentrations are generally required for a satisfactory titration curve and for accurately determining the dissociation constant, K_d. Using such experiments, it is possible to identify the residues that constitute the binding site(s), identify the number of binding sites, determine binding affinities, and, if a structural model exists, delineate the binding sites in the 3D structure.

2.3.1   CHEMICAL SHIFT MAPPING IN THE FAST EXCHANGE REGIME: MANα(1-2)MAN BINDING TO LKAMG

LKAMG is a designed, chimeric protein that was created by combining domain A of TbCVNH and domain B of NcCVNH, with both proteins belonging to the CV-N

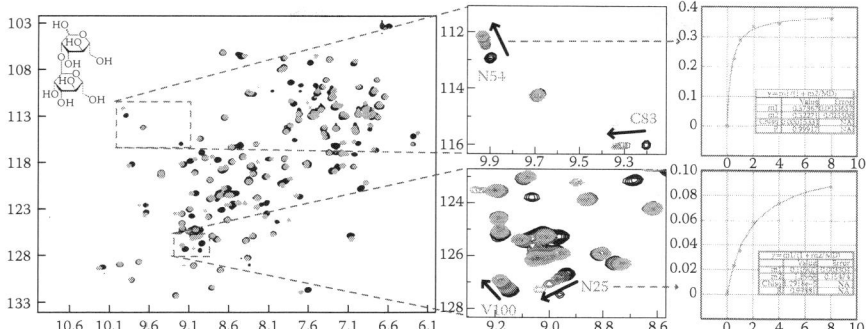

FIGURE 2.2   (See color insert.) Superposition of the ${}^{1}H$-${}^{15}N$ 2D HSQC spectra of free (black) and Manα(1-2)Man-bound (cyan) LKAMG. Two expanded regions, highlighting several shifting resonances, are shown in the middle panels. Binding is in fast exchange, and titration curves for residues N25 and N54 for increasing amounts of ligand are provided in the right-hand side panels.

homolog (CVNH) family (Koharudin et al. 2008, 2009). Similar to the parental proteins, LKAMG binds Manα(1-2)Man on both domains, A and B. Manα(1-2)Man binding to LKAMG was followed in NMR titration experiments using ${}^{15}N$-labeled protein (0.1 mM) in 20 mM Na phosphate buffer, 0.02% NaN_3, 90/10% H_2O/D_2O (pH 6.0) at 25°C at 700 MHz. A series of spectra for protein to sugar molar ratios of 1:0 (free), 1:0.5, 1:1.0, 1:2.0, 1:4.0, and 1:8.0 were recorded.

As illustrated in Figure 2.2, binding of Manα(1-2)Man to LKAMG is in fast exchange. The two binding sites identified via the titration experiment are located at opposite sides of the protein. From the titration curves for resonances that exhibited sizable, saturable shifts and no resonance overlap, apparent K_d values of 1.8 ± 0.2 and 0.3 ± 0.1 mM were determined for the sites on domains A and B, respectively. This illustrates that even extremely weak binding with millimolar binding constants can be detected and characterized by NMR.

2.3.2   Chemical shift mapping in the slow exchange regime: hexaacetyl chitohexaose binding to MoCVNH-LysM

The LysM domain is a carbohydrate-binding module that recognizes the *N*-acetylglucosamine (*N*-GlcNAc) building blocks of complex sugars, as found in chitin and peptidoglycan. Chitin is a fungal cell-wall polymer of *N*-GlcNAc units connected by β1,4-linkages, and peptidoglycans, a bacterial cell-wall polymer, contains alternating β1,4-linked *N*-GlcNAc and *N*-acetylmuramic acid (*N*-MurNAc) units (Buist et al. 2008). The LysM domain is present in a large number of proteins in higher eukaryotes, including plants and humans (de Jonge and Thomma 2009; Ponting et al. 1999). For example, in fungi alone, more than 400 putative LysM domain-containing proteins have been identified, with most of them solely containing strings of LysMs (Buist et al. 2008).

We investigated chitin binding by the LysM domain of MoCVNH-LysM, a type-III CVNH embedded within the hypothetical MGG_03307 protein from *Magnaporthe oryzae* (Koharudin et al. 2011). The binding of hexaacetyl chitohexaose, an oligomer of six *N*-GlcNAc units (*N*-GlcNAc$_6$), and MoCVNH-LysM was followed by NMR titration experiments using ^{15}N-labeled protein. Increasing amounts of *N*-GlcNAc$_6$ were added to the protein and spectra were recorded for each addition of sugar at protein:*N*-GlcNAc$_6$ molar ratios of 1:0 (free protein), 1:0.22, 1:0.45, 1:0.89, and 1:2.4. 2D ^{1}H-^{15}N HSQC spectra for the free and *N*-GlcNAc$_6$-bound states are provided in Figure 2.3. Binding is in slow exchange; that is, new bound resonances appear and increase in intensity, while the free resonances get smaller and disappear (Figure 2.3, left insets). Binding isotherms were plotted on the basis of peak intensities of four bound resonances (right inset), yielding an apparent K_d value of 21 ± 4 µM. Mapping of residues with strongly affected resonances after the addition of *N*-GlcNAc$_6$ revealed that the binding site in the LysM domain comprises the loop region between helix α1 and strand β1, helix α1, and the loop region between helix α2 and strand β2.

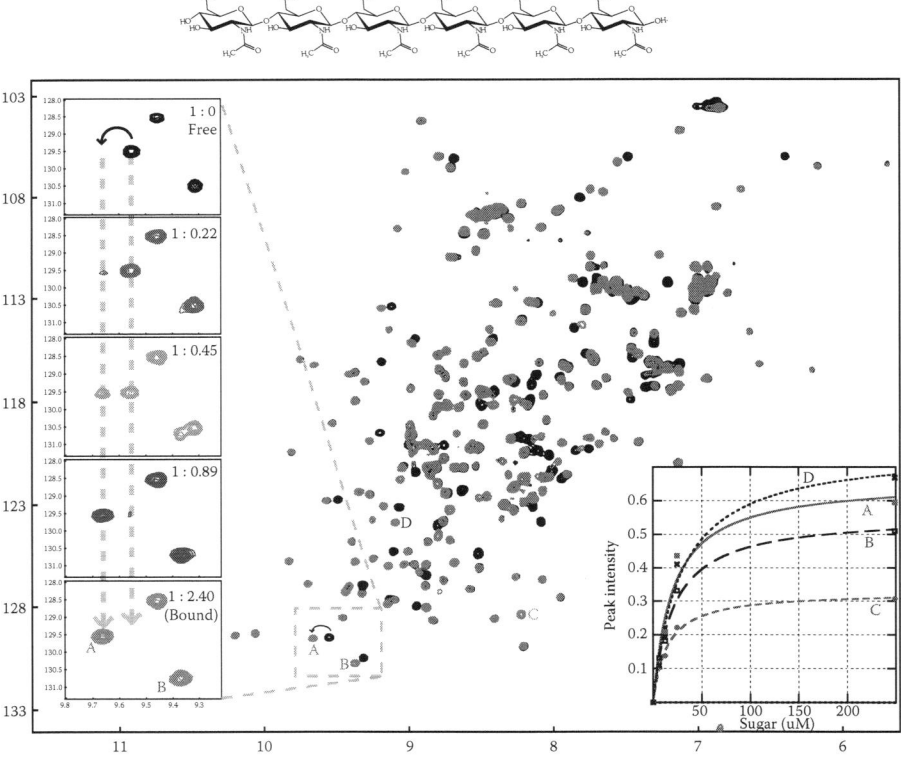

FIGURE 2.3 **(See color insert.)** Superposition of the 2D ^{1}H-^{15}N HSQC spectra of free (black) and hexaacetyl chitohexaose-bound (magenta) MoCVNH-LysM. Expanded regions highlighting an interacting residue in the slow exchange regime on the NMR timescale for a series of glycan additions are shown in the inset panels at the left-hand side. Binding isotherms for several resonances are provided in the inset at the bottom right.

2.4   SATURATION TRANSFER DIFFERENCE NMR

STD-NMR is a ligand-based detection method for determining the binding epitope on the ligand (Haselhorst et al. 2009; Mayer and Meyer 1999, 2001). STD-NMR is not limited to a particular protein-to-ligand ratio since background signal intensity that does not result from the saturation transfer is eliminated by difference spectroscopy. In practice, a protein-to-carbohydrate ratio of greater than 1:50 is employed to ensure sizable STD-NMR signals. The large molar excess of ligand over the protein minimizes the probability of carbohydrate that has already received saturation from reentering the protein binding site. Furthermore, STD-NMR can be applied over a wide range of K_d values, permitting the investigation of nanomolar to high millimolar binding. One prerequisite, however, needs to be fulfilled under all circumstances: binding has to be in the fast exchange regime. Indeed, high-affinity ligands that generally bind in the slow exchange regime reside too long in their protein-binding site(s) and are therefore not detectable by STD-NMR spectroscopy.

The basic principle of STD-NMR is depicted in Figure 2.4. In essence, protein resonances are saturated and magnetization transfer from protein protons via spin diffusion is allowed to proceed. In the *on* resonance experiment, saturation will propagate from the protein to the bound carbohydrate at the complex interface, with the closest sugar atoms receiving the largest magnetization transfer. The *off* resonance spectrum is the control spectrum, in which the same radio frequency power is applied at a frequency outside the protein's spectrum envelope. The difference spectrum between *on* and *off* experiments contains only resonances that experience the saturation transfer. The observed resonances in the difference spectrum will almost exclusively belong to the ligand, since the protein concentration is small, with protein resonances not above the noise level. If they are present, they can be eliminated by a relaxation filter experiment.

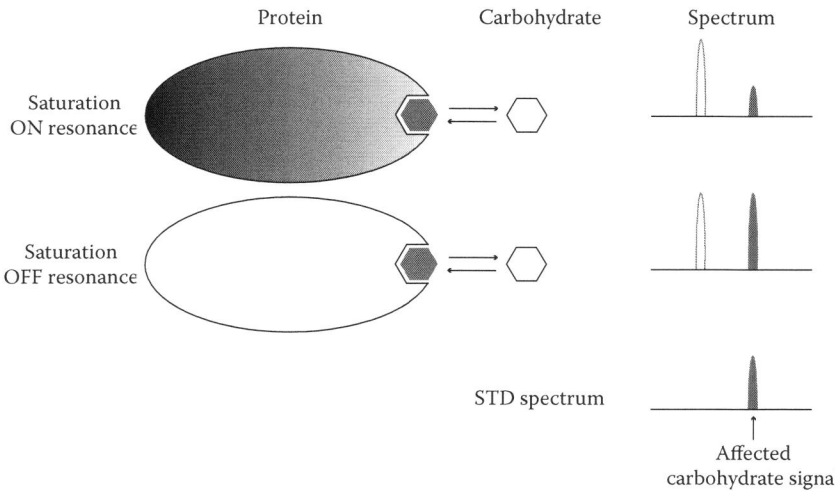

FIGURE 2.4   Illustration of the basic STD-NMR principle for determining the binding epitope on a ligand. A schematic difference spectrum between *on* and *off* experiments is also shown.

Setting up an STD-NMR experiment is relatively straightforward. Briefly, two samples are prepared; one containing the carbohydrate alone (Sample 1) and the other containing the carbohydrate in the presence of protein, with the carbohydrate in large excess over protein (Sample 2). For both samples, it is advisable to use D_2O as the solvent, thereby reducing the otherwise intense H_2O signal. The typical STD-NMR experiment is a 1D ^1H NMR spectrum. Before carrying out the STD-NMR experiments, standard 1D ^1H NMR spectra for both samples are recorded to select the optimal frequency positions for the *on* and *off* experiments. The *on* frequency has to be set to a value where only protein resonances are found and no carbohydrate signals are present. Typically, carbohydrates do not possess any resonances in the aromatic region (6–8 ppm), and this is usually a good region to induce protein saturation. The *off* frequency can be placed anywhere, as long as no protein signals or carbohydrate signals are close, for example, at ~30 ppm. Once both *on* and *off* frequencies have been selected, STD-NMR experiments are performed on Sample 2. Naturally, it is important to have complete assignments for the sugar spectrum. Saturation times of 1.5 to 2 s work generally well. However, for confident and complete epitope mapping, it is recommended to acquire STD-NMR spectra for several saturation times, ranging from 0.5 to 4.0 s. As controls, identical STD-NMR experiments are performed on Sample 1. In principle, perfect subtraction of the on- and off-saturation spectra for the control experiment should occur.

It should also be pointed out that temperature can have a significant effect on STD-NMR signal intensities. It is therefore advisable that a series of STD-NMR spectra at different temperatures is recorded. At lower temperatures, carbohydrates tend to have less conformational flexibility, resulting in stronger STD signals compared with spectra acquired at higher temperature (Groves et al. 2007; Haselhorst et al. 2004). Furthermore, since the quantification of the STD-NMR signal intensities can be influenced by relaxation time, correlation time, exchange rates, and the molecular topology of the bound ligand, it has been suggested that saturation times shorter than the T_1 relaxation time of the complex should be used (Yan et al. 2003).

2.4.1 Mapping of the CV-N-binding epitope on oligomannosides by STD-NMR

STD-NMR was initially employed to screen a carbohydrate library for wheat-germ agglutinin (WGA) binding (Mayer and Meyer 1999). Since then, STD-NMR has been applied in numerous studies to characterize protein–carbohydrate interactions (Benie et al. 2003; Haselhorst et al. 2004; Jayalakshmi et al. 2004). Here we present STD-NMR data from our own laboratory where we mapped the interaction between different oligomannosides and the HIV-inactivating protein Cyanovirin-N (CV-N) (Sandström et al. 2004).

We used a 42- and 100-fold excess of oligosaccharide over 78 µM CV-N. The STD-NMR spectra were acquired at two different temperatures (10°C and 25°C). It was noted that all STD-NMR intensities were larger at 10°C. The WATERGATE pulse sequence was used for water suppression (Piotto et al. 1992) and saturation of the protein magnetization was carried out using a pulse train of 40 Gaussians of 50 ms duration, each separated by a 1 ms delay. Spectra were acquired with *on* resonance frequencies set at 0.25, 0.4, 0.6, 1.3, 2, and 7 ppm, and the reference STD spectra were recorded with the *off* resonance set at 30 ppm. Saturation times of 0.5 and 1 to 6 s incremented in 1 s steps were used, and control experiments were conducted with oligosaccharide-only samples (at the same concentration as used in the presence of CV-N).

The results of this study established that of the three dimannoses, Manα(1-2)Manα(1-OMe), Manα(1-3)Manα(1-OMe), and Manα(1-6)Manα(1-OMe), only the Manα(1-2)Manα(1-OMe) disaccharide yielded large STD-NMR signals when mixed with CV-N (Figure 2.5). The largest STD effects were observed for the H2′, H3′,

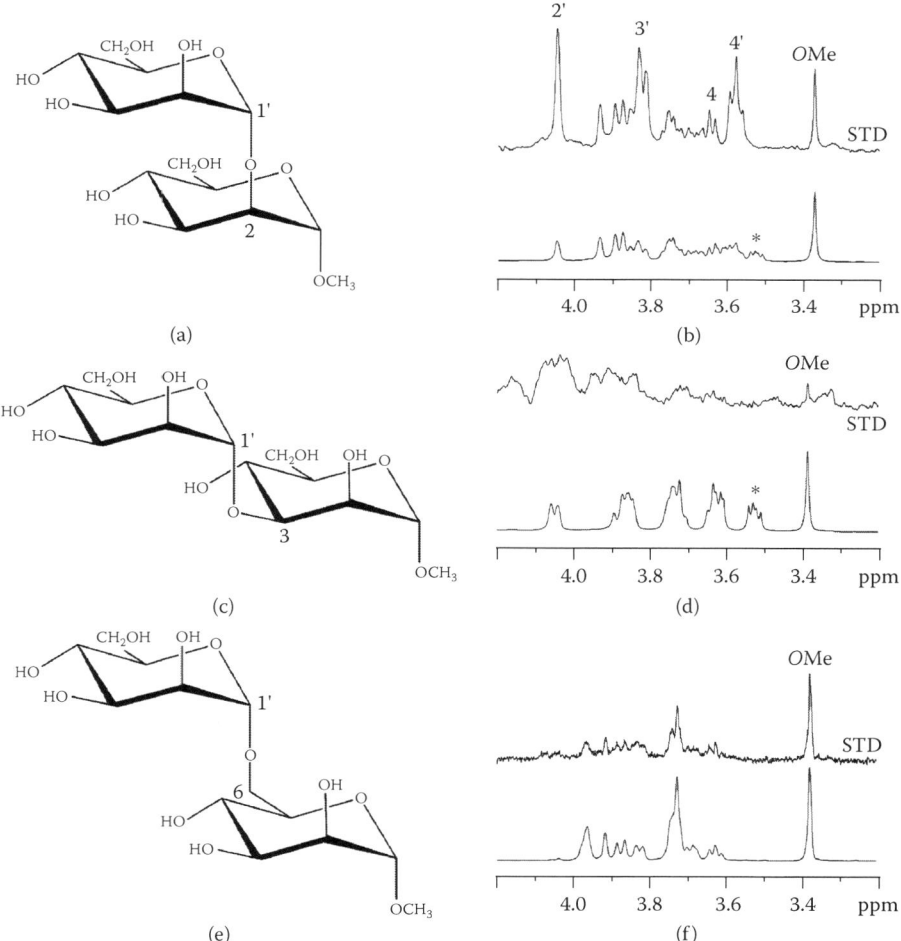

FIGURE 2.5   STD-NMR spectra for three different dimannoses in the presence of CV-N. (a and b) Only the Manα(1-2)Manα(1-OMe) disaccharide yielded large STD-NMR signals when binding to CV-N. Manα(1-3)Manα(1-OMe) (c) and Manα(1-6)Manα(1-OMe) (e) did not show any appreciable signals (d and f), demonstrating that no binding occurred. (Reprinted with permission from Sandström, C. et al., *Biochemistry*, 43, 13926–13931. Copyright 2004 American Chemical Society.)

and H4′protons at the nonreducing end of the Manα(1-2)Manα(1-OMe) disaccharide, indicating that these protons make the most intimate contact with CV-N. The absence of STD-NMR signals for the Manα(1-3)Manα(1-OMe) disaccharide and extremely weak signals for Manα(1-6)Manα(1-OMe) indicated that these two compounds did not interact significantly with CV-N.

2.5 TRANSFERRED NUCLEAR OVERHAUSER EFFECT

Like STD-NMR, the TrNOE method also uses ligand-based detection and allows determination of the ligand structure. This method has the remarkable advantage that very little protein is required and that the experiment works best for very large molecular mass systems. Therefore, no size limitation for proteins (e.g., lectins) exists. Furthermore, since the majority of protein–carbohydrate interactions are in μM to mM range, most systems are in fast exchange on the NMR timescale. Such low-affinity interactions render crystallographic studies extremely challenging, making the TrNOE experiment a very valuable method for investigating protein–carbohydrate complexes (Jiménez-Barbero and Peters 2003; Siebert et al. 2003).

The theoretical underpinnings of the NOE and TrNOE have been described previously (Balaram et al. 1972; Clore and Gronenborn 1983) and here only a brief summary will be provided. In the TrNOE experiment, exchange between the free and bound ligands occurs during the mixing time (illustrated in Figure 2.6) and the observed NOE depends on the weighted average of the cross-relaxation rates in the free and bound states. The experiment is set up such that the bound-state NOE is sizable and negative and after transfer to the free ligand can be easily measured on the free ligand spectrum. Translating the size of the NOEs into distances between the involved protons allows one to determine the bound conformation of the ligand. Since the molecular mass of the protein–ligand complex is usually large, TrNOE signals are generally negative. It also should be pointed out that the TrNOE experiment

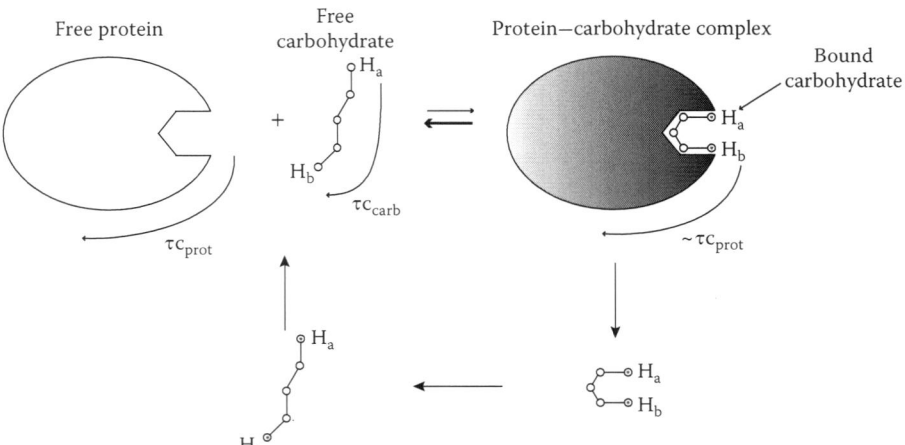

FIGURE 2.6   The basic principle of TrNOE. NOEs present between protons of the bound ligand are transferred to and recorded on the free ligand.

works extremely well for large systems and the larger the size of the complex, the easier it becomes.

The experimental setup for a TrNOE experiment is essentially identical to the common 2D NOESY experiment. The major difference pertains to the sample: for NOESY experiments, a reasonable protein concentration (>100 μM) is required, while in the TrNOE experiment, a large excess of ligand over protein is employed, permitting the use of very small protein concentrations. The essential prerequisite for the TrNOE experiment is suitable exchange kinetics: the off-rate of the ligand has to be fast. This is frequently the case for low-affinity systems (but not necessarily so). In practice, a 10- to 50-fold excess of ligand over the protein is used, resulting in an easily assignable spectrum since it mainly contains free ligand resonances. For carbohydrates, D_2O can be used as the solvent, since no exchangeable protons of the carbohydrate are monitored. Or, alternatively, the WATERGATE pulse sequence can be used for water suppression. In general, several 2D NOESY experiments will be recorded as controls and for evaluating the best conditions. 2D NOESY spectra of the free protein and the free carbohydrate are recorded, as well as several spectra of carbohydrate–protein mixtures at different carbohydrate/protein ratios. It should be pointed out that only the assignments for the free ligand (glycan) need to be available, since all NOEs will be measured on the free resonances (or correctly, the weighted average, but with a 100-fold excess these signals are essentially at the free position). In the ideal case, the free carbohydrate will exhibit positive or zero NOEs, while the carbohydrate/protein mixture will exhibit negative NOEs.

At this point, a few cautionary notes should be interjected. An important source of errors in TrNOE experiments is spin diffusion, and indirect NOEs, especially between very close protons, can lead to erroneous internuclear distances and therefore the TrNOE-derived conformation. In addition, it has to be ascertained that a single binding site for the ligand on the protein is involved; averaging between two or more sites will result in average TrNOEs and an "averaged" conformation that is physically meaningless. Several approaches are available to address these shortcomings, such as TrROE, MINSY, or QUIET experiments (Arepalli et al. 1995; Vincent et al. 1997).

2.5.1 THE CONFORMATION OF THE TRISACCHARIDE RHAα(1-2)[GLCNACβ(1-3)] RHAα(1-OPR) BOUND TO A MONOCLONAL ANTIBODY

Protein–carbohydrate interactions have been studied extensively by TrNOE experiments (for reviews, see Jiménez-Barbero et al. 1999; Peters and Pinto 1996). Here we present as an example the structure determination of the branched Rhaα(1-2) [GlcNAcβ(1-3)]Rhaα(1-OPr) trisaccharide of the cell-wall polysaccharide of Group A Streptococcus (GAS) when bound to a GAS-directed monoclonal antibody (Weimar et al. 1995). The experiments were carried out in D_2O solutions and the conditions were evaluated by titration of the Ab with carbohydrate, collecting 1D transient NOE experiments for each titration step. At the optimized carbohydrate/protein ratio of 14:1 (2.2 mM carbohydrate: 0.08 mM Ab), two 2D TrNOE experiments were recorded, one normal NOESY with pre-saturation for water suppression and a second one that contained a $T_{1\rho}$ filter in the NOESY sequence. In addition, a set of TrROESY experiments was also recorded to remove any TrNOE cross-peaks

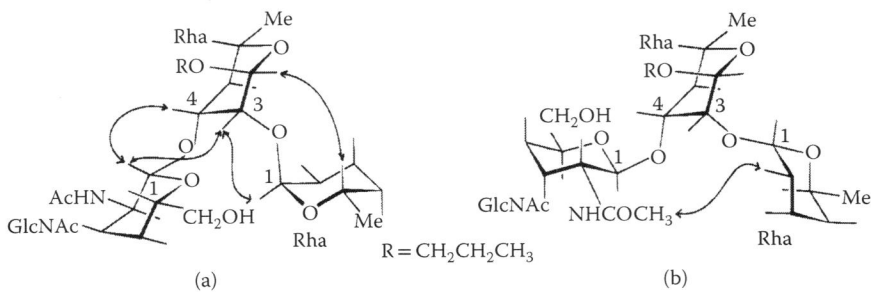

FIGURE 2.7   A comparison between (a) the bound Rhaα(1-2)[GlcNAcβ(1-3)]Rhaα(1-OPr) tri-saccharide conformation, calculated based on TrNOE data, and (b) the lowest energy structure from an ensemble of calculated free conformations. (Reprinted with permission from Weimar, T. et al., *Biochemistry*, 34, 13672–13681. Copyright 1995 American Chemical Society.)

originating from spin diffusion. The final, unambiguous TrNOE signals were converted into distances and used as constraints to determine the conformation of the trisaccharide by simulated annealing.

The bound and free trisaccharide structures are depicted in Figure 2.7. Interestingly, the bound conformation of the trisaccharide is of slightly higher energy (~2 kcal mol^{-1}) and exhibits a different torsion angle around the Rhaα(1-2)Rha linkage compared to the conformation at the global minimum.

2.6   DETERMINATION OF CARBOHYDRATE–LECTIN COMPLEX STRUCTURES BY STANDARD NOESY EXPERIMENTS

In order to obtain a complete structural description of a protein–carbohydrate complex at the atomic level, standard NOESY experiments are conducted to determine the conformation of the bound glycan in the protein. Interatomic distances for as many protons as possible are measured and translated into distances based on the well-known relationship $I_{NOE} \approx \ < 1/r^6 > f(\tau_c)$. The factor $f(\tau_c)$ describes any modulation of the dipole–dipole interaction by stochastic rate processes with an effective correlation time τ_c due to global and local motions of the protein.

Both intramolecular and intermolecular NOEs are required to define the 3D structure of the complex. Both types of NOEs will be present in homonuclear 2D or 3D NOESY spectra and their unambiguous identification is difficult and time-consuming. Therefore, in practice, the proteins are generally ^{15}N or ^{13}C labeled, and protein-only NOEs are identified in heteronuclear separated NOESY experiments. Intermolecular NOEs across the binding interface are derived by combining heteronuclear-filtered and/or heteronuclear-separated NOESY experiments (Burgering et al. 1993; Folkers et al. 1993; Zwahlen et al. 1997).

2.6.1   THE STRUCTURE OF THE CV-N–MANα(1-2)MAN COMPLEX

The solution structure of the complex between Cyanovirin-N (CV-N) and Manα(1-2) Man was determined using standard NOESY experiments (Bewley 2001). CV-N is a

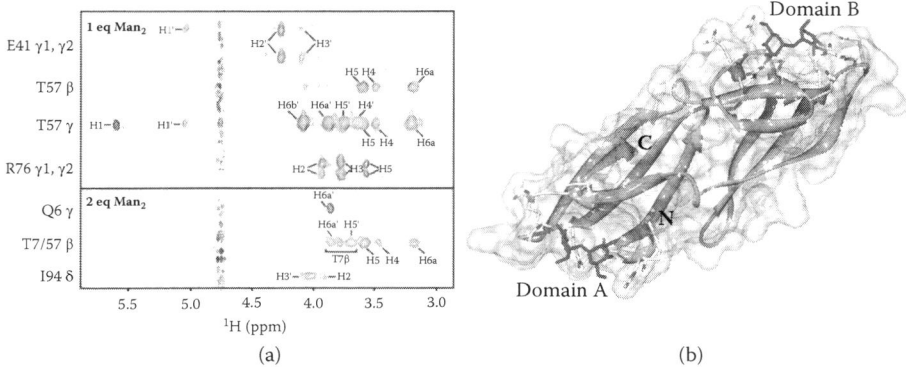

FIGURE 2.8   (See color insert.) (a) Selected intermolecular NOEs between CV-N side chains of E41 Hγ1/γ2, T57 Hβ, T57 Hγ, R76 Hγ1/γ2, Q6 Hγ, T57 Hβ, and I94 Hδ and Manα(1–2) Man. (b) The calculated bound conformation of the dimannose in the protein binding sites. (Reprinted from *Structure*, 9, Bewley, C. A., Solution structure of a cyanovirin-N:Man alpha 1–2Man alpha complex: Structural basis for high-affinity carbohydrate-mediated binding to gp120, pp. 931–940, Copyright 2001, with permission from Elsevier.)

small (11 kDa) virucidal lectin originally identified in aqueous extracts from the cyanobacterium *Nostoc ellipsosporum* during screening for anti-HIV activities (Boyd et al. 1997). The sequence of CV-N comprises two tandem repeats, each containing ~50 amino acids and two pairs of disulfide-bonded cysteines (Bewley et al. 1998; Yang et al. 1999). NMR titration experiments with Manα(1-2)Man (or Man2α) demonstrated the presence of two binding sites, which mapped to opposite ends of the molecule (Barrientos and Gronenborn 2002; Botos et al. 2002; Shenoy et al. 2002).

The solution structure of the CV-N–Manα(1-2)Man complex was solved using ^{15}N, ^{13}C-labeled protein and carbohydrate at 1:1 and 1:2 molar ratios. Intramolecular NOEs for the protein alone and for Manα(1-2)Man in the complex were extracted from 3D ^{15}N-separated and ^{12}C-filtered NOESY experiments, respectively. The intermolecular NOEs between protein and Manα(1-2)Man were obtained using 3D ^{12}C-filtered/^{13}C-separated NOESY.

The structure of the complex was determined using conjoined rigid-body/torsion-angle dynamics (Schwieters and Clore 2001). In this procedure, backbone and non-interfacial side chains of CV-N are fixed, while the interfacial side chains of the protein are free to move, and Manα(1-2)Man is free to translate and rotate subject to the intermolecular restraints. A few examples of the intermolecular NOEs are provided in Figure 2.8a and the bound Manα(1-2)Man conformation in the complex is shown in Figure 2.8b.

2.7   OTHER NMR METHODS

In addition to the four major NMR methods described previously, other NMR techniques have also been developed to monitor interactions between protein and carbohydrate. These include approaches using magnetization transfer from bulk water to the carbohydrate via labile protein protons, commonly called WaterLOGSY

(Dalvit et al. 2000) and measuring changes in diffusion constants, termed as diffusion-edited screening (Hajduk et al. 1997). For details of these two experiments, the reader is referred to the respective references, and only a brief outline is provided here. The first method is basically a variant of STD-NMR in which saturation transfer involves bound water molecules instead of protein atoms. The second method exploits the fact that the decrease of carbohydrate resonance intensities when bound to protein is proportional to the rate of diffusion and thus the strength/length of the gradient pulse. Both methods can be employed to evaluate binding of ligands (sugars) to protein, but so far have not gained widespread popularity.

2.8 SUMMARY

NMR has been applied extensively to study important interactions in glycobiology and has provided invaluable information about sugar-lectin structure and function. The development of advanced pulse sequences, higher magnetic field magnets, and cryoprobe technology now permits NMR studies of interesting biological systems of large molecular sizes. The unique ability of NMR to address systems with weak and transient interactions, as well as tight complexes, renders it extremely powerful for studying protein–glycan systems.

Here, we have summarized those NMR approaches that are most often used in the glycobiology field. Naturally, a full and comprehensive description of these is beyond the scope of this chapter and we simply aimed to provide the basics that are at the heart of each method. The examples are provided to illustrate the nature of the information that can be extracted when using the different approaches and should peak the interest of glycobiologists who are not experts in NMR spectroscopy to delve further into NMR. To that end, a list of available NMR methods is provided in Table 2.1.

TABLE 2.1
NMR Methods Used for the Characterization of Protein–Carbohydrate Interactions

Scheme	Detection Site	Affinity Range	Exchange Regime	Carbohydrate to Protein Ratio	Protein Concentration (µM)	Protein Size (kDa)	Isotope Labeling
CSP	Protein	nM–mM	Fast and Slow	>1	>50	<40	Yes
TrNOE	Carbohydrate	µM–mM	Fast	10–50	<10	The larger, the better	No
STD-NMR	Carbohydrate	nM–mM	Fast	>50	<1	—	No
NOESY	Both	nM–µM	Slow	1–2	>500	<40	Yes
Water LOGSY	Carbohydrate	µM–mM	Fast	10–50	1–10	—	No
Diffusion	Carbohydrate	µM–mM	Fast	>5	50–100	<40	No

REFERENCES

Arepalli, S. R., Glaudemans, C. P., Daves, G. D., Jr., Kovac, P., and Bax, A. 1995. Identification of protein-mediated indirect NOE effects in a disaccharide-Fab′ complex by transferred ROESY. *J Magn Reson B*, 106, 195–198.

Balaram, P., Bothner-By, A. A., and Dadok, J. 1972. Negative nuclear Overhauser effects as probes of macromolecular structure. *J Am Chem Soc*, 94, 4015–4017.

Barrientos, L. G. and Gronenborn, A. M. 2002. The domain-swapped dimer of cyanovirin-N contains two sets of oligosaccharide binding sites in solution. *Biochem Biophys Res Commun*, 298, 598–602.

Benie, A. J., Moser, R., Bauml, E., Blaas, D., and Peters, T. 2003. Virus-ligand interactions: identification and characterization of ligand binding by NMR spectroscopy. *J Am Chem Soc*, 125, 14–15.

Bewley, C. A. 2001. Solution structure of a cyanovirin-N:Man alpha 1–2Man alpha complex: structural basis for high-affinity carbohydrate-mediated binding to gp120. *Structure*, 9, 931–940.

Bewley, C. A., Gustafson, K. R., Boyd, M. R. et al. 1998. Solution structure of cyanovirin-N, a potent HIV-inactivating protein. *Nat Struct Biol*, 5, 571–578.

Botos, I., O'Keefe, B. R., Shenoy, S. R. et al. 2002. Structures of the complexes of a potent anti-HIV protein cyanovirin-N and high mannose oligosaccharides. *J Biol Chem*, 277, 34336–34342.

Boyd, M. R., Gustafson, K. R., McMahon, J. B. et al. 1997. Discovery of cyanovirin-N, a novel human immunodeficiency virus-inactivating protein that binds viral surface envelope glycoprotein gp120: potential applications to microbicide development. *Antimicrob Agents Chemother*, 41, 1521–1530.

Buist, G., Steen, A., Kok, J., and Kuipers, O. P. 2008. LysM, a widely distributed protein motif for binding to (peptido)glycans. *Mol Microbiol*, 68, 838–847.

Burgering, M. J., Boelens, R., Caffrey, M., Breg, J. N., and Kaptein, R. 1993. Observation of inter-subunit nuclear Overhauser effects in a dimeric protein. Application to the Arc repressor. *FEBS Lett*, 330, 105–109.

Clarkson, J. and Campbell, I. D. 2003. Studies of protein–ligand interactions by NMR. *Biochem Soc Trans*, 31, 1006–1009.

Clore, G. M. and Gronenborn, A. M. 1983. Theory of the time-dependent transferred nuclear Overhauser effect: applications to structural analysis of ligand protein complexes in solution. *J Magn Reson*, 53, 423–442.

Dalvit, C., Pevarello, P., Tato, M. et al. 2000. Identification of compounds with binding affinity to proteins via magnetization transfer from bulk water. *J Biomol NMR*, 18, 65–68.

de Jonge, R. and Thomma, B. P. 2009. Fungal LysM effectors: extinguishers of host immunity? *Trends Microbiol*, 17, 151–157.

Folkers, P. J. M., Folmer, R. H. A., Konings, R. N. H., and Hilbers, C. W. 1993. Overcoming the ambiguity problem encountered in the analysis of nuclear Overhauser magnetic resonance spectra of symmetrical dimer proteins. *J Am Chem Soc*, 115, 3798–3799.

Gardner, K. H. and Kay, L. E. 1998. The use of 2H, 13C, 15N multidimensional NMR to study the structure and dynamics of proteins. *Annu Rev Biophys Biomol Struct*, 27, 357–406.

Groves, P., Kover, K. E., Andre, S. et al. 2007. Temperature dependence of ligand-protein complex formation as reflected by saturation transfer difference NMR experiments. *Magn Reson Chem*, 45, 745–748.

Hajduk, P. J., Meadows, R. P., and Fesik, S. W. 1997. Discovering high-affinity ligands for proteins. *Science*, 278, 497–499.

Haselhorst, T., Lamerz, A.-C., and von Itzstein, M. 2009. Saturation transfer difference NMR spectroscopy as a technique to investigate protein–carbohydrate interactions in solution.

In *Methods in Molecular Biology, Glycomics: Methods and Protocols*, Packer, N. H. and Karlsson, N. G. (eds), Totowa, NJ: Humana Press.

Haselhorst, T., Wilson, J. C., Thomson, R. J. et al. 2004. Saturation transfer difference (STD) 1H-NMR experiments and in silico docking experiments to probe the binding of N-acetylneuraminic acid and derivatives to Vibrio cholerae sialidase. *Proteins*, 56, 346–353.

Hassan, H., Bennett, E. P., Mandel, U., Hollingsworth, M. A., and Clausen, H. 2000. O-glycan occupancy is directed by substrate specificities of polypeptide GalNAc-transferases. In *Carbohydrates in Chemistry and Biology*, Ernst, B., Hart, B. W. and Sina, P. (eds), New York: Wiley-VCH.

Herget, S., Toukach, P. V., Ranzinger, R. et al. 2008. Statistical analysis of the Bacterial Carbohydrate Structure Data Base (BCSDB): characteristics and diversity of bacterial carbohydrates in comparison with mammalian glycans. *BMC Struct Biol*, 8, 35.

Jayalakshmi, V., Biet, T., Peters, T., and Krishna, N. R. 2004. Refinement of the conformation of UDP-galactose bound to galactosyltransferase using the STD NMR intensity-restrained CORCEMA optimization. *J Am Chem Soc*, 126, 8610–8611.

Jiménez-Barbero, J., Asensio, J. L., Canada, F. J., and Poveda, A. 1999. Free and protein-bound carbohydrate structures. *Curr Opin Struct Biol*, 9, 549–555.

Jiménez-Barbero, J. and Peters, T. 2003. TR-NOE experiments to study carbohydrate-protein interactions. In *NMR Spectroscopy of Glycoconjugates*, Jiménez-Barbero, J. and Peters, T. (eds), Weinheim, FRG: Wiley-VCH Verlag GmbH & Co. KGaA.

Koharudin, L. M. I., Furey, W., and Gronenborn, A. M. 2009. A designed chimeric cyanovirin-N homolog lectin: structure and molecular basis of sucrose binding. *Proteins*, 77, 904–915.

Koharudin, L. M. I., Viscomi, A. R., Jee, J. G., Ottonello, S., and Gronenborn, A. M. 2008. The evolutionarily conserved family of cyanovirin-N homologs: structures and carbohydrate specificity. *Structure*, 16, 570–584.

Koharudin, L. M. I., Viscomi, A. R., Montanini, B. et al. 2011. Structure-function analysis of a CVNH-LysM lectin expressed during plant infection by the rice blast fungus Magnaporthe oryzae. *Structure*, 19, 662–674.

Marino, K., Bones, J., Kattla, J. J., and Rudd, P. M. 2010. A systematic approach to protein glycosylation analysis: a path through the maze. *Nat Chem Biol*, 6, 713–723.

Mayer, M. and Meyer, B. 1999. Characterization of ligand binding by saturation transfer difference NMR spectroscopy. *Angew Chem Int Ed Engl*, 38, 1784–1788.

Mayer, M. and Meyer, B. 2001. Group epitope mapping by saturation transfer difference NMR to identify segments of a ligand in direct contact with a protein receptor. *J Am Chem Soc*, 123, 6108–6117.

Pervushin, K., Fernandez, C., Riek, R. et al. 2000. Determination of h2J(NN) and h1J(HN) coupling constants across Watson-Crick base pairs in the Antennapedia homeodomain-DNA complex using TROSY. *J Biomol NMR*, 16, 39–46.

Pervushin, K., Riek, R., Wider, G., and Wuthrich, K. 1997. Attenuated T2 relaxation by mutual cancellation of dipole-dipole coupling and chemical shift anisotropy indicates an avenue to NMR structures of very large biological macromolecules in solution. *Proc Natl Acad Sci USA*, 94, 12366–12371.

Peters, T. and Pinto, B. M. 1996. Structure and dynamics of oligosaccharides: NMR and modeling studies. *Curr Opin Struct Biol*, 6, 710–720.

Piotto, M., Saudek, V. and Sklenar, V. 1992. Gradient-tailored excitation for single-quantum NMR spectroscopy of aqueous solutions. *J Biomol NMR*, 2, 661–665.

Ponting, C. P., Aravind, L., Schultz, J., Bork, P., and Koonin, E. V. 1999. Eukaryotic signalling domain homologues in archaea and bacteria. Ancient ancestry and horizontal gene transfer. *J Mol Biol*, 289, 729–745.

Richards, M. R. and Lowary, T. L. 2009. Chemistry and biology of galactofuranose-containing polysaccharides. *Chembiochem*, 10, 1920–1938.

Riek, R., Pervushin, K., and Wuthrich, K. 2000. TROSY and CRINEPT: NMR with large molecular and supramolecular structures in solution. *Trends Biochem Sci*, 25, 462–468.

Sandström, C., Berteau, O., Gemma, E. et al. 2004. Atomic mapping of the interactions between the antiviral agent cyanovirin-N and oligomannosides by saturation-transfer difference NMR. *Biochemistry*, 43, 13926–13931.

Schwieters, C. D. and Clore, G. M. 2001. Internal coordinates for molecular dynamics and minimization in structure determination and refinement. *J Magn Reson*, 152, 288–302.

Sharon, N. 2008. Lectins: past, present and future. *Biochem Soc Trans*, 36, 1457–1460.

Sharon, N. and Lis, H. 1989. Lectins as cell recognition molecules. *Science*, 246, 227–234.

Sharon, N. and Lis, H. 2004. History of lectins: from hemagglutinins to biological recognition molecules. *Glycobiology*, 14, 53R–62R.

Shenoy, S. R., Barrientos, L. G., Ratner, D. M. et al. 2002. Multisite and multivalent binding between cyanovirin-N and branched oligomannosides: calorimetric and NMR characterization. *Chem Biol*, 9, 1109–1118.

Siebert, H. C., Jimenez-Barbero, J., Andre, S., Kaltner, H. and Gabius, H. J. 2003. Describing topology of bound ligand by transferred nuclear Overhauser effect spectroscopy and molecular modeling. *Methods Enzymol*, 362, 417–434.

Ten Hagen, K. G., Fritz, T. A., and Tabak, L. A. 2003. All in the family: the UDP-GalNAc:polypeptide N-acetylgalactosaminyltransferases. *Glycobiology*, 13, 1R–16R.

Tjandra, N. and Bax, A. 1997. Direct measurement of distances and angles in biomolecules by NMR in a dilute liquid crystalline medium. *Science*, 278, 1111–1114.

Tjandra, N., Garrett, D. S., Gronenborn, A. M., Bax, A., and Clore, G. M. 1997. Defining long range order in NMR structure determination from the dependence of heteronuclear relaxation times on rotational diffusion anisotropy. *Nat Struct Biol*, 4, 443–449.

Varki, A. 2009. Multiple changes in sialic acid biology during human evolution. *Glycoconj J*, 26, 231–245.

Vincent, S. J., Zwahlen, C., Post, C. B., Burgner, J. W., and Bodenhausen, G. 1997. The conformation of NAD+ bound to lactate dehydrogenase determined by nuclear magnetic resonance with suppression of spin diffusion. *Proc Natl Acad Sci USA*, 94, 4383–4388.

Weimar, T., Harris, S. L., Pitner, J. B., Bock, K., and Pinto, B. M. 1995. Transferred nuclear Overhauser enhancement experiments show that the monoclonal antibody strep 9 selects a local minimum conformation of a Streptococcus group A trisaccharide-hapten. *Biochemistry*, 34, 13672–13681.

Yan, J., Kline, A. D., Mo, H., Shapiro, M. J., and Zartler, E. R. 2003. The effect of relaxation on the epitope mapping by saturation transfer difference NMR. *J Magn Reson*, 163, 270–276.

Yang, F., Bewley, C. A., Louis, J. M. et al. 1999. Crystal structure of cyanovirin-N, a potent HIV-inactivating protein, shows unexpected domain swapping. *J Mol Biol*, 288, 403–412.

Zuiderweg, E. R. P. 2002. Mapping protein–protein interactions in solution by NMR spectroscopy. *Biochemistry*, 41, 1–7.

Zwahlen, C., Legault, P., Vincent, S. J. F. et al. 1997. Methods for measurement of intermolecular NOEs by multinuclear NMR spectrosocpy: application to a bacteriophage lambda N-peptide/boxB RNA complex. *J Am Chem Soc*, 119, 6711–6721.

3 Electron Cryomicroscopy of Large Glycoprotein Assemblies

Tzviya Zeev-Ben-Mordehai and Kay Grünewald

CONTENTS

3.1 INTRODUCTION

In recent years, the applications of electron cryomicroscopy (cryoEM) for structural studies of macromolecules were greatly enhanced. These applications now include the structural characterization of purified macromolecules, helical filaments, two-dimensional (2D) arrays, and even whole cells. All of this became possible due to the improvements in sample preparation, which preserve a close to native environment for the macromolecules (Dubochet et al. 1981), as well as developments of the microscopes including computer-controlled data acquisition and dedicated data processing (DeRosier and Klug 1968; Crowther et al. 1970a).

In this chapter, the different applications will be introduced and will be followed by recent examples of how they were applied for the study of glycoproteins.

3.2   EXPERIMENTAL BACKGROUND

3.2.1   Electron Cryomicroscopy

The essence of cryoEM is visualizing biological specimens embedded in a thin film of vitreous ice (noncrystalline, amorphous, glass-like; i.e., similar to a liquid with very high viscosity), the thickness of which is only slightly greater than the diameter of the specimen (Taylor and Glaeser 2008). Imaging in the frozen-hydrated state assures a close to native environment for the biological specimen; this is in clear distinction to classical electron microscopy (EM) methods that involve fixation, staining, dehydration, and plastic embedment of the specimen that frequently lead to distortion of the biological specimen.

Methods for preserving specimen hydration within the high vacuum of the electron microscope were initially developed to preserve crystalline order in protein crystals for structure determination at atomic resolution using electron crystallography. Kenneth Taylor and Robert Glaeser were the first to develop a device that allowed freezing crystals in thin films by plunging into liquid nitrogen (Taylor and Glaeser 1973, 1974, 1976). The method seemed successful for preserving the crystal lattice, however, not for freezing particle suspension. The main problem being that crystalline ice did not appear to propagate through the crystal lattice while crystalline hexagonal ice did form in the suspension.

The pioneering work of Jacques Dubochet and colleagues on vitrifying thin water films paved the way for the whole field of cryoEM (Dubochet et al. 1981; Adrian et al. 1984; Dubochet et al. 1988). The idea behind vitrification is to cool the liquid so rapidly that its molecules become immobilized before they have time to crystallize. Despite the previous widespread notion that vitrification of pure water or dilute solutions is fundamentally impossible (Johari 1977), Dubochet and McDowell were able to vitrify thin water layers by immersing them in liquid ethane (Dubochet and McDowall 1981). They were successful owing to their observation that cooling speed increases with decreasing sample size.

Ethane and propane are the preferred cryogens due to their high cooling conduction (Dubochet et al. 1988). Biological specimens are often prepared in buffered aqueous solutions with low salt and can thus be considered dilute solutions. To a first approximation, dilute solutions when vitrified can be treated as water and hence will be referred to as such in all subsequent discussions. In practice, the sample is applied onto an EM grid coated with a thin perforated carbon film (Figure 3.1). Then, the excess of liquid is blotted away with a filter paper resulting in a thin water layer with suspended particles spanning the carbon holes. The thin film of water is then instantly plunged into liquid ethane cooled by liquid nitrogen to a temperature of approximately −180°C (93 K) (Figure 3.1). This approach works well for biological specimens <10 μm in thickness. High cooling rates for thicker specimens (up to few hundred μm) are achieved by high-pressure freezing (Moor 1987; Studer et al. 2008; McDonald et al. 2010). For this, high pressure is applied to the sample while a jet of liquid nitrogen cools it rapidly. The application of the high pressure impedes the formation of crystalline ice by preventing volume expansion. Phase transitions from vitreous water to cubic ice and from cubic ice to hexagonal ice are triggered

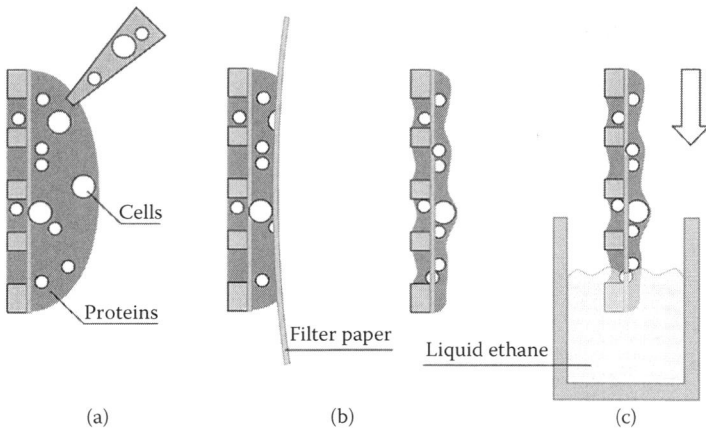

FIGURE 3.1 Vitrification of thin suspension of particles by plunge freezing. (a) The sample is applied to an EM grid coated with a thin perforated carbon film. (b) The excess of liquid is blotted away with a filter paper resulting in a thin aqueous film spanning the carbon holes. (c) This film is vitrified by rapid plunging into liquid ethane that is cooled by liquid nitrogen. (With kind permission from Springer Science+Business Media: *Science of Microscopy*, Cryo-electron tomography (cet), 2006, pp. 535–604, Plitzko, J. M. and Baumeister, W.)

by increased temperature. To avoid devitrification, samples need to be kept at low temperature ($<-135°C$ [138 K]) at all times following the freezing (Dubochet 2007). Important instrumental developments, namely the construction of a cryotransfer mechanism and high-resolution cryostages now allow the visualization of the sample at $-183°C$ (90 K) achieved by liquid nitrogen cooling or at between $-269°C$ (4 K) and $-261°C$ (12 K) achieved by liquid helium cooling (Fujiyoshi et al. 1991; Zemlin et al. 1996; van Heel et al. 2000).

Early on it was noted that macromolecules are highly sensitive to the electron beam, and thus the dose to which the macromolecules are exposed to needs to be kept to a minimum (Hart 1968; Glaeser 1971). Low dose procedures, in which the focusing and astigmatism correction are done on an area adjacent to the one to be recorded, were developed and are dependent on fast electronic shutters just below the electron gun chamber and combination of deflection coils (Unwin and Henderson 1975). CryoEM images collected under low-dose mode minimize the radiation damage but result in very noisy images. Furthermore, macromolecules consist of light elements (H, C, N, and O), which weakly scatter electrons and hence show poor contrast.

Contrast in an image is the variation in intensity displayed between individual features of an object against its background. Contrast arises from differences in the way different parts of a specimen interact with the incoming electrons. The specimen can attenuate the amplitude (intensity) of the incident wave, its phase (velocity), or both. Thin biological specimens, to a first approximation, behave as weak phase objects and produce only small phase shifts without significant amplitude changes in the transmitted electron wave. However, it is not possible to directly detect the phase change in a wave, just to record its intensity. Thus, the ideal image of a weak phase object contains no contrast. How can pure phase objects then be imaged? The answer is that objective lens aberrations destroy the perfect, featureless image of a phase object and provide contrast by introducing a phase shift. In practice, the defocus of the objective lens is changed, which causes a phase shift that introduces contrast. The change in the contrast caused by this phase shift is formulated in the phase contrast transfer function (CTF):

$$\text{CTF}(q) = \sin\left(\frac{\pi}{2}\left(C_{\text{s}} - 2\Delta F \lambda q^2\right)\right) \tag{3.1}$$

where q represents the frequency, C_{s} is the spherical aberration constant of the microscope, λ the wavelength of the electrons, and ΔF the applied defocus (Figure 3.2). This method to introduce contrast, however, is imperfect because the CTF periodically changes sign, which means that for some frequencies there is no contrast (zeroes of the function), and for some there is contrast reversal (Figure 3.2). A way for interpreting the images, or computing structures from them, must either restrict data to the high-contrast region up to the first zero, thus limiting the resolution of the study, or it must invert the contrast in regions of contrast reversal in higher resolution bands.

Transmission electron microscopes (TEMs) provide high energy with short wavelength electron flux, thus imaging of thin samples (<200 nm) in TEM seems nearly transparent when at perfect focus. If the sample is thinner than the microscope's depth of focus, the image can be approximate as a projection of the

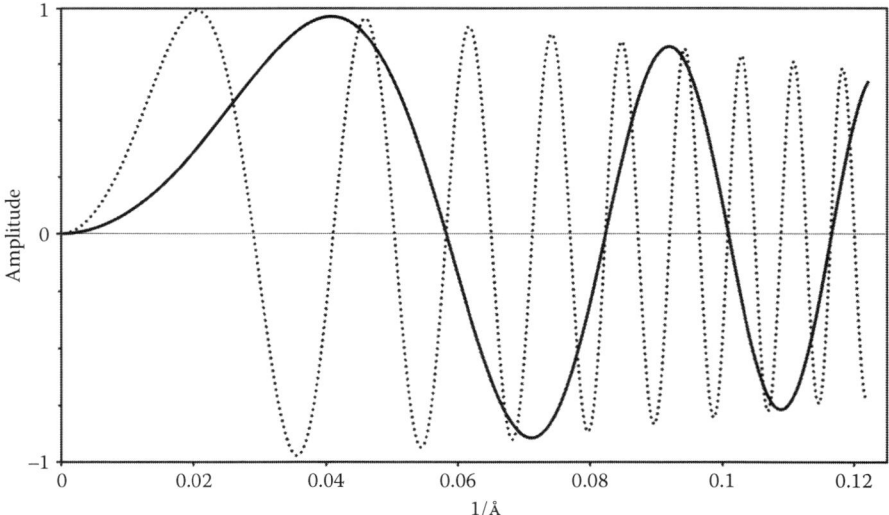

FIGURE 3.2 Phase contrast transfer as a function of frequency. Curves presenting simulated contrast transfer function for 300 keV acceleration voltage, a spherical aberration (C_{s}) value of 2 mm, and defocus of −1.5 µm (solid curve) and −6 µm (dotted curve). Frequencies where the function crosses zero amplitude denote points of zero contrast and contrast reversal. The attenuation of the function is caused by the modulation transfer function of the detector.

three-dimensional (3D) object along the beam direction onto a 2D detector. All 3D image reconstruction techniques are based on the assumption that if we have a gallery of 2D images that are taken of the same sample or of identical samples from different angular views, then we can reconstruct the original 3D structure (DeRosier and Klug 1968).

Due to the aforementioned inherent challenges when working with biological specimens, the maturation of cryoEM of unstained, frozen-hydrated specimens as a tool for structural studies became only possible with the development of 3D image reconstruction techniques. In a general sense, two major modalities are available to obtain 3D structure information of a biological object from 2D projection images; each has a different data collection and data-processing strategy. In one, an object that exists in multiple copies of identical structure is reconstructed from a large number of projections originating from any of these multiple copies (Section 3.2.2) and in the other one, an object is reconstructed from multiple projections obtained by tilting the object of interest in the EM (Section 3.2.3). The former modality is further divided into two subtechniques depending if the objects are ordered in a lattice (Section 3.2.2.1) or are isolated/separated (Section 3.2.2.2). These three approaches will be described in the following in more detail.

3.2.2 3D IMAGE RECONSTRUCTION OF RECURRING MACROMOLECULAR ASSEMBLIES

Although electron microscopes can provide very high-resolution information due to the short wavelength of the electrons, this is not achievable for a biological specimen due to high energy of the electrons and thus the radiation damage they cause to the specimen. A way around this problem is to distribute the radiation dose over many identical particles, which allow dose reduction without sacrificing resolution.

3.2.2.1 Objects Arranged in an Ordered Lattice

This technique is based on having the biological objects arranged in an ordered lattice, either in a 2D crystal or a helical array. The basics of 2D electron crystallography (2DEX) will be described in the following and for helical array reconstruction the reader is referred to the structure of acetylcholine receptor and related literature (Amos et al. 1982; Toyoshima and Unwin 1990; Carragher et al. 1996; DeRosier et al. 1999).

2DEX is mostly used for the crystallographic analysis of membrane proteins, which are packed into a 2D crystal within a lipid bilayer (Fujiyoshi and Unwin 2008). 2D crystals of membrane proteins are often grown from solutions of detergent-solubilized purified protein, the concentration of the detergent is reduced below its critical micellar concentration, where detergent is released from proteins thus exposing the hydrophobic patches of the proteins, which force them to assemble and reconstitute into a lipid monolayer (Jap et al. 1992).

The structural information from 2D crystals is acquired either in direct imaging or in diffraction mode (Figure 3.3a). Electron diffraction records the intensity distribution of the diffracted electron beam (Amos et al. 1982). Since it is insensitive to sample drift or vibration, and not dependent on the microscope's beam coherence,

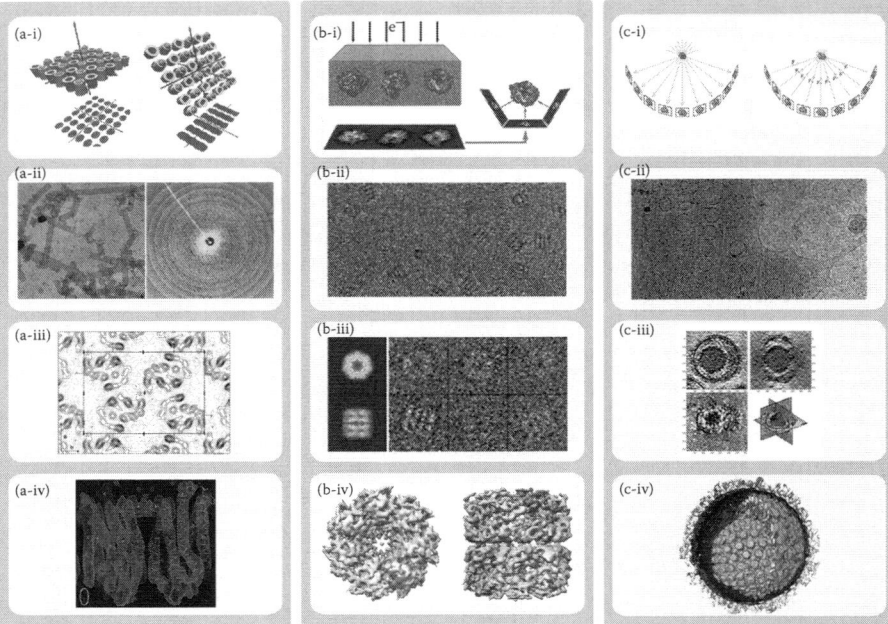

FIGURE 3.3 The three modalities for 3D image reconstruction from 2D projection images (presented in three columns). Column (a): 3D image reconstruction of objects arranged in an ordered lattice, described in detail in Section 3.2.2.1. Column (b): 3D image reconstruction of objects arranged in random orientations, described in detail in Section 3.2.2.2. Column (c): 3D image reconstruction of unique biological objects, described in detail in Section 3.2.3. For each column, four rows present the following aspects: Row (i): Schematic overview of the data collection and reconstruction principle. Row (ii): Example of a raw micrograph. Row (iii): Resulting electron density map. Row (iv): Annotated electron density map. (Reprinted from Schenk, A. D. et al., *Methods Enzymol*, 482, 101–129, 2010 (a-i); Krebs, A. et al., *J Mol Biol*, 282, 991–1003, 1998 (a-ii and a-iii); Ruprecht, J. J. et al., *EMBO J*, 23, 3609–3620, 2004 (a-iv); Mitra, K. and Frank, J., *Annu Rev Biophys Biomol Struct*, 35, 299–317, 2006 (b-i); Ludtke, S. J. et al., *Structure*, 12, 1129–1136, 2004 (b-ii–b-iv); Grünewald, K. et al., *Biophys Chem*, 100, 577–591, 2003b (c-i); Grünewald, K. et al., *Science*, 302, 1396–1398, 2003a (c-iv). With permission.)

atomic resolution information can be recorded. See in Section 3.3.1 the example of the structure of aquaporin-0 at 1.9 Å resolution (Gonen et al. 2005). However, as for x-ray diffraction, the pattern contains very good amplitude values but lacks all phase information. In contrast, direct imaging provides both the amplitude and the phases but is sensitive to specimen drift and vibration. Therefore, if phases of the particle are not available from any other source (e.g., molecular replacement of a homologous particle), direct imaging is the only option to get the structure of the particle.

To obtain the 3D structure of a molecule forming a 2D crystal, data must be collected from many crystals imaged at different tilt angles (Figure 3.3a-i) since all the molecules in the lattice are presented in the same view. The data from the different tilts are processed and merged to give the reflections since these reflections have both amplitude and phase. A Fourier back transformation yields an electron density (Henderson et al. 1990).

Despite the high-resolution information 2DEX can provide, it is not a commonly used technique. One reason for the limited use is that growing 2D crystals of membrane proteins is not easy while competitive approaches for growing 3D crystals of membrane proteins have been developed and already yielded several x-ray structures (Long et al. 2007). Another reason for the limited use of 2DEX is the tedious data collection that requires many crystals.

3.2.2.2 Objects Arranged in Random Orientations

The 3D structure of biological samples that do not form any regular arrays can be determined by the "single-particle" analysis approach (Figure 3.3b). While in 2DEX the particles are in the same orientation relative to each other, that is, prealigned in the crystal lattice, in the "single-particle" analysis, random orientation is crucial (Frank et al. 1991). The particles need to be imaged as isolated and separated from each other, and at a wide distribution of orientation, for example, see Figure 3.3b-ii. Projection images are collected at low dose (~10 electrons/$Å^2$), and computational alignment of images of the particles results in the formation of a "virtual crystal" (Figure 3.3b-iii). If the particles are homogeneous and present in a distribution of different orientations, a set of untilted images can provide all the information needed for 3D reconstruction. The major challenge in recovering the 3D structure is to determine the relative orientation of the 2D projection images. There are two general approaches for determining the relative orientation: collection of two images of the same object at different tilt angles (random conical tilt) (Radermacher et al. 1987; Saito et al. 1988) or common lines analysis of a population of identical particles with a wide distribution of orientations. In the random conical tilt approach, two data sets are collected at a known angular difference. The common lines approach is based on the observation that two projections of the same 3D structure will share at least one 1D (line) projection (Crowther et al. 1970a; van Heel 1987). Once the relative orientations of the common lines between a set of images is determined, an initial 3D model can be reconstructed. Refinement of the initial model proceeds iteratively by projection matching and/or by further common lines analysis (Baker et al. 1999). In projection matching, the current 3D map is reprojected at different orientations to provide reference images for improved alignment. Finally, the aligned images are averaged to improve the signal-to-noise ratio (SNR). The data set must contain a complete set of projections at least around a single axis to give a uniformly determined 3D reconstruction. The angular step size needed depends on the size of the object and the resolution of the analysis. The resolution of the reconstruction depends mainly on the number of particles as long as they are of very similar conformation.

In the most favorable cases, the "single-particle" analysis yields structures at atomic resolution (close to 3 Å, Zhang et al. 2010a) and commonly produces structures at a resolution better than 10 Å. Most of the structures at atomic resolution are of viruses (Yu et al. 2008; Zhang et al. 2010a). The high symmetry of spherical viruses (typically 60-fold) makes them very advantageous objects for structural analysis. In practice, the symmetry increases the number of particles available for averaging. The first approximate structure can be generated using only one image that can be, as soon as its orientation has been determined, positioned in space using symmetry operations. Obtaining electron density maps at atomic resolution allows ab initio model building of novel structures (Zhang et al. 2010a). Near atomic resolution structures of particles with lower symmetry or no symmetry are now emerging as well. For example, see the structure of the chaperonin GroEL at 6 Å (Ludtke et al. 2004) or the structure of the ribosome at 5.5 Å resolution (Armache et al. 2010). Structures at near atomic resolution allow for confident fitting of high-resolution structures from x-ray crystallography or nuclear magnetic resonance (NMR) into the EM map.

In recent years, it has become clear that proteins exist in multiple conformations that are highly related to their function. In combination with dedicated data processing, cryoEM has the ability to capture different conformational states of a protein (Fischer et al. 2010). The projection images can be classified according to the conformational state before averaging using multivariate statistical analysis (van Heel and Frank 1981; van Heel 1984; van Heel et al. 2000). Each class is composed of images representing the same orientation of the particles, and averaging is done within the same class, resulting in several distinct 3D reconstructions.

Combining information from biochemical and kinetic studies allows sorting the distinct structures along a reaction coordinate and thus revealing the dynamics of the reaction at molecular resolution. For example, see the ribosome dynamics and tRNA movement that was recently reported (Fischer et al. 2010).

3.2.3 3D IMAGE RECONSTRUCTION OF UNIQUE BIOLOGICAL OBJECTS

In electron tomography, the 3D structure of an object is reconstructed from data collected from only one copy of that object (Figure 3.3c). As such, tomography is well suited for the structural studies of pleomorphic objects with a unique shape without symmetry, namely cells, organelles, and subcellular structures with the only limitation being the thickness of the specimen. Moreover, electron cryotomography (cryoET) is the only method that can provide molecular resolution information on macromolecular complexes in their cellular environment.

The basic principle behind tomography is the Hegerl–Hoppe dose fractionation theorem (Hegerl and Hoppe 1976; McEwen et al. 1995). The basis of this theorem is that more information can be obtained from a tomogram than from a single projection by using the same electron dose (Lucic et al. 2005). Accordingly, during the data collection the total dose is spread over a series of projection images. As vitrified biological samples are very sensitive to radiation, the total dose available for fractionation is limited. In practice, it is typically kept below 100 electrons/$Å^2$, which is still much higher than the dose used in the other modalities of cryoEM (10–40 electrons/$Å^2$). Full tomographic data sets require collecting projection images at angles ±90° at fine increments. According to the Crowther criterion, the resolution of the reconstruction depends on the number of images collected:

$$R_{max} \cong \frac{\pi \cdot d}{N} \qquad (3.2)$$

where R_{max} is the maximum resolution, d is the thickness of the specimens, and N is the number of images (Crowther et al. 1970b). Due to sample holder geometry, the aforementioned dose limitation, and also the thickening of the sample slab at higher tilt angles, data are usually collected at angles $\pm 60°$ and at increments of 1.5–2°. Taken a total dose of <100 electrons/$\mathring{A}^2$ spread over ~120 images means that any tilt image gets very low electron dose (less than 1 electron/$\mathring{A}^2$) and thus has very low SNR. In fact, the (raw) resolution obtainable in cryoET (~40 $\mathring{A}$) is mainly limited by the low SNR and not the Crowther criterion. Spreading the total dose further over more projections as the Crowther formulation implies will not help as the SNR goes then into a range that makes the signal undetectable in the individual images and the reconstruction. The resolution of the reconstructed tomogram is not isotropic but is lower along the axis parallel to the beam (Figure 3.3c-iii) due to the missing information from the high-angle tilts (the "missing wedge") (Lucic et al. 2008).

The ability of cryoET to visualize unique biological objects makes its application to image macromolecules in their cellular and subcellular context very attractive. The electron beam penetration limit in TEM is about 1 µm. Many of the organelles and subcellular structures can be imaged directly in plunged intact cells, for example, the plasma membrane, the secretory machinery, the mitochondria, to mention few (e.g., see Figure 3.3c-ii). Cell protrusions, like filopodia or dendrites and axons, are also accessible to TEM. The cell nucleus, however, and tissues are rarely thin enough to directly image them by cryoEM. Hence, electron cryomicroscopy of vitreous sections (CEMOVIS) has been developed for those cases (Al-Amoudi et al. 2004a). Thicker samples (up to few hundred micrometers) need to be vitrified by high-pressure freezing as the depth of vitrification by plunge freezing is only in the 10 µm range. The vitrified sample is then sliced at a temperature below the devitrification point into ultrathin sections that are subsequently analyzed by cryoEM or cryoET in the vitrified state. Although several studies reported remarkable results using CEMOVIS (Al-Amoudi et al. 2007; Bouchet-Marquis et al. 2007; Hagen and Grünewald 2008), there are certain limitations that prevent it from becoming a more commonly used approach. One limitation is that the sectioning procedure and handling of vitrified sections is an art in itself and requires substantial training and experience. Other limitations are the cutting artifacts, in particular, knife marks, compression, crevasses, and chatter (Al-Amoudi et al. 2005; Han et al. 2008; Norlen et al. 2009; Bouchet-Marquis and Hoenger 2011). An alternative method for sample thinning, free of compression, using focused ion beam (FIB) is emerging, discussed in more detail in Section 3.4.

Complex molecular assemblies often occur in hundreds or even thousands of identical copies within a single cell. In such cases, their volumes can be extracted from the tomogram volume and further aligned in 3D, classified and averaged, in analogy to the "single-particle" analysis but in three dimensions. This approach is referred to as "subtomogram averaging" and results in a better SNR, discussed in more detail in Section 3.3.2. This way the above-described SNR limitation to the resolution in tomography is overcome, and in parallel, by filling the missing wedge from particles of different orientations, a more isotropic resolution achieved for the resulting averaged structure. For example, see the studies on the nuclear pore complex (Beck et al. 2004, 2007).

3.3   EXAMPLES OF GLYCOPROTEIN STRUCTURES DETERMINED BY ELECTRON CRYOMICROSCOPY

Glycoproteins constitute a significant subclass of the proteome, however, their structure determination by high-resolution techniques was hampered due to the carbohydrate conjugate. The carbohydrate chains tend to introduce biochemical heterogeneity to the protein preparation and local flexibly, thus sometimes averting crystallization (see Chapter 1). Additionally, the carbohydrate chains can dramatically increase the mass of the protein and that causes problems in structure determination by NMR (see Chapter 2). Due to the aforementioned difficulties with the carbohydrate chains, ways to remove them either enzymatically or nonenzymatically were developed. However, those difficulties do not hamper structure determination by cryoEM. The application of any of the approaches described before (Section 3.2) is feasible for macromolecules and macromolecular assemblies with molecular masses ranging from few hundred kilodaltons to hundred megadaltons. Moreover, cryoEM is less sensitive to biochemical heterogeneity and local flexibility and with the appropriate classification can provide structural insight to the different conformations.

As it will become obvious in the following examples, glycoproteins do not require special treatment at the sample preparation for cryoEM. The protein and the carbohydrate chains are treated as one entity, and thus the carbohydrate chains are an integral part of the electron density map obtained. However, the carbohydrate chains are rarely interpreted, perhaps because of the lack of dedicated tools available to do so.

3.3.1   AQUAPORIN

Aquaporins are a family of water-conducting channels (Agre et al. 1997). More than 10 different mammalian aquaporins have been identified, each with an essentially unique pattern of expression among tissues and during development (Ishibashi et al. 2009). Different aquaporins have different patterns of glycosylation. Aquaporin-0 (AQP0) is the most abundant protein in lens fiber cell membranes, where it forms not only water pores but also the 11–13 nm thin lens junctions that assemble after proteolytic cleavage of the cytoplasmic termini (Chepelinsky 2009). AQP0 undergoes glycation, a process that has been implicated in lens opacification (Swamy-Mruthinti and Schey 1997; Swamy-Mruthinti 2001).

The structure of the AQP0-mediated membrane junction was determined at 1.9 Å resolution by electron diffraction from a 2D crystal (Figure 3.4) of protein purified directly from the core of sheep lenses (Gonen et al. 2005). The crystals were obtained by addition of lipids to the protein mixture with subsequent detergent removal by dialysis. The AQP0 structure was solved by molecular replacement using the AQP1 structure as a search model, and the final structure had six transmembrane-spanning helical regions with intracellular N- and C-termini (Figure 3.4). The resolution of the structure was good enough to resolve individual water molecules (Figure 3.4d). Three noninteracting water molecules were observed in the water path forming a hydrogen-bonding network with polar residues of the pore. The structure also revealed the detailed interaction between lipids and protein

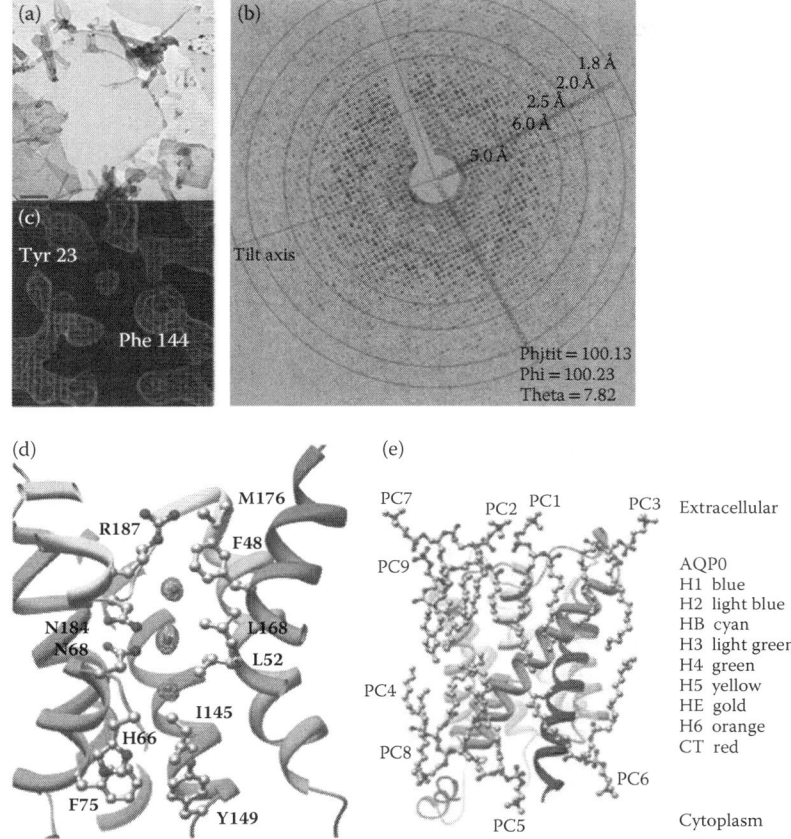

FIGURE 3.4 **(See color insert.)** Structure determination of aquaporin-0 by electron crystallography. (a) 2D crystals of aquaporin-0. (b) Electron diffraction pattern showing diffraction spots to a resolution beyond 2 Å. (c) Region of the final 2Fo-Fc map. (d) The water pore of aquaporin-0. A mesh outlines the density of the water molecules. (e) Lipid–protein interactions in the 2D crystal. (Reprinted from Gonen, T. et al., *Nature*, 438, 633–638, 2005. With permission.)

(Figure 3.4e). AQP0 forms tetramers that are separated by lipids, laterally the tetramers do not interact directly with each other and only the protein–lipid–protein interactions hold tetramers together in a crystalline arrangement. A network of hydrogen bonds and salt bridges holds the annular lipids together, and lipids interact with the protein mainly via their phosphate headgroups. Although the protein was purified directly from the tissue, carbohydrate chains were not resolved in this structure, either because they were disordered or not present. However, this structure is still worthwhile discussing in this context as it remarkably demonstrates the value and power of electron crystallography, in reaching high resolution and allowing the study of the membrane-embedded conformation and the detailed protein–lipid interaction.

3.3.2 Viral Glycoproteins

Viruses were the first vitrified biological specimens to be studied and paved the way for the field of structural studies of biological samples with cryoEM (Adrian et al. 1984). The icosahedral symmetry in spherical viruses makes them very advantageous objects for structural analysis by the "single-particle" approach (Section 3.2.2.2), which now results in structures of atomic resolution (Yu et al. 2008; Zhang et al. 2008; Liu et al. 2010; Zhang et al. 2010a). CryoET, however, is the emerging approach for the study of pleomorphic viruses and their viral glycoproteins, in particular (Grünewald and Cyrklaff 2006). Herpes Simplex virus 1 (HSV-1) was the first virus structure to be solved by this method (Grünewald et al. 2003a) followed by many other viruses, including the retroviruses moloney murine leukemia virus (MoMuLV) (Förster et al. 2005), human immunodeficiency virus (HIV) (Benjamin et al. 2005; Briggs et al. 2006), simian immunodeficiency virus (SIV) (Zanetti et al. 2006; Zhu et al. 2006), and the poxvirus vaccinia (Cyrklaff et al. 2005), just to mention the earliest. This approach is now applied routinely and has yielded many pleomorphic virus structures. The structure of HSV-1 with its surrounding membrane envelope allowed, for the first time, the inspection of the different glycoproteins in their native environment (Grünewald et al. 2003a). Spikes of variable length, thickness, and spacing were observed, and their distribution seemed nonrandom on the membrane, suggesting functional clustering.

A combined approach of cryoET and subsequent averaging of tomographic sub-volumes in 3D can be used to yield structures of viral glycoproteins at molecular resolution (<3 nm). The first proof of principle and also technical landmark paper for this was the 3D structure of the MoMuLV Env glycoprotein spike obtained by cryoET and subtomogram averaging (Förster et al. 2005). Volumes containing spikes protruding from the viral membrane were extracted, aligned, and averaged and produced the structure of Env in a prefusion state at 2.7 nm resolution (Figure 3.5). Importantly, Förster et al. compensated for the "missing wedge" effect, a crucial step at the alignment from tomographic data (see Section 3.2.3). The structure obtained, following alignment and averaging, had isotropic resolution (Figure 3.5). The Env of MoMuLV was shown to be a trimer and consisting of a large domain located ~5 nm from the surface of the viral membrane that is connected to the membrane by three

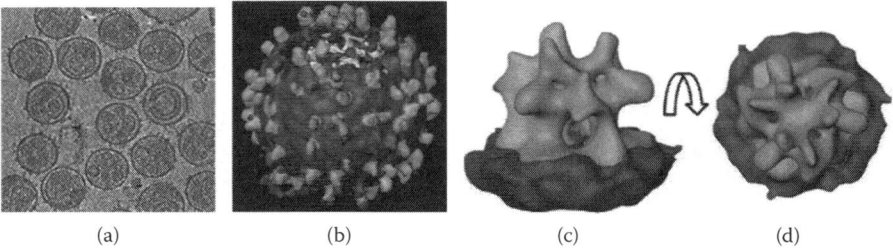

(a)   (b)   (c)   (d)

FIGURE 3.5 **(See color insert.)** Analysis of intact moloney murine leukemia virus and its Env glycoprotein by cryoET. (a) A slice from a tomogram of MoMuLV particles. (b) Surface-rendered tomogram of an individual MoMuLV virion. (c,d) Isosurface representation of the Env complex resulting from subtomogram averaging ([c], side view; [d], top view). (Reprinted from Förster, F. et al., *Proc Natl Acad Sci USA*, 102, 4729–4734, 2005. With permission.)

legs (Figure 3.5c). The significance of this structure emphasized the strength of using cryoET for the study of large assemblies of glycoproteins as the structure of intact Env is not available from x-ray crystallography. This is most likely because Env is highly glycosylated, the glycosylation can exceed 50% of the mass of the protein, but also because Env is a single-pass transmembrane protein and large quantities of soluble Env necessary for crystallization attempts are difficult to prepare, and the material is often unstable and heterogeneous (Elliott 1996). The cryoET Env reconstruction allowed the tentative docking of the available x-ray structure of the receptor-binding domain and was used to propose a fusion mechanism.

The Env of SIV and HIV were also studied by cryoET followed by iterative alignment and averaging of spike subtomograms by several research groups. The resulting structures from the different groups yielded different structures, mainly in the stem region of the complex. In one structure, the stem has a tripod organization as of the Env of MoMuLV (Zhu et al. 2006, 2008; Wu et al. 2010), while in the other there is a single stalk (Zanetti et al. 2006; White et al. 2010). This variation might be in part caused by difference in the alignment procedures used but is still under intensive research to understand the differences.

The N-linked glycans of the HIV and SIV Env protein comprise about 50% of the mass of the protein (Leonard et al. 1990), as in the case of the Env of MoMuLV. The carbohydrates believed to form a "glycan shield," which is one of the major mechanisms for blocking or minimizing the virus neutralizing antibody response (Wei et al. 2003). Using the carbohydrate chains as a constraint, Zanetti et al. proposed two models to fit the available crystal structure of gp120 into the cryoEM density map. The resulting models demonstrated the glycan shield (Figure 3.6b and c).

The approach of cryoET followed by subtomogram averaging was applied not only for the study of retrovirus glycoproteins. The *Bunyaviridae* are a family of negative-sense RNA viruses with three-part segmented genomes (Elliott 1996). Virions are enveloped and decorated with spikes derived from a pair of

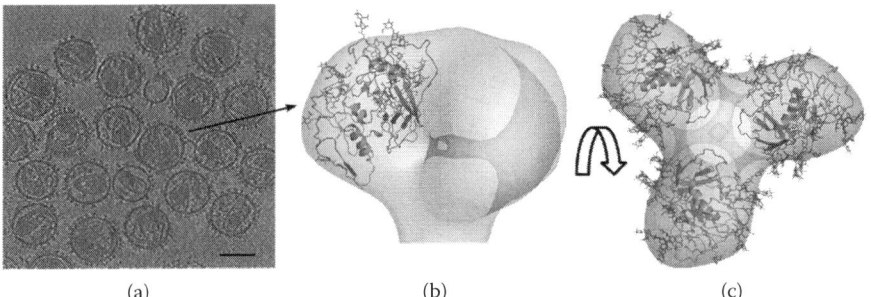

(a)   (b)   (c)

FIGURE 3.6 **(See color insert.)** The structure of the simian immunodeficiency virus (SIV) Env complex by cryoET. (a) A slice from a tomogram of SIV revealing the Env spikes on the virions surface. Scale bar represents 100 nm. (b,c) Isosurface representation of the subtomogram averaged Env complex ([b], side view; [c], top view). The crystal structure of gp120 (shown in cartoon representation) was fitted into the averaged EM map. Additionally, the carbohydrate chains (shown in ball-and-stick representation) were added on the predicted N-glycosylation sites. (Reprinted from Zanetti, G. et al., *PLoS Pathogens*, 2, e83, 2006. With permission.)

glycoproteins (GN and GC). Huiskonen et al. determined the structure of the GN–GC spike complex from Tula Hantavirus of the *Bunyaviridae* to 3.6 nm resolution by cryoET and subsequent averaging of tomographic subvolumes (Huiskonen et al. 2010). Each spike complex was shown to have a square-shaped assembly with fourfold symmetry. The spikes formed ordered patches on the viral membrane by means of specific lateral interactions, and the authors suggested that these interactions might be sufficient for creating membrane curvature during virus budding. Notably, the GN–GC complex from Hantaan virus (another member of the Hantavirus genus) was determined by "single-particle" analysis (Section 3.2.2.2) from side views of the glycoprotein spikes and yielded a tetrameric structure to 2.5 nm resolution with a very good agreement to the Tula virus (Battisti et al. 2011). Large cytoplasmic extensions associated with each GN–GC spike forming a lattice on the inner surface of the viral membrane could be observed.

3.3.3 Desmosomes

Desmosomes are cadherin-based intercellular junctions that primarily provide mechanical stability to tissues such as epithelia and cardiac muscle but also play an important role in tissue morphogenesis (Vleminckx and Kemler 1999). A complex network of lateral interactions between glycoproteins on the extracellular side and intracellular proteins is believed to account for the strength and durability inherent to desmosomes (Stokes 2007). Extracellularly, desmosomes use physical associations between glycoproteins from the cadherin family, viz., desmocollins (Dsc) and desmoglein (Dsg). Intracellularly, desmosomes are made up of a set of proteins from the armadillo signaling family, namely plakoglobin (PG) and plakophilin (PP) that constitute an electron-dense plaque, which is ultimately connected to the cell cytoskeleton. The cytoplasmic domains of Dsg and Dsc bind to PG and PP, which, in turn, associate with the N-termini of the elongated desmoplakin (DP) dimers, which are ultimately linked, through their C-termini, to intermediate filaments (IF).

Combining state-of-the-art cryoEM techniques with the available x-ray crystal structures of C-cadherins, Al-Amoudi et al. derived a pseudoatomic model of the architecture of a desmosome from human epidermis (Figure 3.7), under close to native conditions (Al-Amoudi et al. 2007). This is particularly remarkable as they study the desmosome at the tissue level. Due to the nature of the specimen and its thickness, CEMOVIS (cf. Section 3.2.3) was vital to achieve best specimen preservation. 2D projection images of human epidermis recorded with CEMOVIS revealed the periodic nature of the desmosome (Figure 3.7b) (Al-Amoudi et al. 2004b). The 3D reconstruction of electron tomograms from vitrified sections verified the quasiperiodic arrangement of the cadherins at ~7 nm intervals along the desmosome's midline with a curved shape resembling the x-ray structure of C-cadherin (Al-Amoudi et al. 2007). Subtomogram averaging yielded an EM map of the cadherin organization at 3.4 nm resolution. It is important to note that averaging was possible only within the same section. The reason for this is mainly the compression of the sections that occur at different orientations relative to the cadherin particles for each section. Fitting of the cadherin atomic structure indicated the periodic arrangement of a *trans* W-like and a *cis* V-like interaction corresponding

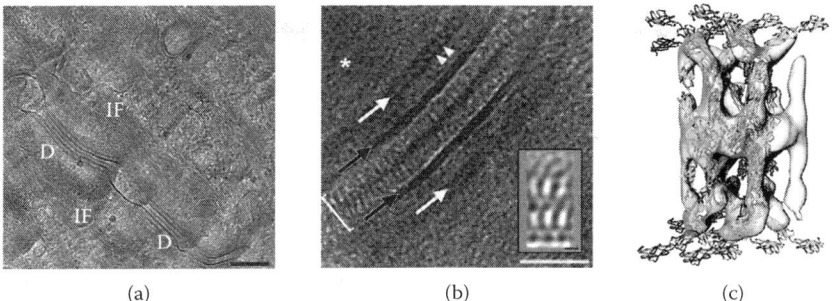

(a)   (b)   (c)

FIGURE 3.7 **(See color insert.)** Molecular architecture of cadherins in native epidermal desmosomes. (a) Projection image of a desmosomal region in a vitreous section of human epidermis (D, desmosomes; IF, filament networks). (b) Higher magnification projection image of a desmosome from an ultrathin vitreous section. (Inset) A slice through the desmosome extracellular space density resulting from subtomogram averaging. (c) An isosurface representation of the desmosomal cadherins with the C-cadherin crystal structure fitted. Scale bars: (a), 200 nm; (b), 50 nm; inset in (b), 7 nm. (Reprinted from Al-Amoudi, A. et al., *EMBO J*, 23, 3583–3588, 2004a; Al-Amoudi, A. et al., *Nature*, 450, 832–837, 2007. With permission.)

to molecules from opposing membranes and the same cell membrane, respectively. The resulting structure of cadherin organization allowed the authors to explain the mechanism of cadherin-based cell adhesion and to suggest a conceivable model for the assembly and disassembly of desmosomes. Unfortunately, the authors did not attempt to model the carbohydrate chains of the cadherins. Recently, a more complete model of the protein–protein interactions of the desmosomal plaque was reported from the same group by fitting the high-resolution structures of the major plaque proteins (Al-Amoudi et al. 2011).

3.4 OUTLOOK AND PROSPECTS

Electron cryomicroscopy has emerged as a prominent technique to study the structure of glycoproteins, in particular, in their native state and environment. The increase of mass and heterogeneity that often accompanies the conjugation of carbohydrate chains to the protein itself are not posing an obstacle for structure analysis by cryoEM. As such, glycoproteins studied by this technique are characterized together with their carbohydrate chains as a single entity.

The structures obtained by cryoEM, however, are usually of lower resolution compared to crystallography and NMR. Instrumental improvements, such as the highly coherent illumination provided by field emission gun microscopes, and the stability of specimen stages at cryogenic temperatures had significant impact on the improvements of resolution now achieved by cryoEM. The low SNR in images of radiation-sensitive specimens is still, however, limiting the resolution. Ways to improve the poor SNR are actively being pursued and recent instrument developments hold promise for the near future. One of them is the development of thin phase plates called the Zernike phase plate, conceptually similar to the phase plate used for light microscopes. After slow progress in the efforts to develop phase plates for electron microscopes, functional phase plates with thin carbon films have recently been reported, demonstrating benefits for both "single-particle" analysis and cryoET (Chang et al. 2010; Danev et al. 2010). The second upcoming instrument development is the introduction of direct electron detectors (McMullan et al. 2009). These detectors promise to provide superior detector quantum efficiency (DQE) over the current indirect detectors, that is, the charge couple device (CCD) cameras, that have <10% DQE (Mooney 2007). Consequently, the direct detectors should significantly improve the SNR, and hence, the contrast and resolution.

Image reconstruction from cryoEM data is readily providing structural information at molecular resolution, where high-resolution structures can confidently be fitted into the EM map (Topf et al. 2008; Zhang et al. 2010b). Tools to model the carbohydrate chains are now available and allow the fitting into the EM density map (Bohne-Lang and von der Lieth 2005) but unfortunately still not used routinely. Integrating proteomics information available from mass spectrometry on protein–protein interaction can further constrain fitting into EM maps of protein complexes or assemblies (Förster and Villa 2010). The field is still, however, missing robust methods for structure validation (equivalent to the crystallographic R_{factor} and R_{free} values) that wait to be developed.

The functionality of many complexes is revealed by their different conformational states often not provided by crystallographic studies. The classification algorithms routinely applied for "single-particle" reconstructions provide the tools to characterize this conformational heterogeneity. Different new approaches have now been developed for classification in three dimensions to expand this possibility of dealing with sample heterogeneity of subtomogram volumes, for example, algorithms based on maximum likelihood (Scheres et al. 2009) or neural networks (Yu and Frangakis 2011).

CryoET is the only technique capable of providing structural information at nanometer resolution in the native environment of the cell or even tissue. The inherent limitation of application of cellular cryoET is the thickness of the sample. Although the application of CEMOVIS (Section 3.2.3) provides remarkable results (e.g., see Section 3.3.3), it suffers from sectioning artifacts and being an art in itself, both limiting its wider use (Al-Amoudi et al. 2005; Han et al. 2008; Norlen et al. 2009; Bouchet-Marquis and Hoenger 2011). Focused ion beam (FIB) milling, previously mainly used in materials science, is an alternative for sample thinning. In this method, a finely focused beam of ions (usually gallium) is used at high beam currents for site-specific milling in scanning electron microscope (SEM). Recently, effort has been made to adopt this technique for biological specimens and for cryospecimens. Efforts are concentrating on minimizing the impact of local heating by milling in shallow angles almost parallel to the support surface and the developments for transfer of the lamella obtained after the milling from the SEM/FIB dual beam instrument into the TEM for data acquisition (Marko et al. 2007; Hayles et al. 2010; Rigort et al. 2010). Specimens obtained by this method do not suffer from compression and will allow particle averaging from different lamellas.

Identification of a specific macromolecule within the crowded environment of the cell or tissue is vital for following the macromolecules fate. Clonable labeling techniques providing electron dense markers with a versatility and power similar to that of green fluorescence protein (GFP) labeling for fluorescence microscopy are still being sought. One suggestion is the fusion to ferritin that can be detected in EM when charged with iron and was shown feasible in *E. coli* (Wang et al. 2011). However, in its current form this will probably not be useful for labeling in mammalian cells due to toxicity of the excess of iron. Labeling glycoproteins on the extracellular side of the plasma membrane is potentially easier and usually done by immunogold labeling. Using small gold beads with Fab fragments will allow detecting glycoproteins with higher precision and efficiency.

CryoEM is a powerful technique that has the potential to bridge the gap between the high-resolution methods (discussed in chapters 1 and 2) and advanced light microscopy. Methods and instrumentations have been developed to correlate data acquisition and processing between the light and fluorescence microscopes and the electron microscope. An integrated approach of the high-resolution structure determination methods and correlative light and electron microscopy will allow looking at dynamic biological processes at different resolution and complexity and ultimately will lead to a better perception of those processes.

REFERENCES

Adrian, M., Dubochet, J., Lepault, J., and Mcdowall, A. W. 1984. Cryo-electron microscopy of viruses. *Nature*, 308, 32–36.

Agre, P., Lee, M. D., Devidas, S., and Guggino, W. B. 1997. Aquaporins and ion conductance. *Science*, 275, 1490–1492.

Al-Amoudi, A., Castano-Diez, D., Devos, D. P. et al. 2011. The three-dimensional molecular structure of the desmosomal plaque. *Proc Natl Acad Sci USA*, 108, 6480–6485.

Al-Amoudi, A., Chang, J. J., Leforestier, A. et al. 2004a. Cryo-electron microscopy of vitreous sections. *EMBO J*, 23, 3583–3588.

Al-Amoudi, A., Diez, D. C., Betts, M. J., and Frangakis, A. S. 2007. The molecular architecture of cadherins in native epidermal desmosomes. *Nature*, 450, 832–837.

Al-Amoudi, A., Norlen, L. P. O., and Dubochet, J. 2004b. Cryo-electron microscopy of vitreous sections of native biological cells and tissues. *J Struct Biol*, 148, 131–135.

Al-Amoudi, A., Studer, D., and Dubochet, J. 2005. Cutting artefacts and cutting process in vitreous sections for cryo-electron microscopy. *J Struct Biol*, 150, 109–121.

Amos, L. A., Henderson, R., and Unwin, P. N. T. 1982. Three-dimensional structure determination by electron-microscopy of two-dimensional crystals. *Prog Biophys Mol Biol*, 39, 183–231.

Armache, J. P., Jarasch, A., Anger, A. M. et al. 2010. Cryo-EM structure and rRNA model of a translating eukaryotic 80s ribosome at 5.5-angstrom resolution. *Proc Natl Acad Sci USA*, 107, 19748–19753.

Baker, T. S., Olson, N. H., and Fuller, S. D. 1999. Adding the third dimension to virus life cycles: Three-dimensional reconstruction of icosahedral viruses from cryo-electron micrographs. *Microbiol Mol Biol Rev*, 63, 862–922.

Battisti, A. J., Chu, Y. K., Chipman, P. R. et al. 2011. Structural studies of hantaan virus. *J Virol*, 85, 835–841.

Beck, M., Förster, F., Ecke, M. et al. 2004. Nuclear pore complex structure and dynamics revealed by cryoelectron tomography. *Science*, 306, 1387–1390.

Beck, M., Lucic, V., Förster, F., Baumeister, W., and Medalia, O. 2007. Snapshots of nuclear pore complexes in action captured by cryo-electron tomography. *Nature*, 449, 611–615.

Benjamin, J., Ganser-Pornillos, B. K., Tivol, W. F., Sundquist, W. I., and Jensen, G. J. 2005. Three-dimensional structure of HIV-1 virus-like particles by electron cryotomography. *J Mol Biol*, 346, 577–588.

Bohne-Lang, A. and Von Der Lieth, C. W. 2005. Glyprot: In silico glycosylation of proteins. *Nucleic Acids Res*, 33, W214–219.

Bouchet-Marquis, C. and Hoenger, A. 2011. Cryo-electron tomography on vitrified sections: A critical analysis of benefits and limitations for structural cell biology. *Micron*, 42, 152–162.

Bouchet-Marquis, C., Zuber, B., Glynn, A. M. et al. 2007. Visualization of cell microtubules in their native state. *Biol Cell*, 99, 45–53.

Briggs, J. a. G., Grünewald, K., Glass, B. et al. 2006. The mechanism of HIV-1 core assembly: Insights from three-dimensional reconstructions of authentic virions. *Structure*, 14, 15–20.

Carragher, B., Whittaker, M., and Milligan, R. A. 1996. Helical processing using phoelix. *J Struct Biol*, 116, 107–112.

Chang, W. H., Chiu, M. T. K., Chen, C. Y. et al. 2010. Zernike phase plate cryoelectron microscopy facilitates single particle analysis of unstained asymmetric protein complexes. *Structure*, 18, 17–27.

Chepelinsky, A. B. 2009. Structural function of mip/aquaporin 0 in the eye lens; genetic defects lead to congenital inherited cataracts. *Handb Exp Pharmacol*, 190, 265–297.

Crowther, R. A., Amos, L. A., Finch, J. T., Derosier, D. J., and Klug, A. 1970a. Three dimensional reconstructions of spherical viruses by Fourier synthesis from electron micrographs. *Nature*, 226, 421–425.

Crowther, R. A., Derosier, D. J., and Klug, A. 1970b. The reconstruction of a three-dimensional structure from projections and its application to electron microscopy. *Proc R Soc London, A*, 317, 319–355.

Cyrklaff, M., Risco, C., Fernandez, J. J. et al. 2005. Cryo-electron tomography of vaccinia virus. *Proc Natl Acad Sci USA*, 102, 2772–2777.

Danev, R., Kanamaru, S., Marko, M., and Nagayama, K. 2010. Zernike phase contrast cryo-electron tomography. *J Struct Biol*, 171, 174–181.

Derosier, D. J. and Klug, A. 1968. Reconstruction of three dimensional structures from electron micrographs. *Nature*, 217, 130–134.

Derosier, D., Stokes, D. L., and Darst, S. A. 1999. Averaging data derived from images of helical structures with different symmetries. *J Mol Biol*, 289, 159–165.

Dubochet, J. 2007. The physics of rapid cooling and its implications for cryoimmobilization of cells. In *Cellular Electron Microscopy*, Mcintosh, J. R. (ed), pp. 7–21. San Diego: Academic Press.

Dubochet, J., Adrian, M., Chang, J. J. et al. 1988. Cryo-electron microscopy of vitrified specimens. *Q Rev Biophys*, 21, 129–228.

Dubochet, J., Booy, F. P., Freeman, R., Jones, A. V., and Walter, C. A. 1981. Low-temperature electron-microscopy. *Annu Rev Biophys Bioeng*, 10, 133–149.

Dubochet, J. and Mcdowall, A. W. 1981. Vitrification of pure water for electron microscopy. *J Microsc*, 124, 3–4.

Elliott, R. M. 1996. *The Bunyaviridae*, New York: Plenum Press.

Fischer, N., Konevega, A. L., Wintermeyer, W., Rodnina, M. V., and Stark, H. 2010. Ribosome dynamics and tRNA movement by time-resolved electron cryomicroscopy. *Nature*, 466, 329–333.

Förster, F., Medalia, O., Zauberman, N., Baumeister, W., and Fass, D. 2005. Retrovirus envelope protein complex structure in situ studied by cryo-electron tomography. *Proc Natl Acad Sci USA*, 102, 4729–4734.

Förster, F. and Villa, E. 2010. Integration of cryo-em with atomic and protein-protein interaction data. *Methods Enzymol*, 483: 47–72.

Frank, J., Penczek, P., Grassucci, R., and Srivastava, S. 1991. 3-dimensional reconstruction of the 70S *Escherichia coli* ribosome in ice—The distribution of ribosomal-RNA. *J Cell Biol*, 115, 597–605.

Fujiyoshi, Y., Mizusaki, T., Morikawa, K. et al. 1991. Development of a superfluid-helium stage for high-resolution electron-microscopy. *Ultramicroscopy*, 38, 241–251.

Fujiyoshi, Y. and Unwin, N. 2008. Electron crystallography of proteins in membranes. *Curr Opin Struct Biol*, 18, 587–592.

Glaeser, R. M. 1971. Low temperature electron microscopy—Radiation damage in crystalline biological materials. *J Microsc-Oxf*, 12, 133–138.

Gonen, T., Cheng, Y. F., Sliz, P. et al. 2005. Lipid-protein interactions in double-layered two-dimensional AQPO crystals. *Nature*, 438, 633–638.

Grünewald, K. and Cyrklaff, M. 2006. Structure of complex viruses and virus-infected cells by electron cryo tomography. *Curr Opin Microbiol*, 9, 437–442.

Grünewald, K., Desai, P., Winkler, D. C. et al. 2003a. Three-dimensional structure of herpes simplex virus from cryo-electron tomography. *Science*, 302, 1396–1398.

Grünewald, K., Medalia, O., Gross, A., Steven, A. C., and Baumeister, W. 2003b. Prospects of electron cryotomography to visualize macromolecular complexes inside cellular compartments: Implications of crowding. *Biophys Chem*, 100, 577–591.

Hagen, C. and Grünewald, K. 2008. Microcarriers for high-pressure freezing and cryosectioning of adherent cells. *J Microsc-Oxf*, 230, 288–296.

Han, H. M., Zuber, B., and Dubochet, J. 2008. Compression and crevasses in vitreous sections under different cutting conditions. *J Microsc-Oxf*, 230, 167–171.

Hart, R. G. 1968. Electron microscopy of unstained biological material: The polytropic montage. *Science*, 159, 1464–1467.

Hayles, M. F., De Winter, D. a. M., Schneijdenberg, C. et al. 2010. The making of frozen-hydrated, vitreous lamellas from cells for cryo-electron microscopy. *J Struct Biol*, 172, 180–190.

Hegerl, R. and Hoppe, W. 1976. Influence of electron noise on three-dimensional image reconstruction. *Z Naturforsch*, 31, 1717–1721.

Henderson, R., Baldwin, J. M., Ceska, T. A. et al. 1990. Model for the structure of bacteriorhodopsin based on high-resolution electron cryomicroscopy. *J Mol Biol*, 213, 899–929.

Huiskonen, J. T., Hepojoki, J., Laurinmaki, P. et al. 2010. Electron cryotomography of Tula hantavirus suggests a unique assembly paradigm for enveloped viruses. *J Virol*, 84, 4889–4897.

Ishibashi, K., Hara, S., and Kondo, S. 2009. Aquaporin water channels in mammals. *Clin Exp Nephrol*, 13, 107–117.

Jap, B. K., Zulauf, M., Scheybani, T. et al. 1992. 2D crystallization: From art to science. *Ultramicroscopy*, 46, 45–84.

Johari, G. P. 1977. On the heat capacity, entropy and glass transition of vitreous ice. *Philos Mag*, 35, 1077–1090.

Krebs, A., Villa, C., Edwards, P. C., and Schertler, G. F. X. 1998. Characterisation of an improved two-dimensional p22121 crystal from bovine rhodopsin. *J Mol Biol*, 282, 991–1003.

Leonard, C. K., Spellman, M. W., Riddle, L. et al. 1990. Assignment of intrachain disulfide bonds and characterization of potential glycosylation sites of the type 1 recombinant human immunodeficiency virus envelope glycoprotein (gp120) expressed in Chinese hamster ovary cells. *J Biol Chem*, 265, 10373–10382.

Liu, H. R., Jin, L., Koh, S. B. S. et al. 2010. Atomic structure of human adenovirus by cryo-em reveals interactions among protein networks. *Science*, 329, 1038–1043.

Long, S. B., Tao, X., Campbell, E. B., and Mackinnon, R. 2007. Atomic structure of a voltage-dependent K+ channel in a lipid membrane-like environment. *Nature*, 450, 376–382.

Lucic, V., Förster, F., and Baumeister, W. 2005. Structural studies by electron tomography: From cells to molecules. *Annu Rev Biochem*, 74, 833–865.

Lucic, V., Leis, A., and Baumeister, W. 2008. Cryo-electron tomography of cells: Connecting structure and function. *Histochem Cell Biol*, 130, 185–196.

Ludtke, S. J., Chen, D. H., Song, J. L., Chuang, D. T., and Chiu, W. 2004. Seeing GroEL at 6 Ångstrom resolution by single particle electron cryomicroscopy. *Structure*, 12, 1129–1136.

Marko, M., Hsieh, C., Schalek, R., Frank, J., and Mannella, C. 2007. Focused-ion-beam thinning of frozen-hydrated biological specimens for cryo-electron microscopy. *Nat Methods*, 4, 215–217.

Mcdonald, K., Schwarz, H., Muller-Reichert, T. et al. 2010. "Tips and tricks" for high-pressure freezing of model systems. In *Electron Microscopy of Model Systems*, Mueller-Reichert, T. (ed), pp. 671–693. Oxford: Elsevier.

Mcewen, B. F., Downing, K. H., and Glaeser, R. M. 1995. The relevance of dose-fractionation in tomography of radiation-sensitive specimens. *Ultramicroscopy*, 60, 357–373.

Mcmullan, G., Faruqi, A. R., Henderson, R. et al. 2009. Experimental observation of the improvement in MTF from backthinning a CMOS direct electron detector. *Ultramicroscopy*, 109, 1144–1147.

Mitra, K. and Frank, J. 2006. Ribosome dynamics: Insights from atomic structure modeling into cryo-electron microscopy maps. *Annu Rev Biophys Biomol Struct*, 35, 299–317.

Mooney, P. 2007. Optimization of image collection for cellular electron microscopy. In *Cellular Electron Microscopy*, Mcintosh, J. R. (ed), pp. 661–719. San Diego: Academic Press.

Moor, H. 1987. Theory and practice of high pressure freezing. In *Cryotechniques in Biological Electron Microscopy*, Steinbrecht, R. and Zierold, K. (eds), pp. 175–191. Berlin: Springer.

Norlen, L., Oktem, O., and Skoglund, U. 2009. Molecular cryo-electron tomography of vitreous tissue sections: Current challenges. *J Microsc-Oxf*, 235, 293–307.

Plitzko, J. M. and Baumeister, W. 2006. Cryo-electron tomography (cet). In *Science of Microscopy*, Hawkes, P. and Spence, J. (eds), pp. 535–604. New York: Springer.

Radermacher, M., Wagenknecht, T., Verschoor, A., and Frank, J. 1987. Three-dimensional reconstruction from a single-exposure, random conical tilt series applied to the 50S ribosomal subunit of *Escherichia coli*. *J Microsc*, 146, 113–136.

Rigort, A., Bäuerlein, F. J. B., Leis, A. et al. 2010. Micromachining tools and correlative approaches for cellular cryo-electron tomography. *J Struct Biol*, 172, 169–179.

Ruprecht, J. J., Mielke, T., Vogel, R., Villa, C., and Schertler, G. F. 2004. Electron crystallography reveals the structure of metarhodopsin i. *EMBO J*, 23, 3609–3620.

Saito, A., Inui, M., Radermacher, M., Frank, J., and Fleischer, S. 1988. Ultrastructure of the calcium release channel of sarcoplasmic reticulum. *J Cell Biol*, 107, 211–219.

Schenk, A. D., Castano-Diez, D., Gipson, B. et al. 2010. 3D reconstruction from 2D crystal image and diffraction data. *Methods Enzymol*, 482, 101–129

Scheres, S. H. W., Melero, R., Valle, M., and Carazo, J. M. 2009. Averaging of electron sub-tomograms and random conical tilt reconstructions through likelihood optimization. *Structure*, 17, 1563–1572.

Stokes, D. L. 2007. Desmosomes from a structural perspective. *Curr Opin Cell Biol*, 19, 565–571.

Studer, D., Humbel, B. M., and Chiquet, M. 2008. Electron microscopy of high pressure frozen samples: Bridging the gap between cellular ultrastructure and atomic resolution. *Histochem Cell Biol*, 130, 877–889.

Swamy-Mruthinti, S. 2001. Glycation decreases calmodulin binding to lens transmembrane protein, mip. *Biochim Biophys Acta*, 1536, 64–72.

Swamy-Mruthinti, S. and Schey, K. L. 1997. Mass spectroscopic identification of in vitro glycated sites of mip. *Curr Eye Res*, 16, 936–941.

Taylor, K. A. and Glaeser, R. M. 1973. Hydrophilic support films of controlled thickness and composition. *Rev Sci Instrum*, 44, 1546–1547.

Taylor, K. A. and Glaeser, R. M. 1974. Electron diffraction of frozen, hydrated protein crystals. *Science*, 186, 1036–1037.

Taylor, K. A. and Glaeser, R. M. 1976. Electron-microscopy of frozen hydrated biological specimens. *J Ultrastruct Res*, 55, 448–456.

Taylor, K. A. and Glaeser, R. M. 2008. Retrospective on the early development of cryoelectron microscopy of macromolecules and a prospective on opportunities for the future. *J Struct Biol*, 163, 214–223.

Topf, M., Lasker, K., Webb, B. et al. 2008. Protein structure fitting and refinement guided by cryo-em density. *Structure*, 16, 295–307.

Toyoshima, C. and Unwin, N. 1990. 3-dimensional structure of the acetylcholine-receptor by cryoelectron microscopy and helical image-reconstruction. *J Cell Biol*, 111, 2623–2635.

Unwin, P. N. and Henderson, R. 1975. Molecular structure determination by electron microscopy of unstained crystalline specimens. *J Mol Biol*, 94, 425–440.

Van Heel, M. 1984. Multivariate statistical classification of noisy images (randomly oriented biological macromolecules). *Ultramicroscopy*, 13, 165–183.

Van Heel, M. 1987. Angular reconstitution: A posteriori assignment of projection directions for 3D reconstruction. *Ultramicroscopy*, 21, 111–123.

Van Heel, M. and Frank, J. 1981. Use of multivariate statistics in analysing the images of biological macromolecules. *Ultramicroscopy*, 6, 187–194.

Van Heel, M., Gowen, B., Matadeen, R. et al. 2000. Single-particle electron cryo-microscopy: Towards atomic resolution. *Q Rev Biophys*, 33, 307–369.

Vleminckx, K. and Kemler, R. 1999. Cadherins and tissue formation: Integrating adhesion and signaling. *Bioessays*, 21, 211–220.

Wang, Q., Mercogliano, C. P., and Lowe, J. 2011. A ferritin-based label for cellular electron cryotomography. *Structure*, 19, 147–154.

Wei, X., Decker, J. M., Wang, S. et al. 2003. Antibody neutralization and escape by HIV-1. *Nature*, 422, 307–312.

White, T. A., Bartesaghi, A., Borgnia, M. J. et al. 2010. Molecular architectures of trimeric SIV and HIV-1 envelope glycoproteins on intact viruses: Strain-dependent variation in quaternary structure. *PLoS Pathogens*, 6, e1001249.

Wu, S. R., Loving, R., Lindqvist, B. et al. 2010. Single-particle cryoelectron microscopy analysis reveals the HIV-1 spike as a tripod structure. *Proc Natl Acad Sci USA*, 107, 18844–18849.

Yu, Z. and Frangakis, A. S. 2011. Classification of electron sub-tomograms with neural networks and its application to template-matching. *J Struct Biol*, 174, 494–504.

Yu, X. K., Jin, L., and Zhou, Z. H. 2008. 3.88 angstrom structure of cytoplasmic polyhedrosis virus by cryo-electron microscopy. *Nature*, 453, 415–419.

Zanetti, G., Briggs, J. A. G., Grünewald, K., Sattentau, Q. J., and Fuller, S. D. 2006. Cryoelectron tomographic structure of an immunodeficiency virus envelope complex in situ. *PLoS Pathogens*, 2, e83.

Zemlin, F., Beckmann, E., and Vandermast, K. D. 1996. A 200 kv electron microscope with Schottky field emitter and a helium-cooled superconducting objective lens. *Ultramicroscopy*, 63, 227–238.

Zhang, X., Jin, L., Fang, Q., Hui, W. H., and Zhou, Z. H. 2010a. 3.3 Ångstrom cryo-em structure of a nonenveloped virus reveals a priming mechanism for cell entry. *Cell*, 141, 472–482.

Zhang, X., Settembre, E., Xu, C. et al. 2008. Near-atomic resolution using electron cryomicroscopy and single-particle reconstruction. *Proc Natl Acad Sci USA*, 105, 1867–1872.

Zhang, S. H., Vasishtan, D., Xu, M., Topf, M., and Alber, F. 2010b. A fast mathematical programming procedure for simultaneous fitting of assembly components into cryoEM density maps. *Bioinformatics*, 26, i261–i268.

Zhu, P., Liu, J., Bess, J. et al. 2006. Distribution and three-dimensional structure of aids virus envelope spikes. *Nature*, 441, 847–852.

Zhu, P., Winkler, H., Chertova, E., Taylor, K. A., and Roux, K. H. 2008. Cryoelectron tomography of HIV-1 envelope spikes: Further evidence for tripod-like legs. *PLoS Pathogens*, 4, e1000203.

Section II

Theortical/Modeling Techniques to Predict Three-Dimensional Structures

4 Protein–Carbohydrate Interactions

Computational Aspects

Anita Sarkar and Serge Pérez

CONTENTS

4.1 INTRODUCTION

In nature, carbohydrates form an important family of biomolecules, as simple or complex carbohydrates, either alone or covalently linked to proteins or lipids. Most of the earlier studies on carbohydrates focused on plant polysaccharides, such as cellulose, starch, pectins, and so on, largely because of their wide range of applications. More recently, the role of carbohydrates in biological events has been recognized, and glycobiology has emerged as a new and challenging research area at the interface of biology and chemistry. Of special interest are the carbohydrate-mediated recognition events that are important in biological phenomena, which give a pivotal role to the study of protein–carbohydrate interactions. Actually, the binding protein partners of carbohydrates encompass a wide variety of macromolecules involved in functions such as recognition, biosynthesis, modification, hydrolysis, and so on (Figure 4.1).

Determination of the three-dimensional (3D) structural and dynamical features of complex carbohydrates, carbohydrate polymers, and glycoconjugates, along with the understanding of the molecular basis of their associations and interactions, represent the main challenges in structural glycoscience (Woods and Tessier 2010).

Elucidation of the 3D structures and the dynamical properties of oligosaccharides is a prerequisite for a better understanding of the relationships between structures and functions, involving the biochemistry of recognition processes and the subsequent rational design of carbohydrate-derived drugs. Seemingly, the elucidation and the understanding of the different structural levels of polysaccharides are required to relate structure to properties. Ultimately, some polysaccharides are also carriers of biological information that can only be deciphered if their interactions with other biological macromolecules are understood. Unfortunately, oligosaccharides, either in their free form or as part of glycoconjugates, are inherently difficult to crystallize, and structural data from x-ray studies are sparse (Pérez et al. 2000). In solution, the flexibility of certain glycosidic linkages produces multiple conformations that coexist in equilibrium. The use of several spectroscopic methods, with appropriate time resolution, is necessary for analysis of the conformational behavior of such molecules (Rice et al. 1993; Peters and Pinto 1996). As for polysaccharides, they differ from other biological macromolecules because the diffraction data that can be obtained are not sufficient to permit crystal structure determination based on the data alone. Hence, procedures for molecular modeling of carbohydrates and carbohydrate polymers have been devised as an important tool for structural studies of these compounds. Various molecular modeling methods have been developed (Woods 1996) and have been widely used for the determination of oligosaccharide and polysaccharide conformations (Pérez and Kouwijzer 1999). The progress made in algorithms and computational power allows for the simulation of carbohydrates in their natural environment, that is, solvated in water, in organic solvent, or in concentrated solution. These developments along with their applications have been thoroughly reviewed in a previously published chapter (Pérez 2007).

Carbohydrates, along with proteins and nucleic acids, constitute one of the central building blocks of life. The interactions between proteins and carbohydrates

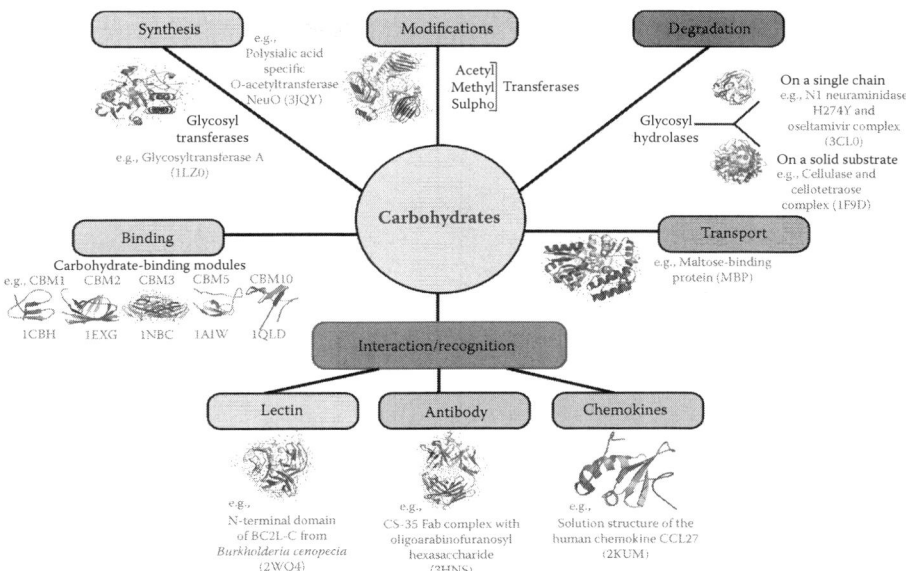

FIGURE 4.1 (See color insert.) Synopsis of the families of proteins interacting with carbohydrates, illustrated with examples from PDB for synthesis (Parthasarathy et al. 2002), modifications (acetyltransferase) (Jansma et al. 2010), degradation by glycosyl hydrolases on a single chain (Collins et al. 2008) or on a solid substrate (Parsiegla et al. 2000), binding by carbohydrate-binding modules (CBMs) (Brun et al. 1997; Kraulis et al. 1989; Raghothama et al. 2000; Tormo et al. 1996; Xu et al. 1995), transport (Cuneo et al. 2009), interaction/recognition by lectins (Sulak et al. 2010), antibodies (Murase et al. 2009), and chemokines (Schulz et al. 2011).

play a role in numerous biological processes such as protein specificity in antibody–antigen recognition, cell–cell adhesion, enzyme–substrate specificity, molecular transport, and so on. They are critical to the onset, detection, and, potentially, also the prevention of human diseases such as cancer. The interactions between proteins and complex carbohydrates such as polysaccharides are also involved in the biosynthesis and biodegradation of the major raw materials on Earth. Experimental assessment of carbohydrate recognition by x-ray crystallography is impeded by difficulties of cocrystallizing proteins and carbohydrates. Nevertheless, highly resolved protein–carbohydrate complexes gathered from x-ray synchrotron investigations have accumulated to the point where it has been possible to compare the experimentally derived structures with those predicted from computational methods. Some general features governing protein–carbohydrate interactions have been derived, and computational tools have evolved and improved accordingly. These tools provide efficient ways to increase our understanding of the different contributions to the binding energy. These developments allow searching conformational space efficiently and yield reliable estimates of binding free energy. They allow exploring *in silico* cases where the experimental data are lacking, and provide sound structural information for a rational design of bioactive carbohydrates or carbohydrate mimetics.

In this chapter, we aim to review the significant contributions and the present status of the application of computational methods to the characterization and prediction of protein–carbohydrate interactions.

4.2 SPECIFIC FEATURES OF CARBOHYDRATE MODELING

Carbohydrates have a potential information content that is several orders of magnitude higher than any other biological macromolecule. The diversity of carbohydrate structures results from the broad range of monomers (>100) of which they are composed and the different ways in which these monomers are joined (glycosidic bonds). Thus, even a small number of monosaccharide units can provide a large number of different oligosaccharides (also referred to as glycans), including branched structures, a unique feature among biomolecules. For example, the number of all possible linear and branched isomers of a hexasaccharide exceeds 10^{12} (Laine 1994).

The carbohydrate recognition mechanism depends on (1) the sequence of the monosaccharides in the glycan (i.e., glucose vs. mannose), (2) the anomeric centers (i.e., α or β), (3) the linkage positions (i.e., 1–3 vs. 1–4), and (4) the chemical modifications to the core glycan (i.e., sulfation, phosphorylation, methylation, acetylation, etc.). The strength of this interaction is also determined by the carbohydrate conformation and orientation with respect to the binding site.

Carbohydrates and their derivatives possess many hydroxyl groups and thus a large number of rotatable bonds. Due to the many hydroxyl groups, these compounds are usually highly water soluble, and their logP is often negative. The surface of carbohydrates and their derivatives is composed of hydrophobic and hydrophilic patches formed by nonpolar aliphatic protons and polar hydroxyl groups. This leads to anisotropic solvent densities around carbohydrate molecules. In aqueous environments, favorable interactions of water molecules with the hydrophilic patches result from electrostatic interactions and hydrogen bonding. Conversely, the interaction of water with hydrophobic surface patches is unfavorable. Such equilibrium between hydrophobic and hydrophilic patches forms the basis for such properties as carbohydrate solubility in water, or such functions as molecular recognition.

Another essential feature of carbohydrates is their conformational flexibility (Tvaroska and Pérez 1986). Compared to drug-like molecules, carbohydrates are typically much more flexible. The relative orientations of two consecutive monosaccharide units in a disaccharide moiety are expressed in terms of the glycosidic linkage torsional angles Φ and Ψ around the glycosidic bonds, which are defined as $\Phi = \text{O5–C1–O–C}_x$ and $\Psi = \text{C1–O–C}_x\text{–C}_{(x+1)}$ for a $(1 \rightarrow x)$ linkage (Figure 4.2). The energetically favorable conformations of a carbohydrate dimer may be easily shown on energy plots called (Φ, Ψ) maps, which are somewhat similar to the Ramachandran plots used to visualize the backbone dihedral angles of the constituent amino acids in proteins. These plots feature multiple minima with the separating energy barriers being over 10–15 kcal/mol.

However, carbohydrates in complex were found to adopt conformations belonging to different minima. These observations underline the necessity for thoroughly

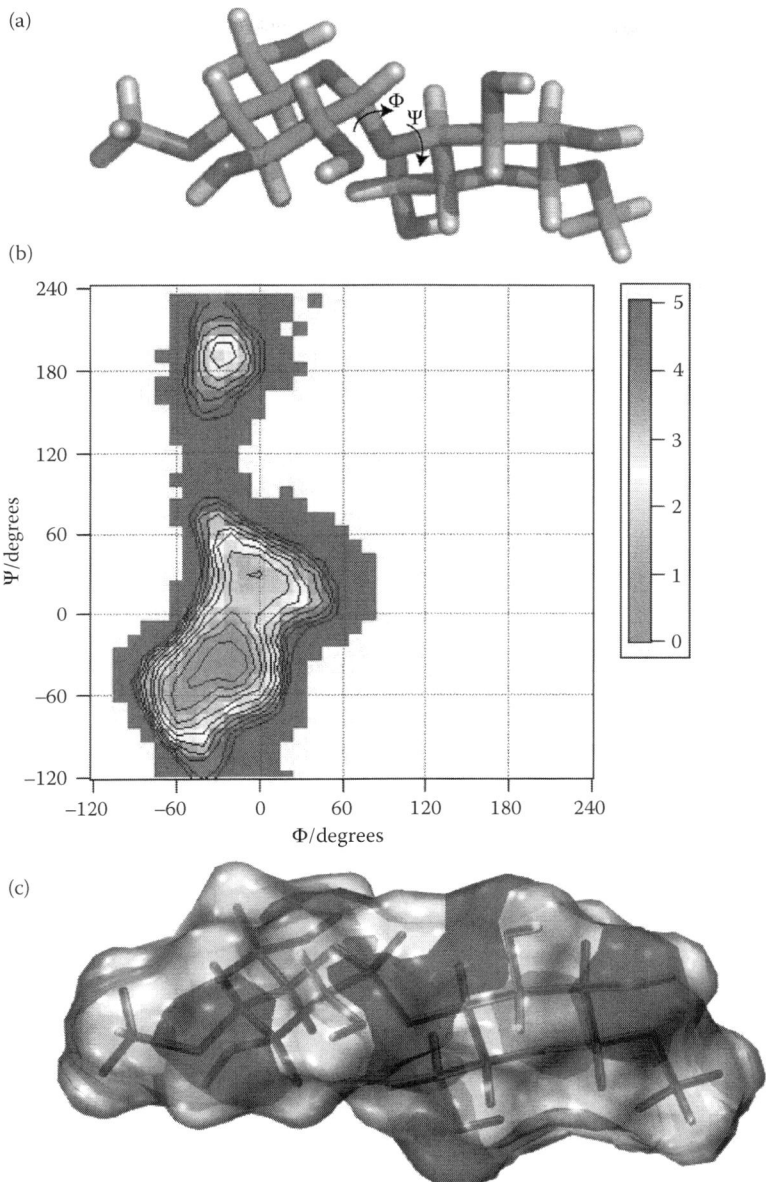

FIGURE 4.2 **(See color insert.)** (a) Molecular representation of the disaccharide (Glcα(1-4) Glcβ) with the Φ and Ψ torsion angles shown on the glycosidic linkage. The potential energy surface shows conformational energy with respect to the Φ and Ψ torsion angles. (b) The favored low-energy Φ/Ψ combinations are shown in light color, while the high-energy regions are shown in red and the inaccessible regions are shown in white. (c) The surface of the disaccharide is composed of hydrophobic (green) and hydrophilic (red) patches, formed by nonpolar aliphatic protons and polar hydroxyl groups.

sampling the conformational space of carbohydrate oligomers during docking. While this may be feasible for glycosidic bonds, the number of degrees of freedom increases rapidly when, in addition to this, we take into account the orientation of the hydroxyl groups.

4.3 PROTEIN–CARBOHYDRATE INTERACTIONS

As with other types of macromolecular interactions, the formation of the complex is driven by favorable changes in enthalpy (ΔH) and entropy (ΔS). Thermodynamic measurements have indicated that the binding free energy (ΔG) of monosaccharides to proteins is quite small. ΔG increases in a significant manner whenever disaccharides or higher oligosaccharides are interacting with proteins. Whenever such proteins are interacting with carbohydrates, a high "avidity" is observed as a result of a multivalent effect. The binding free energy between a carbohydrate molecule and a protein partner is indeed the variable of interest to be assessed. It is assumed to be composed of independent contributions in terms of van der Waals forces, electrostatic interactions with or without encompassing hydrogen bonding, the hydrophobic effect, and so on.

4.3.1 Van der Waals and Electrostatic Interactions

From the large number of hydrogen bond donors and acceptors present in carbohydrates, complex and dense hydrogen bonding networks with proteins arise. The complexity of such networks is enhanced by the competition occurring with the water molecules for hydrogen bonds. The overall enthalpic gain from hydrogen bonding may be counterbalanced by some entropic cost.

4.3.2 CH–π Interactions

These characterize the enthalpy of binding of carbohydrates to protein. It is defined as a type of hydrogen bond occurring between a hydrogen atom attached to a carbon and the π systems of arenes. Typically, this is a weak effect. Despite the full recognition of this effect, its computation requires a high level of theory and is not fully taken into account in the computational procedures (Spiwok et al. 2004).

As observed in many crystal structures of protein–carbohydrate complexes, aromatic residues of the proteins are often stacked against some faces of the carbohydrates. Such an arrangement results from the hydrophobic effect wherein small hydrophobic moieties of the solute induce an ordering of the water molecules at the solvent interface. The resulting decrease of the hydrophobic surface area induces a decrease in solvent ordering and a consequent favorable change in entropy. Alternatively, a nonclassical hydrophobic effect has also been documented to occur in lectin–carbohydrate complexes, where the complex formation is driven by enthalpy due to favorable interactions between the solute molecules forming the complex as well as favorable interactions between the solvent molecules (Poveda et al. 1997; Chervenak and Toone 1994).

4.3.3 SOLVATION–DESOLVATION

As a result of docking carbohydrates into proteins, the number of atomic contacts between the ligand and the protein is maximized, and the subsequent structure is such that the carbohydrate lies more or less flat on the protein surface (Tschampel and Woods 2003). However, x-ray crystal structures show contradictory features, with carbohydrate residues extending into the surrounding solvent. These structures might be correctly computed if the impact of solvation and desolvation on the binding free energy were properly taken into account.

4.4 FORCE FIELDS DESIGNED FOR CARBOHYDRATES

To study carbohydrate structures and properties using molecular modeling techniques, molecular mechanics potential energy functions and parameters specific for this class of molecules are required. Appropriate force fields for carbohydrate systems have been created, with the aim of reproducing the particular effects that influence their global structural properties in solution (Sorin and Pande 2005). The exocyclic hydroxymethyl group behavior is defined by the ω-angle (O5–C5–C6–O6), and its preference for *gauche* states can be reproduced by introducing scaling factors that slightly modify the 1–4 nonbonded interactions (Kirschner and Woods 2001). 1–4 nonbonded interactions define the influence, in terms of electrostatic and van der Waals potentials. 1–4 nonbonded interactions are not treated in the same manner in all force fields (Figure 4.3), and this could be a problem in simulating complex

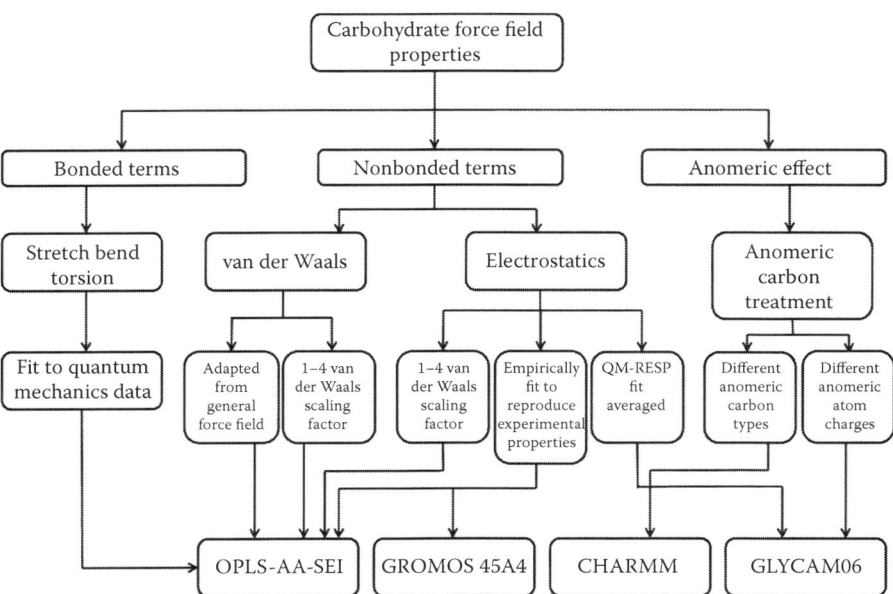

FIGURE 4.3 Parameterization protocol comparison between the carbohydrate force fields: GLYCAM06; GROMOS 45A4, CHARMM, and OPLS-AA-SEI.

systems in which two different force fields have to be used. In these cases, the separate treatment of 1–4 nonbonded interactions can assure a full compatibility among the force fields. The potential impact of choosing the 1–4 scaling factors often becomes irrelevant when glycans bind to proteins because generally their freedom in the binding site is reduced. In literature, several reviews describe and compare the performance of carbohydrate force fields used in glycomodeling (Imberty and Pérez 2000; Pérez et al. 1998).

To simulate the behavior of carbohydrates *in vacuo* or in solution (e.g., to study ring puckering (Dowd et al. 1994) or rotational barriers of oligosaccharides), either established force fields or special parameterizations may be used (Glennon and Merz 1997; Homans 1990; Kirschner et al. 2008; Momany and Willett 2000; Woods et al. 1995). Such force fields allow investigation and prediction of the deformation of carbohydrate rings. These special force fields (as well as previously established ones) have been employed repeatedly for molecular dynamics (MD) simulations of protein–carbohydrate complexes (Bradbrook et al. 2000; Tempel et al. 2002). In some cases, the simulations were successfully used for estimating binding free energies (Bryce et al. 2001; Clarke et al. 2001; Liang et al. 1996; Pathiaseril and Woods 2000).

Despite the many possible advantages of established force fields, they were not designed to predict binding free energies or enthalpies in protein–ligand docking. Since solvent molecules are usually modeled explicitly, force fields do not need to include extra terms for hydrophobic effects. The special CH–π interactions are not taken into account (Laughrey et al. 2008; Ramirez-Gualito et al. 2009).

Some force fields do model hydrogen bonds explicitly, while others regard it as part of the electrostatic interaction. Irrespective of the approach, displacement of water molecules competing for hydrogen bonds is not accounted for.

Some force fields correlate well with *ab initio* calculations for *ab initio* optimized geometries (Spiwok et al. 2005). A recent comparison of the results of *ab initio* and force field calculations underlines the difficulties in predicting binding enthalpies in protein–carbohydrate complexes using existing force fields; for example, the stabilizing interaction energy for the interaction between fucose and tryptophan is heavily overestimated by the AMBER* force field (Vandenbussche et al. 2008).

GLYCAM06 is a widely used force field for modeling carbohydrates, glycoproteins, and glycolipids, as well as for protein–carbohydrate complexes (Kirschner et al. 2008; Tessier et al. 2008). It can be used for describing the physicochemical properties of complex carbohydrate derivatives, and it is fully compatible with the AMBER force field. Parameters have been developed taking into account a test set of 100 molecules from the chemical families of hydrocarbons, alcohols, ethers, amides, esters, carboxylates, molecules of mixed functional groups, and simple ring systems related to cyclic carbohydrates and fit to quantum mechanical data. GLYCAM06 may be used in simulation packages other than AMBER through the employment of appropriate file conversion tools.

To facilitate the parameter transferability, all atomic sequences have an explicitly defined set of torsion terms, with no generic terms, and PARM94 parameters,

---

* As implemented in the Maestro Program (1995 version).

the same used in AMBER, are used for modeling the carbohydrate van der Waals terms (Cornell et al. 1995). No scaling factors for treating 1–4 interactions are introduced for reproducing the *gauche* effect on ω angle rotamers (Kirschner and Woods 2001).

In GLYCAM06, the stereoelectronic effects that influence bond and angle variations at the anomeric carbon atom are included in a unique anomeric atom type. This feature permits to mimic the ring flipping observed in glycosidic monomers that occur, for example, during catalytic events (Biarnes et al. 2006). Comparison with experimental data confirmed that the force field is able to reproduce rotational energies and carbohydrate features quite well if combined with an appropriate charge set, except for highly polar molecules for which empirical terms have been introduced to correct energetic torsion errors (Kirschner et al. 2008). The atomic partial charges are calculated residue by residue. For each residue, 50–100 ns MD simulation is performed, 100–200 snapshots are extracted, and charges are calculated by fitting to the averaging quantum mechanics molecular electrostatic potential (ESP). This strategy is adopted for incorporating the dependence of molecular conformations on partial charges. Restraints are employed in the ESP-fitting procedure (RESP) to ensure that the charges on all aliphatic hydrogen atoms are zero since C–H aliphatic hydrogen atoms are not significant for reproducing dipole moments (Basma et al. 2001; Woods et al. 1990). An optimal RESP charge restraint weight of 0.01 is applied, based on simulations of carbohydrate crystal lattices (Woods and Chappelle 2000).

GROMOS 53A6(CARBO), CHARMM, and OPLS-AA are alternative carbohydrate force fields used, together with GLYCAM06, to describe conformational carbohydrate properties in computational chemistry. The GROMOS force field was earlier developed for MD simulations of proteins, nucleotides, or sugars in aqueous or apolar solutions or in crystalline form, but it has been modified and upgraded to include the anomeric effects for mono- and oligopyranoses (Lins and Hünenberger 2005; Hansen and Hünenberger 2011). As in GLYCAM06, quantum mechanics methods are used for calculating bond and angle force constants, whereas dihedral parameters derivation and van der Waals terms are directly taken from previous GROMOS versions (Schuler et al. 2001; Schuler and Van Gunsteren 2000). An ESP-fitting procedure, with restraints on aliphatic hydrogen atoms and averaging over atom types, is chosen for reproducing the electrostatic potential, using a trisaccharide as a model for charge development (Lins and Hünenberger 2005). No distinction is done between α and β monomers in terms of charges and anomeric atom type, and electrostatic–van der Waals 1–4 scaling factors are not introduced so as to correctly reproduce the *gauche* effects on ω angles. Twenty-nanosecond-long MD simulation in explicit water (Berendsen et al. 1981) was used for validating the force field, showing the capability to correctly predict the stereoelectronic effects and the most stable ring conformations but sometimes failing to reproduce their correct energies. GROMOS was proposed as the more adapted force field to mimic the transition from $^4C_1$ to skew boat conformations of the iduronic acid residues in heparin MD simulations (Gandhi and Mancera 2009).

The CHARMM force field was extended to glucopyranose and its diastereomers (Guvench et al. 2008). Several revisions for carbohydrates have been proposed in order to extend this force field to five member sugar rings and oligosaccharides

(Guvench et al. 2009; Hatcher et al. 2009). The same hierarchical parameterization procedure and treatment of 1–4 nonbonded interactions are used to ensure a full compatibility with other CHARMM biomolecular force fields (MacKerell et al. 2000, 1998, 2004). Preliminary parameter sets are created using small-molecule models corresponding to fragments of pyranose rings and then successively applied to complete pyranose monosaccharide structures. Missing dihedral parameters are developed by fitting over 1800 quantum mechanical hexopyranose conformational energies. Both partial atomic charges and Lennard–Jones parameter values, taken from previous CHARMM versions, are adjusted to reproduce scaled quantum mechanical carbohydrate–water interaction energies and distances and further refined to reproduce experimental heats of vaporization and molecular volumes for liquids. The force field, with different atom types for α and β anomers, was validated as it reproduces calculated quantum mechanical and experimental properties using MD simulations with TIP3P water models.

The OPLS force field has been expanded to include carbohydrates (Kony et al. 2002). In OPLS-AASEI (scaling electrostatic interactions) force field, 1–4, 1–5, and 1–6 scaling factors are introduced to improve the prediction of Φ/Ψ conformations properties, as well as anomeric effects and relative energies (Kony et al. 2002). Unique charge sets and atom types for α and β anomers are used. All nonbonded parameters are imported directly from the parent force field OPLS-AA (Damm et al. 1997). Charges are derived, as done for previous force field versions (Damm et al. 1997; Jorgensen et al. 1996), from standard alcohols and acetals to simply reproduce consistent energetic properties, and then transferred to carbohydrates.

Other force fields are employed to understand carbohydrate properties *in silico*. In particular, MM3, a force field initially meant for hydrocarbons, is applicable to a wide range of compounds. The MM3 force field for amides, polypeptides, and proteins (Allinger et al. 1990; Lii and Allinger 1991) is widely used for the construction of adiabatic maps of disaccharides. TRIPOS molecular mechanics force field is designed to simulate both peptides and small organic molecules (Clark et al. 1989), but parameter extension for oligosaccharides includes sulfated glycosaminoglycan (GAG) fragments and glycopeptides carbohydrate interactions (Imberty et al. 1991; Pérez et al. 1995). The TRIPOS force field is implemented in the molecular package SYBYL (Tripos Associates, St. Louis, Missouri) and commonly used for geometry optimizations.

4.5 COMPUTATIONAL TOOLS FOR DOCKING CARBOHYDRATES ON PROTEINS

4.5.1 MOLECULAR DOCKING

Molecular docking is a computational procedure that aims at predicting the preferred orientation of a ligand to its target protein, when bound to each other to form a stable complex (Lengauer and Rarey 1996). In order to perform computational protein–ligand docking calculations, the 3D structure of the target protein must be known. Each docking program operates slightly differently, but they share common features that enable them to (1) search for locations on the protein surface that lead to favorable interactions with the ligand, (2) sample the conformational space of the ligand, and (3) compute the interaction energy (or score the binding) between the protein and ligand. The interaction with the ligand relies on both the protein backbone fold in the region of the binding site and the orientation of the side chains in the binding site. One of the most significant limitations in docking is that it is generally performed while keeping the protein surface rigid, which prevents the consideration of the effects of induced fit within the binding site. For an overview of carbohydrate–protein docking, see Chapter 5.

4.5.1.1 Difficulties in Molecular Docking

These difficulties are mostly due to the high number of degrees of freedom characterizing a protein–ligand system that increases the computational cost of the calculations. Thus, several approximations about the flexible states may be introduced in molecular docking experiments. The simplest approximation (rigid docking) considers only the three translational and three rotational degrees of freedom of the protein and of the ligand, treating them as two distinct rigid bodies. The most widely used algorithms at present enable the ligand to fully explore its conformational degree of freedom in a rigid-body receptor (Bultinck et al. 2004; Eklund 2005; Höltje et al. 2008; Kranjc 2009; Walker 2003).

4.5.1.2 Docking Algorithms

The docking algorithms can be grouped into deterministic and stochastic approaches. Deterministic algorithms are reproducible, whereas stochastic algorithms include random factors that do not allow the full reproducibility. The following describes the most widely used algorithms in docking simulations.

4.5.1.2.1 Incremental Construction Algorithms

These algorithms consist of the division of a ligand into rigid fragments. One of the fragments is selected and placed in the protein binding site. The reconstruction of the ligand is then carried out *in situ*, adding the remaining ligand fragments. For example, DOCK (Ewing et al. 2001) uses incremental construction algorithm to treat ligand flexibility. It generates points (sphere centers) that fill the binding site and try to capture the binding site shape properties for identifying favorable regions in which the ligand atoms may be located. The ligand is divided along each flexible bond to generate rigid segments. An anchor fragment is then selected from all the rigid pieces and oriented in the active site by matching ligand atoms with sphere centers. Fragments are then added, and all possible placements are scored on the basis of their interactions with the protein using the energetic scoring function. Best anchor fragments are used for completing the construction of the ligand in the protein binding site. The best scored poses of the complete ligand are selected.

4.5.1.2.2 Genetic Algorithms

Genetic algorithms are stochastic searching approaches that use techniques inspired by evolutionary biology to find reliable results. It mimics the process of evolution by manipulating a collection of data structures called chromosomes.

AutoDock (Morris et al. 1998) uses this algorithm for obtaining reliable docking results. First, the protein is placed inside a cube with a predefined size, characterized by a defined number of points (grid points). In the second step, probes corresponding to the different atom types of the ligand are then moved through the cube and, in particular, at each point, protein–probe interaction energies are calculated and stored in affinity maps. Thirdly, a conformational search of the ligand is performed by applying the Lamarckian genetic algorithm. Its characteristic is that environmental adaptations of an individual's phenotype can become heritable traits, transferred to its genotype. At this stage, a minimization or local search is performed, and the results are taken into account modifying the initial conformation that will enter in a new iteration of crossover and mutation of the genetic algorithm cycle.

4.5.1.2.3   Hierarchical Algorithms

The algorithm used in Glide (Friesner et al. 2004) can be defined as a hierarchical algorithm. It uses an exhaustive systematic search for discovering the most favored ligand conformations in the protein active site, with a screening based on progressively restricted energetic cutoffs. Fields containing information of the protein receptor properties are calculated before the algorithm search. Then a set of initial ligand conformations is produced. Initial screens are performed over the whole phase space available to the ligand to locate promising ligand poses in the respective receptor fields. Ligands are minimized in the field of the receptor using a standard molecular mechanics energy function (Damm et al. 1997). Finally, the lowest-energy poses are subjected to a Monte Carlo procedure that examines torsional minima. A composite scoring function is then used to select the correct docked poses.

A variety of other sampling methods like simulated annealing have been implemented in docking programs.

4.5.1.3   Scoring Functions

Energy scoring functions are necessary to evaluate the free energy of binding between proteins and ligands. The equation below is the Gibbs-Helmholtz equation that describes the ligand-receptor affinity:

$$\Delta G = \Delta H - T\Delta S \qquad (4.1)$$

ΔG gives the free energy of binding that is the measure of energetic changes between two states represented by the bound and unbound state of the receptor and the ligand. ΔH is the enthalpy, T the temperature expressed in Kelvin, and ΔS is the entropy of the system. ΔG is related to the binding constant K_a by the equation:

$$\Delta G = -RT \ln K_a \qquad (4.2)$$

where R is the gas constant.

Some sophisticated techniques for predicting binding free energies are currently too slow to be used in molecular docking of large sets of compounds. Thus, fast scoring functions have been developed. Empirical scoring functions use a set of parameterized terms describing properties known to be important in protein–ligand binding to construct an equation for predicting affinities. Multilinear regression is used to optimize these terms using a set of known protein–ligand complexes. These terms usually describe polar–apolar interactions, loss of ligand flexibility (entropy), and desolvation effects. The Glide score 2.5 (Friesner et al. 2004) is a regression-based scoring function:

$$\Delta G = C_{\text{lipo}} \sum f(r_{\text{lr}}) + C_{\text{Hbond}} \sum g(\Delta r) \, h(\Delta\alpha) + C_{\text{metal}} \sum f(r_{\text{lm}})$$
$$+ C_{\text{polar-phob}} V_{\text{polar-phob}} + C_{\text{coul}} E_{\text{coul}} + C_{\text{vdw}} E_{\text{vdw}} + \text{solvation term} \quad (4.3)$$

The first term describes the lipophilic and aromatic interactions, whereas the polar terms are included in the second (hydrogen bonds, separated into differently weighted components that depend on the electrostatic properties of donor and acceptor atoms) and third (ionic interactions) terms. The fourth term rewards instances in which a polar but nonhydrogen bonding atom is found in a hydrophobic region. Coulomb and van der Waals interaction energies between the ligand and the receptor are evaluated as well as the solvation effect.

Force field-based scoring functions (AutoDock, DOCK) are based on the non-bonded terms of the classical molecular mechanics force fields. A Lennard-Jones potential describes van der Waals interactions, whereas the Coulomb energies describe the electrostatic interactions. In AutoDock (Morris et al. 1998), the implemented scoring function has the following form:

$$\Delta G = \Delta G_{\text{vdw}} \sum_{i,j} \left[\left(A_{ij} / r_{ij}^{12} \right) - B_{ij} / r_{ij}^{6} \right] + \Delta G_{\text{Hbond}} + \Sigma_{i,j} \, E(t) \left[\left(C_{ij} / r_{ij}^{12} \right) - D_{ij} / r_{ij}^{10} \right]$$
$$+ \Delta G_{\text{elec}} \sum_{i,j} q_1 q_2 / \varepsilon \left(r_{ij} \right)^2 + \Delta G_{\text{tor}} N_{\text{tor}} + \Delta G_{\text{sol}} \sum_{i,j} \left(S_i V_j + S_i V_i \right)^{e\left(-r 2ij / 2\sigma 2 \right)}$$

$$(4.4)$$

where the five ΔG terms are coefficients empirically determined using linear regression analysis from a set of protein–ligand complexes with known binding constants. The summations are performed over all pairs of ligand atoms, i, and protein atoms, j. The first three terms describe the Lennard-Jones dispersion, the directional hydrogen bonds, and the Coulomb electrostatic potential taken from the AMBER force field (Cornell et al. 1995). ΔG_{tor} is an empirical measure of the unfavorable entropy of ligand binding due to the restriction of conformational degrees of freedom, whereas N_{tor} is the number of ligand rotatable bonds. In the fifth term, for each atom type within the ligand, fragmental volumes of the surrounding protein atoms V are weighted by an exponential function and summed, evaluating the percentage of volume around the ligand atom that is occupied by protein atoms. This percentage is then weighted by the atomic solvation parameter S of the ligand atom to give the desolvation energy (Morris et al. 1998). Several developed docking approaches use knowledge-based scoring functions based on statistical observations of intermolecular close contacts in protein–ligand x-ray databases, which are used to derive potentials of mean force. This method assumes that the frequency of close intermolecular interactions between certain ligand and protein atoms contributes favorably to the binding affinity. In this approach, no fitting to experimental affinities is required, and solvation and entropic terms are treated implicitly (Muegge and Martin 1999).

4.5.2 MD Simulations

In MD simulations (Figure 4.4), an ensemble of configurations is generated by applying the laws of motion to the atoms of the molecule. The concept behind MD simulation involves calculating the displacement coordinates in time (trajectory) of a molecular system at a given temperature. Finding positions and velocities of a set of particles as a function of time is done classically by integrating Newton's equation of motion in time. Molecular simulations are usually carried out as a micro-canonical (constant-NVE) or canonical (constant-NVT) ensemble. As a consequence, all other thermodynamic quantities must be determined by ensemble averaging. In a classical system, Newton's equations of motion conserve energy and thus provide a suitable scheme for calculating a microcanonical ensemble. However, canonical ensemble can readily be performed

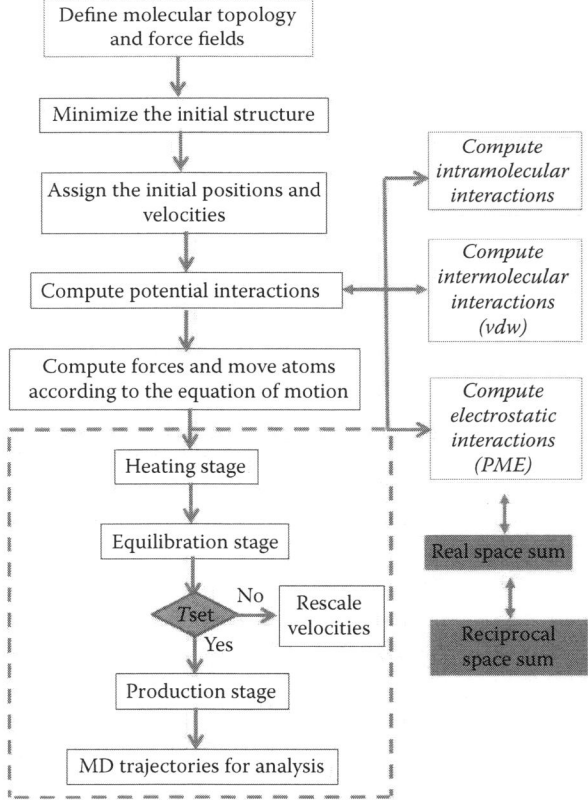

FIGURE 4.4 General scheme and the practical sequential approach of MD simulations.

by coupling the molecular system to a constant-temperature bath, which rescales the atomic velocities according to the desired temperature. In a similar manner, constant-pressure simulations can be performed by scaling through coupling to a constant-temperature position, as the pressure can be calculated from the virial theorem.

Several algorithms have been developed for MD simulations. Such simulations follow a system for a limited time. Physically observed properties are computed as the appropriate time averages through the collective behavior of individual molecules. For the results to be meaningful, the simulations must be sufficiently long so that the important motions are statistically well sampled. Experimentally accessible spectroscopic and thermodynamic quantities can be computed, compared, and related to microscopic interactions. It should be noted that MD is severely limited by the available computer power. With presently available computers, it is feasible to perform a simulation with several thousand explicit atoms for a total time of up to a few microseconds. To explore the conformational space adequately, it is necessary to perform many such simulations. In addition, it may be possible that carbohydrate molecules undergo dynamical events on longer timescales. These motions cannot be investigated with standard MD techniques. Another way is to use high-temperature dynamics to allow the molecule to assume high-energy conformations. This approach has to be used with caution since it can make the molecules acquire "nonphysiological" conformations.

4.5.3 MOLECULAR ROBOTICS

While a number of simulations are performed using MD calculations, it must be recognized that they are usually performed on short time scales and have therefore allowed modeling of the dynamic properties of equilibrium states. Indeed, simulations that are needed to capture an entire conformational event, particularly with explicit simulation of solvents are usually too short, relative to the characteristic time of conformational changes occurring upon binding. Consequently, alternative methods such as essential dynamics or normal-mode analysis have been successfully applied to selectively enhance conformational sampling along specific directions of motions and identify large collective motions that may occur in the protein upon binding to a carbohydrate. Novel methods are being developed with the aim of simulating molecular motions that can occur on large spatial and temporal scales.

In addition, to simplify models, methods alternative to MD simulation can be applied to perform an effective exploration of the conformational space. Algorithms originally developed to compute robot motions have been extended and proposed as alternative methods to compute molecular motions (Figure 4.5). Robotics-based algorithms have been applied to the study of several problems such as ligand docking and accessible pathways in flexible receptors, or conformational changes of proteins, due to loop motions, domains motions, and so on. A methodology named "molecular robotics" has been developed that separates the search for conformational pathways into two stages. The first stage consists of the exploration of geometrically feasible motions, using the robotics-based approach, whereas the second stage uses molecular mechanics for an evaluation of solutions found in the previous stage, while taking

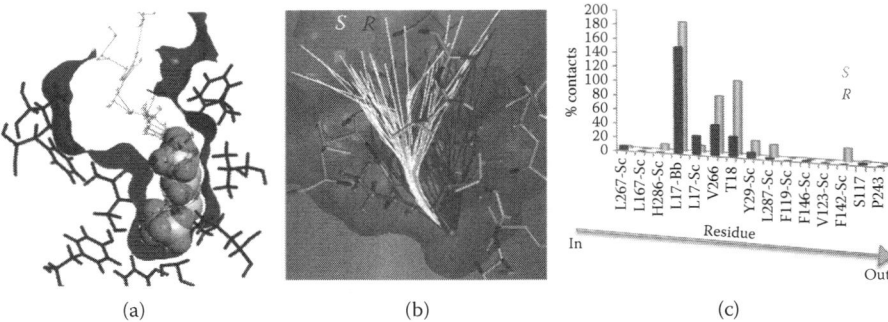

(a)   (b)   (c)

FIGURE 4.5   (See color insert.) Illustration of the molecular robotics approach to investigate the role of substrate accessibility to the active site on *Burkholderia cepacia* lipase enantioselectivity. (a) Conformational exploration of the active-site pocket using path-planning algorithms in order to search exit paths of the ligand from its catalytic position. Exit paths computed for the *R*- and *S*-enantiomers (50 paths for each enantiomer). (b) The distribution obtained for the *R*-enantiomer (blue) appears clearly larger and less constrained than for the *S*-enantiomer (white). (c) Histogram representing for each enantiomer the relative frequency of interatomic contacts (averaged among the 50 paths) with amino acid residues. This automated analysis of ligand–protein contacts enables to highlight amino acid hindering the displacement of enantiomers and thus provide target residues for engineering enantioselectivity (Barbe et al. 2011).

into account explicit simulation of solvents. Such a conformational search method handles large molecular motions in a continuous way and within very short computing times. The key advantage of the robotics-based approach is that it enables fast exploration of high-dimensional conformational spaces thanks to the combination of a geometrical treatment of the main molecular constraints, with the performance of path-planning algorithms.

Based on robotics background, computationally efficient methods have been developed in recent years for sampling and exploring conformational space of biological macromolecules. Combined with methods in computational physics such as normal mode analysis (Kirillova et al. 2008), or using appropriate multiscale molecular models (Haspel et al. 2010), robot path-planning algorithms relying on a mechanistic modeling of (macro)molecules are able to compute large-amplitude conformational transitions in proteins with several orders of magnitude faster than standard simulation methods such as MD (Barbe et al. 2011; Cortés et al. 2011). These robotics-inspired methods have also been developed to simulate ligand displacement inside an active-site pocket of a protein considering both partners as flexible molecular models with very low computational cost (Cortés et al. 2011, 2008, 2010) and provide information about the interactions between the ligand and the protein and about the required conformational changes that are important for understanding the complex biochemical processes. Such methods have already been successfully applied for rational enzyme engineering (Guieysse et al. 2008; Lafaquière et al. 2009), showing the efficiency and the potential of molecular robotics methods to guide the engineering of enzyme mutants with improved activity, selectivity, and specificity.

4.5.4 FREE ENERGY CALCULATIONS

The absolute ligand–receptor interaction energies can be obtained by performing average Molecular Mechanics/Poisson–Boltzmann Surface Area (MM-PBSA) calculations on an ensemble of uncorrelated snapshots in an implicit water environment, collected from an equilibrated MD simulation (Figure 4.6). MM–PBSA is a method that approximates the average free energy of binding ΔG between the ligand L and the receptor R in an implicit aqueous environment:

$$\Delta G = \Delta G_{RL} - \Delta G_R - \Delta G_L \qquad (4.5)$$

Each term of the above equation is further decomposed as follows:

$$\Delta G_{RL} = \Delta E_{MM} + \Delta G_{PBSA} - T\Delta S_{MM} \qquad (4.6)$$

$$\Delta G_R = \Delta E_{MM} + \Delta G_{PBSA} - T\Delta S_{MM} \qquad (4.7)$$

$$\Delta G_L = \Delta E_{MM} + \Delta G_{PBSA} - T\Delta S_{MM} \qquad (4.8)$$

where ΔE_{MM} is the average molecular mechanical energy containing the bond angles, torsion angles, van der Waals, and electrostatic energetic terms described in

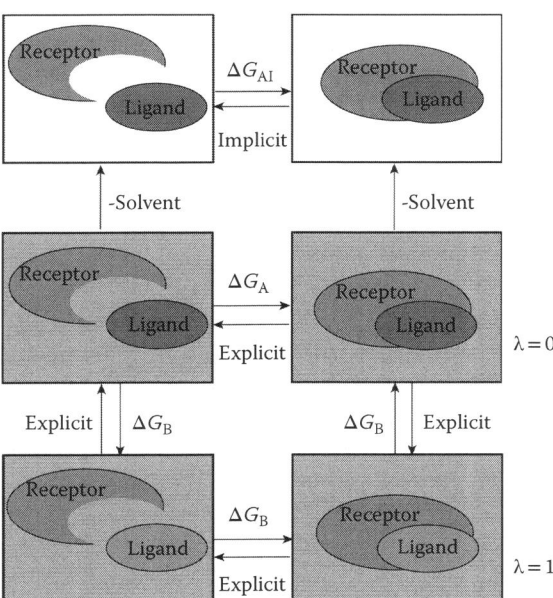

FIGURE 4.6 MM–PBSA calculations determine the absolute free energy of binding of a ligand to a receptor (ΔG_{AI}) in an implicit solvent environment, whereas thermodynamics integration methods calculate the free energy of binding difference between receptor–ligand complexes ($\Delta G = \Delta G_C - \Delta G_D = \Delta G_A - \Delta G_B$), where only the ligand is changed.

the force field. The solvation free energy term ΔG_{PBSA} term contains the electrostatic and nonpolar solvent contributions:

$$\Delta G_{PBSA} = \Delta G_{PB}^{el} + \Delta G_{SA}^{np} \tag{4.9}$$

The Poisson–Boltzmann equation is solved for determining the solvent polar effects ΔG_{PB}^{el} (Gilson and Honig 1988), whereas the solvent accessible surface area is used to determine the nonpolar energetic term ΔG_{SA}^{np} (Sitkoff et al. 1994). Finally, $T\Delta S_{MM}$ represents the entropic term, due to the loss of degrees of freedom upon association. The evaluation of this term represents an issue in computational chemistry, commonly performed by using a quasi-harmonic method or by normal-mode analysis (Srinivasan et al. 1998). The high computational cost combined with a very slow convergence and the approximations introduce significant uncertainty in the result (Basdevant et al. 2006; Gohlke and Case 2004). Thus, the entropy contribution can be neglected in case a comparison of states of similar entropy is desired such as a series of similar ligands binding to the same protein receptor (Kollman et al. 2000).

4.5.4.1 Relative Free Energy of Binding

Thermodynamic integration (TI) calculations compute the free energy difference between two closely related systems A and B by slowly transforming the initial state A to the final state B. The two states are coupled via a parameter λ that serves as an additional, nonspatial coordinate. This parameter describes the transformation from the reference system A to the target system B and allows the free energy difference between the states to be computed as follows:

$$\Delta G(TI) = 1 \int 0 < \delta V(\lambda)/\delta(\lambda) > \lambda \, d\lambda \tag{4.10}$$

In this equation, λ represents the coupling parameter that corresponds to the potential energy $V(A)$ for $\lambda = 0$ and $V(B)$ for $\lambda = 1$. The integration is carried out over the average of the λ derivative of the coupled potential function at given λ values. Thus, MD simulations in explicit water at different discrete λ points are performed, and the value of the integral is calculated numerically. For TI calculations, the system should not undergo significant conformational changes during the transformation, otherwise MD simulations will most likely not sample enough phase space for obtaining converged results (DeMarco and Woods 2008).

4.6 CASE STUDIES

Avoiding the risk of transforming this section into a catalog, only select examples are provided that deal with the relevant classes of the macromolecules for which a range of the conformational features of protein–carbohydrate interactions have been reported by the application of computational methods.

4.6.1 Recognition

4.6.1.1 Lectins

Lectins are oligomeric proteins that can specifically recognize carbohydrates, which per present knowledge act like molecular tools to decipher sugar-encoded messages. They play biologically important roles in recognition processes involved in fertilization, embryogenesis, inflammation, metastasis, and parasite–symbiote recognition from microbes and invertebrates to plants and vertebrates. In the plant kingdom, lectins have been demonstrated to play a role in defense against pathogens or predators and hypothesized to be involved in establishing symbiosis with mushrooms and with bacteria of the *Rhizobia* species. Among the proteins that interact noncovalently with carbohydrates, lectins bind mono- and oligosaccharides reversibly and specifically while displaying no catalytic or immunological activity.

More than 700 crystal structures of lectins have been solved, most of them as complexes with carbohydrate ligands (Krengel and Imberty 2007). At present the 3D lectin database ([Bettler 1998] www.cermav.cnrs.fr/lectines/) makes 922 lectin structures available. The wealth of experimental data obtained from the crystallographic studies of oligosaccharides with lectins provided an essential driving force to the development of molecular modeling methods of complex oligosaccharides in their interactions with proteins. These confirmed the flexible conformational behavior of oligosaccharides that was anticipated from earlier calculations. Studies of molecular recognition of the histo-blood group oligosaccharides by lectins paved the way for the conformational analysis of complex carbohydrate–protein interactions, an area that has been thoroughly reviewed (Imberty and Pérez 2000). Several docking procedures have been developed and tested against the experimental data available.

Increasing crystallographic explorations of oligosaccharide–lectin complexes have made significant progress in the characterization of the binding sites of lectins, which are usually rather shallow, located near the surface, and thus accessible to solvent. This allows predicting the binding mode of complex carbohydrates to proteins (Sulak et al. 2009). In several lectin families of different origins, one or two calcium ions are involved in the carbohydrate-binding site with direct coordination to the sugar hydroxyl groups. Thanks to the availability of well documented 3D structures of lectins in their native and complexed form, they have been considered as a rich playground to develop and test the robustness of docking methods in predicting the binding mode of complex carbohydrates to proteins.

Flexible docking methods of AutoDock, DOCK, and Grid-based Ligand Docking with Energetics (Glide) were compared for a set of bacterial and animal calcium-dependent lectins and their calcium-dependent sites (Nurisso et al. 2008). DOCK represented crystallographic information well, but its lowest energy conformations did not conform to experimental data for all tested cases. Glide results were similar to that of DOCK, but the lowest energy poses were always satisfactory that could mimic the real carbohydrate orientation. AutoDock showed reasonable accuracy in sugar orientation and reported the most accurate distances between calcium ions and sugar hydroxyl groups.

4.6.1.2 Antibodies

The major role of carbohydrates in blood group transfusion and in organ transplants dramatically highlights the importance of carbohydrate–protein interactions as key to major biological processes. The two major histo-blood group carbohydrate determinants (Clausen and Hakomori 1989) are the antigen families, so-called ABH(O) and the Lewis determinants. The majority of the ABO antigens are expressed on human erythrocytes, at the ends of long polylactosaminic chains, while a minority of the epitope is expressed on neutral glycosphingolipids. Despite the key role played by these determinants, the description at the molecular level of the interactions occurring between the antigens and the antibodies is only beginning to be resolved and characterized, for instance, crystal structures of Fab against Lewis determinants (de Geus et al. 2009; Ramsland et al. 2004; van Roon et al. 2004, 2003). The exhaustive investigation of the cross-reaction patterns on nine antibodies against 12 carbohydrate antigens has been conducted through computational methods (Imberty et al. 1995, 1996). 3D descriptors of the molecular properties of the carbohydrate antigens were used in comparative molecular field analysis (CoMFA). Processing of the QSAR data gave indications on the carbohydrate epitopes essential for antibody recognition while yielding insights into the nature of the molecular recognition.

The successful transplantation of pig organs to human (xenotransplantation) is prevented by the occurrence of carbohydrate antigens on the surface of pig organs, which are recognized by xenoreactive antibodies in the human bloodstream. *In silico* protocol aimed at analyzing the interaction between these xenoantigens and antibodies interactions has been developed (Agostino et al. 2009) and applied (Agostino et al. 2010) to the determination of the structures of these terminating carbohydrate antigens in complex with a panel of xenoreactive antibodies.

Cell surface complex carbohydrates and polysaccharides are potent targets for recognizing pathogen infections or cancerous cells. As such, they offer promising or already successful vaccine components against various pathologies. Consequently, their interactions with antibodies are of a significant interest. The elucidation of the molecular basis of the formation of the complexes and also the balance between the enthalpic and the entropic contribution involved in the binding are both required. For the time being, only an appropriate combination of computational and experimental methods may help in establishing these features, in view of developing broad-serotype coverage vaccines.

A majority of life-threatening cases of septicemia, meningitis, and pneumonia occur from the deleterious action of surface capsular polysaccharides of bacteria. Although these polysaccharides may have similar carbohydrate sequences, they may markedly differ in immunogenicity, antigenicity, virulence, and geographical dispersion, for example, the case of Group B *Streptococcus agalactiae* and *Streptococcus pneumomia*.

The generation of the antibody complexed with carbohydrate antigens was performed through a combination of comparative antibody modeling and automated ligand docking. Subsequently, several 10 ns MD simulations were performed using the Molecular Mechanics–Generalized Born Surface Area method with explicit hydration, augmented by conformational entropy estimates. While providing detailed insight into the molecular details and the energy components involved in the formation of the complexes, the analysis offered a comprehensive interpretation of a large body of biochemical and immunological data related to antibody recognition of bacterial polysaccharides (Kadirvelraj et al. 2006).

Shigella flexneri is the main causal agent of the endemic form of bacillary dysentery. The O-antigen is the polysaccharide moiety of the lipopolysaccharide; it is the major target of the serotype-specific protective humoral response elicited upon host infection by *Shigella flexeneri*. The repeating unit of the O antigen is a pentasaccharide. The availability of the x-ray structure of the Fab/[AB(E)CD]$_2$ complex, at a resolution of 1.80 Å (Vulliez-Le Normand et al. 2008), along with a sufficient amount of well-characterized pentasaccharides, and IgG monoclonal antibody allowed a thorough analysis of the complexes by Saturation Transfer Difference (STD) NMR experiments and extensive MD simulations (Figure 4.7). The study brought into light information on the dynamics of the corresponding antibody–carbohydrate complexes that is available neither from the x-ray structure nor from the NMR analysis independently (Theillet et al. 2011). The proposed protocol making use of MD simulations and STD-NMR is likely to facilitate the design of either ligands or carbohydrate recognition domains, according to needed improvements of the natural carbohydrate-receptor properties.

4.6.1.3 Chemokine–GAG Interactions

The GAGs comprise a family of complex anionic polysaccharides including (1) glucosaminoglycans (heparin and heparan sulfate), (2) galactosylaminoglycans (chondroitin sulfate and dermatan sulfate), and (3) hyaluronic acid and keratan sulfate. In addition to their participation in the physicochemical properties of the extracellular matrix, GAG fragments are specifically recognized by protein receptors, and they play a role in the regulation of many processes, such as hemostasis, growth factor control, anticoagulation, and cell adhesion (Kjellen and Lindahl 1991). Given the importance of protein–GAG interactions, oligosaccharide fragments are important targets for drug design.

Docking of GAG oligosaccharide in protein receptor binding sites presents two main difficulties: (1) the binding site does not generally adopt a pocket or crevasse shape that would allow for easy identification and (2) both the ligand and the protein present a high flexibility of side chains. In addition to a simple molecular visualization program, the analysis of the projection of the electrostatic potential on the Connolly surface of the protein, for example, with the MOLCAD program (included in SYBYL, Tripos Associates), has proven to be useful. The GRID program (Goodford 1985), which allows prediction of the most energetically favorable region for binding of small probes on the protein surface, is very successful in identifying sulfate-binding regions. For predicting the orientation of the oligosaccharide on the protein surface, the AutoDock program (Morris et al. 1998), which considers flexibility at glycosidic linkages and pendant groups (hydroxyl groups, hydroxymethyl, etc.), can be used for charged oligosaccharide fragments. It should be kept in mind that such an approach generally yields several families of conformations and that

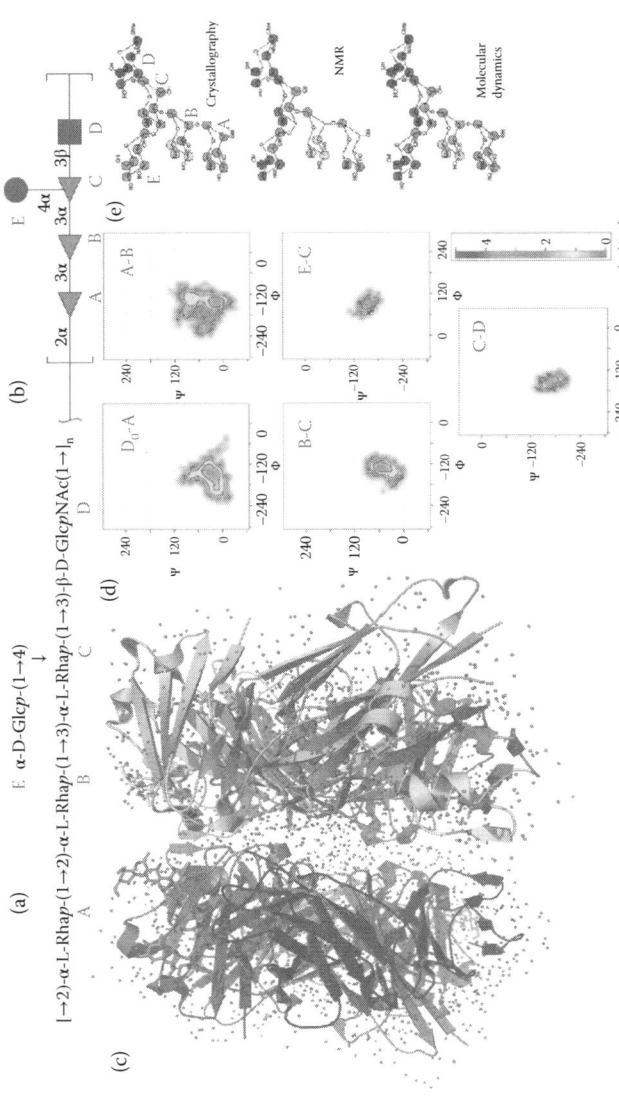

FIGURE 4.7 **(See color insert.)** Features of *Shigella flexneri* O antigen interacting with monoclonal antibody. (a) Primary structure of the *Shigella flexneri* SF2a O antigen (Simmons and Romanowska 1987) common AB(E)CD repeat unit. (b) CFG representation of *Shigella flexneri* common AB(E)CD linear backbone repeat unit, where the green triangles represent rhamnose, the circle denotes glucose, and the blue square denotes *N*-acetylglucosamine. (c) Crystal structure of synthetic O-antigen decasaccharide from serotype 2a *Shigella flexneri* (PDB ID: 3BZ4) in complex with a protective monoclonal antibody Fab F22-4. (d) Φ, Ψ maps of MD simulations for the glycosidic linkages of two repeat units of the bound conformation of the *Shigella flexneri* O antigen D_0 AB(E)CD pentasaccharide. (e) Comparison between the predicted Saturation Transfer Difference (STD) values of the two repeat units of the truncated crystal structure of F22-4 and the measured STD-NMR intensities and the predicted values of the 50 MD simulation snapshots of AB(E)CD.

further simulations, including MD in the presence of explicit water and counter ions, have to be envisaged for a thorough investigation.

The conformational behavior of the heparin pentasaccharide responsible for high affinity to antithrombin III has been the subject of several investigations. This study is complicated by the fact that a conformational change occurs in the protein upon binding (Johnson et al. 2006; Skinner et al. 1997). The first model obtained using homology modeling for the protein and hand-docking of the pentasaccharide allowed the determination of the basic amino acids involved in the recognition of the sulfate and carboxylate groups (Grootenhuis and Vanboeckel 1991). A study making use of several newly developed docking programs arrived at the same prediction for the binding site (Bitomsky and Wade 1999). In the crystal structures of the complex between antithrombin III and pentasaccharide (Jin et al. 1997; Johnson et al. 2006), a cluster of basic amino acids has been demonstrated to interact with the oligosaccharide's sulfate and carboxylate groups. The conformation of the bound pentasaccharide is also subjected to induced fit upon binding (Figure 4.8). At the present time, both x-ray crystallographic studies and NMR data coupled with molecular modeling (Hricovini et al. 2001) agree that the binding is accompanied by dihedral angle variations of two glycosidic linkages and conformational shift of the 2-O-sulfated iduronic residue.

Among the numerous proteins that bind heparin, the fibroblast growth factors (FGFs) have received special attention because they are involved in the control of cell proliferation, migration, and differentiation. Heparin fragments have been cocrystallized as ternary complexes with two FGFs and their receptors (FGFRs), and the minimal binding sequences could be determined (Pellegrini et al. 2000; Schlessinger et al. 2000). Analysis of the crystal structures together with molecular modeling demonstrated that upon binding, the regular helical shape of heparin is kinked at one point by both modification of one glycosidic linkage conformation and one iduronate ring shape (Raman et al. 2003). Such "induced fit" of the ligand in GAG/protein interactions is very likely to happen since it is classically observed in lectin–oligosaccharide interactions.

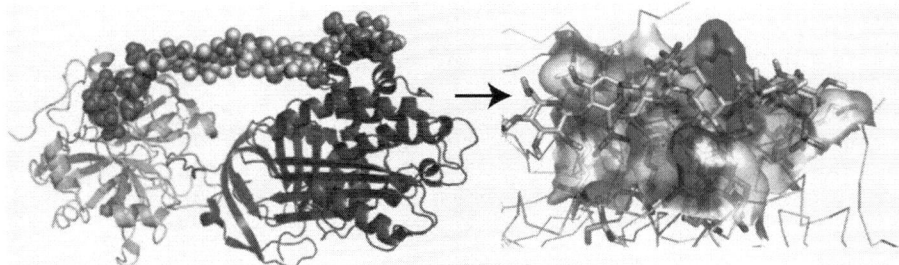

FIGURE 4.8   (See color insert.) General view (left) of the crystal structure of ternary complexes between antithrombin (reddish-brown ribbon), thrombin (green ribbon), and heparin analog (Dementiev et al. 2004; Li et al. 2004); (righ) blow-up of the binding site of antithrombin interacting with the specific heparin fragment.

Chemokines, derived from chemo-attractant cytokines, represent a large family of small proteins which, based on their physiological features, have been classified as "inflammatory" (or inducible) or "homeostatic" (or constitutive) (Rossi and Zlotnik 2000). Their roles include events as diverse as development, angiogenesis, neuronal patterning, hematopoiesis, viral infection, wound healing, and metastasis. Given the importance of protein–GAG interactions, oligosaccharide fragments are important targets for drug design. Chemokines interact with GAGs in general and heparan sulfate in particular. This binding is thought to create a local concentration, or a gradient, of chemokines on tissues where some GAGs are specifically expressed. Modeling studies have therefore been used for describing the interaction between chemokines and heparan sulfate. One interesting structural feature is that chemokines may exist in solution as monomer or dimer (sometimes tetramer), but they bind GAGs in the dimeric or tetrameric state. Depending on the dimerization mode and positions of basic amino acids in the peptide sequences, chemokines will present positively charged clusters on their accessible surfaces that define several possibilities for binding heparan sulfate (Handel et al. 2005; Lortat-Jacob et al. 2002).

4.6.1.4 Transport

Carbohydrates such as the malto-oligosaccharides of lactose, sucrose, raffinose, fructooligosaccharides, L-fucose, trehalose, oligoalginate, oligogalacturonate, and so on constitute a source of carbon for many organisms. These molecules have to be transported across channels and pores, and their motion is critically important for understanding mechanism of many cellular processes. At the protein level, this is achieved by a family of proteins, collectively referred to as transporters. These transmembrane proteins allow permeation of sugars: their structures along with the mechanistic transport model are the subject of intense research. The recent high-resolution structural elucidation of transporters is enabling investigation into the MD of fundamental transport processes.

Transport across the membrane is mediated by channel-forming proteins, of which maltoporin has been most extensively studied. The elucidation of the first high-resolution structure of maltoporin (Schirmer et al. 1995) revealed the general model of specific channel-forming membrane proteins: a β-barrel with 18 antiparallel strands. Like the general diffusion porins, the functional unit of maltoporin is a trimer with long loops exposed to the cell exterior and short turns exposed to the periplasm. A striking feature is a consecutive stretch of aromatic residues in the channel arranged in a left-handed helical path, which has been described as the "greasy slide."

The translocation mechanism of malto-oligosaccharides across the maltoporin membrane channel (Figure 4.9) has been investigated by MD calculations (Dutzler et al. 2002). The first event is the binding of sugar to the first residue of the "greasy slide," which occurs via van der Waals interactions to the hydrophobic face of the glucosyl ring. Deeper penetration into the channel occurs throughout guided diffusion of the oligosaccharide along the "greasy slide." A gradual dehydration of the malto-oligosaccharide favors the establishment of transitory hydrogen bonds between the sugars' hydroxyl groups and the surrounding amino acids. This is made possible by the conformational flexibility occurring at the glycosidic linkages and at the primary hydroxyl groups. The presence of the charged side chains (referred to as

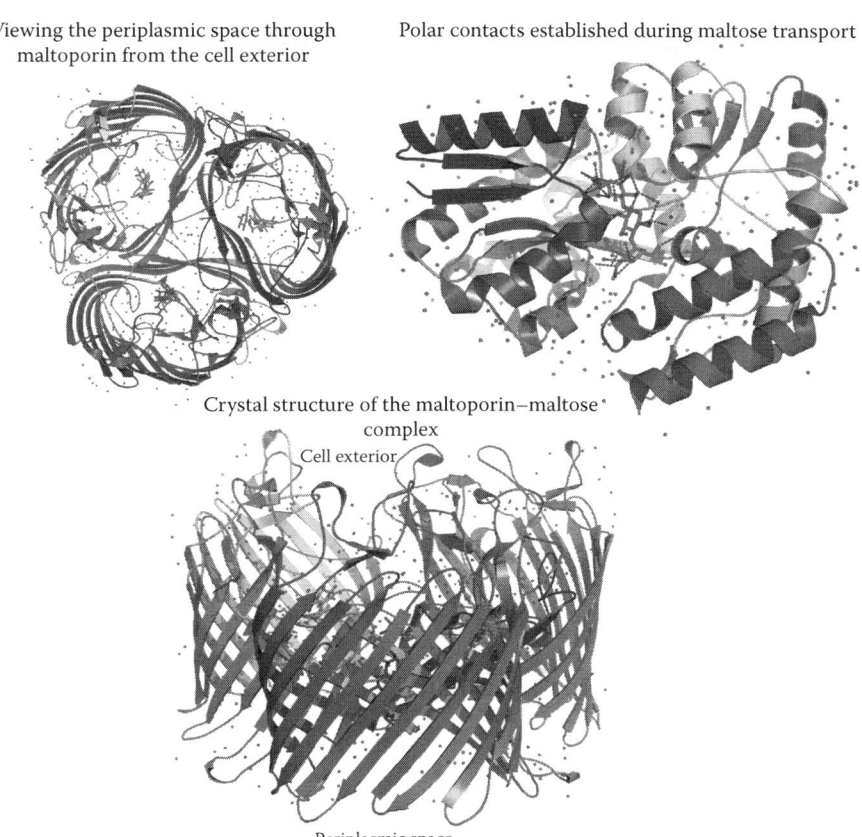

Viewing the periplasmic space through maltoporin from the cell exterior

Polar contacts established during maltose transport

Crystal structure of the maltoporin–maltose complex

Cell exterior

Periplasmic space

FIGURE 4.9 (**See color insert.**) Three-dimensional structure of maltoporin (Dutzler et al. 1996) along with snapshots of the interaction of maltooligosaccharides within the channel.

"polar tracks") mimics the lost hydration shell to the sugar by providing hydrogen bonds to the hydroxyl groups of the carbohydrate. The polar tracks are divided into donor and acceptor lanes all along the greasy slide. The movement of the carbohydrate residues to the next binding site of the greasy slide in combination with a rearrangement of hydrogen bonds is referred to as the "register shift." The continuous making and breaking of hydrogen bonds results in the oligosaccharide moving through the porin in a capillary-like fashion.

Within the super family of carbohydrate transporter exists the MFS family, which exploits the electrochemical potential to shuttle substrates across cell membranes. These transporters are thought to use an alternating-access mechanism to upload and download substrates. The elucidation of the 3D structure of a fucose transporter (Dang et al. 2010) opened the way to MD simulation of the L-fucose residue complexed to the transmembrane protein inserted into a POPE bilayer to mimic the bacterial membrane. Structural, biochemical, and computational analysis provided insights into the function of the transporter, with the identification of key amino acids that play an essential role in the active transport path.

As the structures of other unique transport systems are revealed, the power of computational methods in transporter analysis and prediction will grow exponentially.

4.6.2 Synthesis: Glycosyltransferases

The central process of oligosaccharides, polysaccharides, and glycoconjugate biosynthesis is performed by the action of glycosyltransferases (GTs). These enzymes constitute a large family of proteins, which are present in prokaryotes, eukaryotes, and viruses, and mediate a wide range of functions from structures and storage to signaling. GTs are responsible for the formation of the glycosidic bond by attaching a sugar moiety of an appropriate donor substrate, mainly a nucleotide sugar, to a specific acceptor substrate. These proteins are highly stereo- and regioselective, and they are usually classified by their preferred sugar substrates, acceptor molecules, and the types of glycosidic linkage they generate (Breton et al. 2006; Breton et al. 2009) (www.cermav.cnrs.fr/cgi-bin/rxgt/rxgt.cgi).

Molecular modeling of GTs along with their interactions with the nucleotide sugar and the specific acceptor substrate is difficult. The number of available crystal structures is still limited (Breton et al. 2009); only a limited number of folds have been observed. In these crystal structures, the ratio of loops to secondary elements is high, and many of them do not describe the entire catalytic domain as the electron densities are not clear due to the flexible polypeptide extremities and/or several loops. Flexible loops appear to play an important role in substrate binding. For some of these enzymes, structural and calorimetric binding studies indicate that an obligatory ordered binding of donor and acceptor substrates, linked to a donor substrate-induced conformational change, and the direct participation of UDP in acceptor binding, induces a large conformational change. It has been shown that the open state (free enzyme) has no or little affinity for the oligosaccharide acceptor. Alternatively, the closed active conformation creates a pocket that serves as the binding site for the acceptor. Starting from an available crystal structure of a GT, or using such theoretical approaches as fold recognition (Heissigerova et al. 2003), a 3D model of the GT of interest is first constructed. Further, a combination of methods like fold recognition and molecular modeling might aid the prediction of the acceptor specificities of the putative GT. Docking of substrates also appears to be a difficult task because of the conformational flexibility of the nucleotide sugar and the presence of phosphate and divalent cation. Appropriate energy parameters have been developed based on the AMBER force field interfaced with CICADA for conformational search (Petrova et al. 1999). Interacting with the protein, the nucleotide sugar(s), the metal ions, and the oligosaccharide acceptor(s) are then submitted to a docking procedure, which is followed by energy optimization of the amino acid side chains surrounding the substrates. At present such molecular modeling procedures are aimed at revealing the key catalytic amino acids and the nucleotide-sugar donor specificity and are performed in conjunction with site-specific mutagenesis and biochemical analysis (Botte et al. 2005). The availability of a well-resolved crystal structure of GT(51), a penicillin-binding protein, has opened the way to investigate how computational methods can be used to explore drug targets for antibiotic resistance. Docking and scoring methodology (Surflex-Dock and FlexX-Pharm) have been applied resulting in the discovery of nine novel potential leads for GT(51) inhibition (Yang et al. 2008). Detailed characterization of the mechanisms involved in either the inversion or the retention of stereochemistry can only be interpreted with the use of ab initio molecular orbital study (Andre et al. 2003; Kozmon and Tvaroska 2006; Krupicka and Tvaroska 2009).

4.6.3 GLYCOSYL HYDROLASES/GLYCOSIDASES

The hydrolysis of glycosidic bonds in carbohydrates, polysaccharides, glycoproteins, glycolipids, and so on is performed by glycosidases. These enzymes are classified into endo- and exo-types. Exo-type glycosidases attack and hydrolyze monoglycosides into free sugar and aglycon. When acting on oligo- or polysaccharides, they liberate a monosaccharide unit from the nonreducing end. Endo-type glycosidases act on oligo- and polysaccharides and catalyze the hydrolysis of an internal glycosidic linkage, thereby liberating two carbohydrate moieties or releasing an oligosaccharide (or polysaccharide) and monoglycoside of the reducing end. Some glycosidases are capable of acting as both exo- and endo-types. The reactions resulting from the catalytic action of glycosidases can also be characterized by the anomeric configuration of the glycosidic bond of the substrate that the enzyme attacks, that is, with the retention or inversion of the anomeric configuration.

4.6.3.1 Glycosyl Hydrolases on Single Chain

Computational methods are essentially used to dock the oligosaccharide, polysaccharides, and so on into the active state (which is usually identified by systematic mutations). The glycosidases and their carbohydrate ligands are considered in their energetically stable conformations, and their interaction energies are compared. The resulting docked structures are used to propose a model for substrate and conformer selectivity based on the dimensions of the active site. The docking of substrates and inhibitors indicates the dimensions of the binding site, which are usually large, extending over several monosaccharide units, beyond and toward the cleaving site. The key amino acids that may be involved in the catalytic mechanism can be identified from these results. Such computational protocols have been applied to the study of several classes of glycosidases. Most recent examples incorporating state-of-the-art modeling tools have been used to investigate the features of heparanase interacting with heparin (Gandhi and Mancera 2012; Sapay et al. 2012). These docking ligand–protein complex models can interpret the substrate specificity of heparanase, providing a rationale for the design of polysaccharides that may act as inhibitors of the enzymatic activity of heparanase. Predicted heparin/complexes show that the interactions of the heparin-binding domains in combination with the catalytic domain can be targeted for the design of inhibitors.

Enzyme inhibitors can be classified into substrate analogs and transition-state analogs. Both types of analogs inhibit the enzyme via generally competing with the substrate for binding to the active site of the enzymes but are not affected by the enzyme. Substrate analogs mimic the structural features of the substrates, whereas transition-state analogs have some structural characteristics that are unique to the transition state.

Structural analysis of influenza virus neuraminidase (Varghese et al. 1983) and neuraminidase in complex with sialic acid (Varghese et al. 1992) led to the design of a potent inhibitor of neuraminidase activity: zanamivir (von Itzstein et al. 1993). Based on the efficacy of zanamivir (Relenza™), another neuraminidase inhibitor was also developed: oseltamivir phosphate (Tamiflu™) (Kim et al. 1997). Both Relenza, a carbohydrate-based drug, and Tamiflu, a carbocyclic mimetic, are potent and clinically effective anti-influenza drugs (McCullers et al. 2005). Despite the efficacy of these drugs, major concerns remain regarding the development of resistance to these drugs, which is already occurring. Point mutations in the influenza virus enzyme neuraminidase have been reported that lead to dramatic loss of activity for known neuraminidase inhibitors cited above. A more sound understanding of the molecular basis of such resistance is needed toward developing improved next-generation drugs. Modeling the binding of ligands with neuraminidase has been undertaken using explicit solvent all-atom MD simulations, free energy calculations, and residue-based decomposition. The simulations predicted the effects of a known mutation at one amino acid (R292K) and provided clues as to the origins of resistance to the mutant. The results significantly enhance experimental observations (Chachra and Rizzo 2008).

The likelihood of future influenza pandemics (including the possibility of highly pathogenic H5N1 strains) highlighted the need for additional computational methods. The binding properties of the H5N1 influenza virus neuraminidase have been inferred from molecular modeling (Raab and Tvaroska 2011). They concerned the binding properties between sialic acid, methyl 3′-sialyl lactoside, methyl 6′-sialyllactoside, and H5N1 influenza virus neuraminidase using molecular docking and MD simulations. The obtained results indicate that, in the complex, sialic acid undergoes a conformational transition of the ring. Meanwhile, methyl 3′-sialyl lactoside establishes only weak interactions with a key loop of the neuraminidase, in contrast to what is observed for the complex with methyl 6′-sialyl lactoside. The differences could be attributed to the occurrence of distinct conformations about the glycosidic linkages. As these molecular modeling results are consistent with available experimental data on the specificity of neuraminidase, they provide sound structural information for a rational design of novel and specific inhibitors of H5N1 neuraminidase as potential therapeutics for the treatment of avian flu.

4.6.3.2   Glycosyl Hydrolases on a Solid Substrate

Many polysaccharides occur in the form of highly packed 3D arrangements as a result of extensive inter- and intramolecular hydrogen bonding networks and van der Waals interactions. These features render the structures completely insoluble in water (e.g., cellulose and chitin) and provide them with substantial resistance from attack by most enzymes. The hydrolysis of cellulose in nature is the result of plant cell wall degrading complexes, referred to as cellulosomes (Doi and Kosugi 2004) including cellulases. Cellulases consist of a core of glycoside hydrolases and cellulose-binding modules (also referred to as carbohydrate-binding modules or CBMs) and a linker that binds the two enzymatic components. By playing the dual role of recognizing and adhering to the solid state surface of the polysaccharide, and maintaining the proximity effect, the presence of CBMs is a key factor in the ability of the enzyme to efficiently breakdown insoluble polysaccharides. Once bound to the crystalline substrate, the active center of the core domains of cellulases can attack the cellulose chains. The cellulases are classified into two types: the exo- and endocellulases, depending on whether or not the cellulose can recognize the reducing end of the cellulose chains. The morphology of the native crystals is of course an essential feature in the enzymatic digestion of crystalline cellulose, and this is a major scientific and industrial question (Demain et al. 2005). It is established that the enzymatic breakdown and degradation of cellulose requires a complex of enzymes working together (Figure 4.10). The general picture that has emerged from earlier investigations in this area indicates that the cooperation of at least three types of enzymes is required for efficient digestion of crystalline cellulose into glucose. These are (1) endoglucanases (EC 3.2.1.4), which cleave the chains randomly, (2) cellobiohydrolases (EC 3.2.1.91), which recurrently cleave cellobiose from the chain end of cellulose, and (3) β-glucosidases (EC 3.2.2.21), which hydrolyze cellobiose. As for the cellulose-binding modules, numerous studies have established that three aromatic residues are

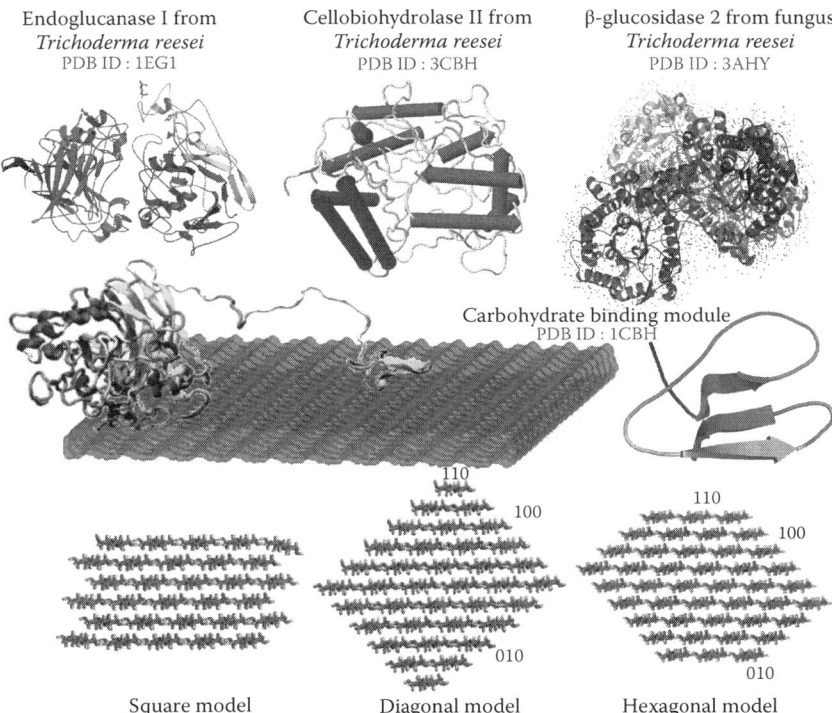

Endoglucanase I from
Trichoderma reesei
PDB ID : 1EG1

Cellobiohydrolase II from
Trichoderma reesei
PDB ID : 3CBH

β-glucosidase 2 from fungus
Trichoderma reesei
PDB ID : 3AHY

Carbohydrate binding module
PDB ID : 1CBH

Square model   Diagonal model   Hexagonal model

FIGURE 4.10 **(See color insert.)** The enzymatic digestion of cellulose. The top panel depicts the 3D structures of the three main categories of enzymes (from *Trichoderma reesei*) that digest crystalline cellulose. The central panel provides a visual as to how cellulose-digesting enzyme (shown here as cellobiohydrolase I or Cel7A) (Crowley 2008) interacts with the cellulose crystalline arrangement and breaks down cellulose into glucose and also illustrates a carbohydrate-binding module (CBM, PDB ID: 1CBH). The bottom panel shows the three faces of the cellulose Iα crystal models in projection with the Miller indices of their constituent crystal planes.

needed for binding onto cellulose crystals and that tryptophan residues contribute to higher binding affinity than tyrosines. However, evidence has accumulated showing that different binding sites for the same cellulose-binding domains could occur.

A systematic study of the CBM protein of Cel7A of *Trichoderma reesei*, with the cellulose Iα crystal model, has been performed using a combined Grid docking search and MD calculations (Yui et al. 2010). Three types of cellulose Iα crystal models with infinite dimensions were constructed, each consisting of different crystallographic faces, that is, (1 1 0), (1 0 0), and (0 1 0). The (1 1 0) complex models exhibited larger affinities at the interface than the other ones. It was found that the CBM was more stably bound to the (1 1 0) surface when it was placed in an antiparallel orientation with respect to the cellulose fiber axis. The predicted directional specificity of the CBM at the optimum positions was consistent with the observed processing direction of the Cel7A (Imai et al. 1998). In the solvated dynamic state, the curved (1 1 0) surface resulting from the fiber twist somewhat assisted a complementary fit with the CBM at the interface.

Much can be learned about the processivity by conducting carefully designed MD simulation of the binding of the catalytic domains of cellulases with various substrate configurations, solvation models, and thermodynamics protocols (Hashimoto 2006).

Computational model of cellobiohydrolase I (Cel7A) from *Trichoderma reesei* on a cellulose (1 0 0) surface displays the large catalytic domain (left), linker (middle single strand), and cellulose-binding module (right small domain). A cellodextrin strand is shown peeled out of the surface of the cellulose and threaded into the catalytic tunnel of Cel7A (Crowley 2008). The investigation requires the consideration of approximately 800,000 atoms. In order to face such a computational challenge, most of the numerical simulations shall require major modifications of existing code and algorithms.

4.7 CONCLUSIONS

In the past few years, there has been an increase in the development and application of computational methods aimed at establishing the molecular features characterizing protein–carbohydrate interactions. Quite naturally, these computational methods are becoming reliant on experimental studies for the elucidation of structural and dynamics feature in the field of glycoscience. Significant steps have been made among which are the developments and implementations of force fields capable of taking into account the specificity of carbohydrates (stereoelectronic effect, gauche effect, etc.) and their compatibility with the computational tools that have been developed for proteins. Recently, methods for handling many rotatable bonds in flexible docking of conformationally flexible carbohydrates have been established. It has been recognized that the surface of carbohydrates and their derivatives that are composed of hydrophobic and hydrophilic patches remain a source of complexity in modeling. Nevertheless, the balance between hydrophobic and hydrophilic patches is essential for carbohydrate solubility and for molecular recognition. The occurrence of such a feature combined with the enhanced conformational flexibility is a unique characteristic that explains how complex oligosaccharides can be transported throughout transmembrane proteins in a capillary-like and yet selective fashion.

Calculations of binding free energies and enthalpies with the required accuracy remain to be improved and tested against well-characterized experimental data. Certainly, calculation of free energy perturbations is a promising approach for the prediction of carbohydrate–receptor binding affinity. Such calculations cannot be performed without a full understanding of solvation. Progresses in this area imply a better handling of hydration and the major role played by solvation and desolvation of both carbohydrates and proteins in their isolated state and during the course of their interactions.

At present these computational tools are considered as useful as the other methods of structural investigation. They can actually help in reconciling the experimental results gathered from separate experiments in different conditions and environments and in extrapolating the results. The wealth of successful applications for many different protein interactions with carbohydrates is a testimony to the maturity of the modeling methods and protocols that have been developed. Nevertheless, these success cases are almost exclusively dealing with cases where proteins interact with carbohydrates, without any further catalytic actions.

Complementary computational methods need to be developed and/or integrated to allow the study of enzymatic reaction and the subsequent optimization of biocatalysts. These methods, based on molecular robotics algorithms, would be used for an efficient virtual screening of configurational and conformational spaces of high dimensions. The ongoing developments of robotics algorithms are likely to provide efficient tools to explore the dynamic functionality of enzymes. These are based on efficient path-planning algorithms and fast geometric operators designed for complex articulated chains. The aim is to reduce, in a significant but relevant way, the exploration of the combinatorial space of the enzyme sequences, based on the geometric feasibility for a ligand either to access or to leave the catalytic site in a "productive" way. The enhancement of the predictive performances of such algorithms will require the use of simplified energy functions to prefilter the conformations that are nonviable to construct the network of concerted motions while the enzyme is interacting with the ligand.

The investigation of the catalytic mechanism of inverting and retaining carbohydrate active enzymes requires high level DFT hybrid QM/MM calculations. The studies of the catalytic reaction and the dynamic motions undergone by the enzymes are being investigated independently at present. Consequently, developments are required to set up MD "hybrid methods" based on the principles of quantum mechanics with the aim of studying the dynamics of electronic effects and charge transfer within the catalytic site. Such hybrid methods would incorporate *ab initio* dynamics as developed by Carr and Parinello (CPMD) and a "classical" MD force field. Applications of these computational methods will allow exploiting further the protein–carbohydrate interactions, especially for therapeutic purposes. Design of transition-state analog inhibitors of glycosyl hydrolases and glycosyl transferases requires knowledge of the mechanism of the enzymatic reaction along with the geometry and charge distribution of transition state.

Extremely challenging cases are being identified. Many of the carbohydrates with biological functions are found at the surfaces of proteins and cells. Some physicochemical principles that underline their associations are being considered to model such systems, for example, patches of glycolipids and glycosurfaces. As the concept of "glycolandscape" is being recognized, new modeling protocols need to be developed. They require novel computational tools capable of constructing the landscape resulting from the side-by-side arrangements of glycoconjugates such as glycolipids. A new paradigm will emerge, and the attention will not only be given to the interaction of a protein with a single carbohydrate unit (the so-called "tree vision") but instead the interaction with glycosurfaces (the so-called "glycocanopy", as an analogy to the crown canopy, that is, the uppermost layer in a forest formed by the crown of the trees.

This concept is likely to become more prevalent as the field of research dealing with the solid-state degradation of plant cell wall polysaccharides by enzymes offers formidable challenges. Plant biomass is an alternative natural source for chemical and feed stocks with a replacement cycle short enough to meet the demand of the world fuel market. The enzymatic hydrolysis of cellulose is still considered a main limiting step of the biological production of biofuels from lignocellulosic biomass. This step involves the action of three types of cellulose-degrading enzymes acting in a synergistic way. In view of designing a functional kinetic model integrating the respective properties of each enzyme along with their synergies, much can be learned by conducting carefully designed computer simulations of the binding of the cellulose-binding domains and the catalytic domains of cellulases with various substrates, solvation models, and thermodynamics protocols. Such an extraordinary computational challenge is delineating the new frontiers of the area of protein–carbohydrate interactions.

ACKNOWLEDGMENTS

The research leading to this chapter has received funding from the European Commission's Seventh Framework Programme FP7/2007-2013 under grant agreement no. 215536. We are grateful to the Marie Curie Initial Training Network as part of the FP7 People Programme for training and funding.

REFERENCES

Agostino, M., Sandrin, M. S., Thompson, P. E., Yuriev, E., and Ramsland, P. A. 2009. In silico analysis of antibody-carbohydrate interactions and its application to xenoreactive antibodies. *Mol Immunol*, 47, 233–246.

Agostino, M., Sandrin, M. S., Thompson, P. E., Yuriev, E., and Ramsland, P. A. 2010. Identification of preferred carbohydrate binding modes in xenoreactive antibodies by combining conformational filters and binding site maps. *Glycobiology*, 20, 724–735.

Allinger, N. L., Li, F., Yan, L., and Tai, J. C. 1990. Molecular mechanics (MM3) calculations on conjugated hydrocarbons. *J Comput Chem*, 11, 868–895.

Andre, I., Tvaroska, I., and Carver, J. P. 2003. On the reaction pathways and determination of transition-state structures for retaining alpha-galactosyltransferases. *Carbohydr Res*, 338, 865–877.

Barbe, S., Cortés, J., Siméon, T. et al. 2011. A mixed molecular modeling-robotics approach to investigate lipase large molecular motions. *Proteins*, 79, 2517–2529.

Basdevant, N., Weinstein, H., and Ceruso, M. 2006. Thermodynamic basis for promiscuity and selectivity in protein-protein interactions: PDZ domains, a case study. *J Am Chem Soc*, 128, 12766–12777.

Basma, M., Sundara, S., Calgan, D., Vernali, T., and Woods, R. J. 2001. Solvated ensemble averaging in the calculation of partial atomic charges. *J Comput Chem*, 22, 1125–1137.

Berendsen, H. J .C., Postma, J. P. M., van Gunsteren, W. F., and Hermans, J. 1981. Interaction models for water in relation to protein hydration. In *Intermolecular Forces*, Pullman, B. (ed) pp. 331–342, Dordrecht: Reidel Publishing Company.

Bettler, E., Imberty, A., Loris, R., and Rivet, A. 1998. Online database of 3 dimensional structures of lectins. http://www.cermav.cnrs.fr/lectines/

Biarnes, X., Nieto, J., Planas, A., and Rovira, C. 2006. Substrate distortion in the Michaelis complex of Bacillus 1,3-1,4-beta-glucanase. Insight from first principles molecular dynamics simulations. *J Biol Chem*, 281, 1432–1441.

Bitomsky, W. and Wade, R. C. 1999. Docking of glycosaminoglycans to heparin-binding proteins: Validation for aFGF, bFGF, and antithrombin and application to IL-8. *J Am Chem Soc*, 121, 3004–3013.

Botte, C., Jeanneau, C., Snajdrova, L. et al. 2005. Molecular modeling and site-directed mutagenesis of plant chloroplast monogalactosyldiacylglycerol synthase reveal critical residues for activity. *J Biol Chem*, 280, 34691–34701.

Bradbrook, G. M., Forshaw, J. R., and Pérez, S. 2000. Structure/thermodynamics relationships of lectin–saccharide complexes. *Eur J Biochem*, 267, 4545–4555.

Breton, C., Audry, M., Fasmer, S., Imberty, A., and Rivet, A. 2009. Online database of 3 dimensional structures of glycosyltransferases. http://www.cermav.cnrs.fr/cgi-bin/rxgt/rxgt.cgi

Breton, C., Šnajdrová, L., Jeanneau, C., Koča, J., and Imberty, A. 2006. Structures and mechanisms of glycosyltransferases. *Glycobiology*, 16, 29R–37R.

Brun, E., Moriaud, F., Gans, P. et al. 1997. Solution structure of the cellulose-binding domain of the endoglucanase Z secreted by *Erwinia chrysanthemi*. *Biochemistry*, 36, 16074–16086.

Bryce, R. A., Hillier, I. H., and Naismith, J. H. 2001. Carbohydrate-protein recognition: Molecular dynamics simulations and free energy analysis of oligosaccharide binding to concanavalin A. *Biophys J*, 81, 1373–1388.

Bultinck, P., Winter, H. D., Langenaecker, W., and Tollenare, J. P. 2004. *Computational Medicinal Chemistry for Drug Discovery*. New York: Marcel Dekker, Inc.

Chachra, R. and Rizzo, R. C. 2008. Origins of resistance conferred by the R292K neuraminidase mutation via molecular dynamics and free energy calculations. *J Chem Theory Comput*, 4, 1526–1540.

Chervenak, M. C. and Toone, E. J. 1994. A direct measure of the contribution of solvent reorganization to the enthalpy of binding. *J Am Chem Soc*, 116, 10533–10539.

Clark, M., Cramer, R. D., and Van Opdenbosch, N. 1989. Validation of the general purpose Tripos 5.2 force field. *J Comput Chem*, 10, 982–1012.

Clarke, C., Woods, R. J., Gluska, J. et al. 2001. Involvement of water in carbohydrate–protein binding. *J Am Chem Soc*, 123, 12238–12247.

Clausen, H. and Hakomori, S.-i. 1989. ABH and related histo-blood group antigens; immunochemical differences in carrier isotypes and their distribution. *Vox Sanguinis*, 56, 1–20.

Collins, P. J., Haire, L. F., Lin, Y. P. et al. 2008. Crystal structures of oseltamivir-resistant influenza virus neuraminidase mutants. *Nature*, 453, 1258–1261.

Cornell, W. D., Cieplak, P., Bayly, C. I. et al. 1995. A second generation force field for the simulation of proteins, nucleic acids, and organic molecules. *J Am Chem Soc*, 117, 5179–5197.

Cortés, J., Barbe, S., Erard, M., and Siméon, T. 2011. Encoding molecular motions in voxel maps. *IEEE-ACM Trans Comput Biol Bioinform*, 8, 557–563.

Cortés, J., Jaillet, L., and Siméon, T. 2008. Disassembly path planning for complex articulated objects. *IEEE Trans Robot*, 24, 475–481.

Cortés, J., Le, D. T., Iehl, R., and Siméon, T. 2010. Simulating ligand-induced conformational changes in proteins using a mechanical disassembly method. *Phys Chem Chem Phys*, 12, 8268–8276.

Crowley, M. F., Uberbacher, E. C., Brooks, C. L. et al. 2008. Developing improved MD codes for understanding processive cellulases. In *Journal of Physics Conference Series*, Stevens, R.L. (ed) V. 125, pp. 12049–12049. Bristol: IOP Publishing Ltd.

Cuneo, M. J., Changela, A., Beese, L. S., and Hellinga, H. W. 2009. Structural adaptations that modulate monosaccharide, disaccharide, and trisaccharide specificities in periplasmic maltose-binding proteins. *J Mol Biol*, 389, 157–166.

Damm, W., Frontera, A., Tirado–Rives, J., and Jorgensen, W. L. 1997. OPLS all-atom force field for carbohydrates. *J Comput Chem*, 18, 1955–1970.

Dang, S., Sun, L., Huang, Y. et al. 2010. Structure of a fucose transporter in an outward-open conformation. *Nature*, 467, 734–738.

de Geus, D. C., van Roon, A. M., Thomassen, E. A. et al. 2009. Characterization of a diagnostic Fab fragment binding trimeric Lewis X. *Proteins*, 76, 439–447.

Demain, A. L., Newcomb, M., and Wu, J. H. D. 2005. Cellulase, clostridia, and ethanol. *Microbiol Mol Biol Rev*, 69, 124–154.

DeMarco, M. L. and Woods, R. J. 2008. Structural glycobiology: A game of snakes and ladders. *Glycobiology*, 18, 426–440.

Dementiev, A., Petitou, M., Herbert, J. M., and Gettins, P. G. W. 2004. The ternary complex of antithrombin-anhydrothrombin heparin reveals the basis of inhibitor specificity. *Nat Struct Mol Biol*, 11, 863–867.

Doi, R. H. and Kosugi, A. 2004. Cellulosomes: Plant-cell-wall-degrading enzyme complexes. *Nat Rev Microbiol*, 2, 541–551.

Dowd, M. K., French, A. D., and Reilly, P. J. 1994. Modeling of aldopyranosyl ring puckering with MM3 (92). *Carbohydr Res*, 264, 1–19.

Dutzler, R., Schirmer, T., Karplus, M., and Fischer, S. 2002. Translocation mechanism of long sugar chains across the maltoporin membrane channel. *Structure*, 10, 1273–1284.

Dutzler, R., Wang, Y. F., Rizkallah, P. J., Rosenbusch, J. P., and Schirmer, T. 1996. Crystal structures of various maltooligosaccharides bound to maltoporin reveal a specific sugar translocation pathway. *Structure*, 4, 127–134.

Eklund, R. 2005. Computational analysis of carbohydrates: Dynamical properties and interactions. Doctoral Thesis, Stockholm University, Sweden.

Ewing, T. J. A., Makino, S., Skillman, A. G., and Kuntz, I. D. 2001. DOCK 4.0: Search strategies for automated molecular docking of flexible molecule databases. *J Comput Aided Mol Des*, 15, 411–428.

Friesner, R. A., Banks, J. L., Murphy, R. B. et al. 2004. Glide: A new approach for rapid, accurate docking and scoring. 1. Method and assessment of docking accuracy. *J Med Chem*, 47, 1739–1749.

Gandhi, N. S. and Mancera, R. L. 2009. Free energy calculations of glycosaminoglycan-protein interactions. *Glycobiology*, 19, 1103–1115.

Gandhi, N. S. and Mancera, R. L. 2012. Computational analyses of the catalytic and heparin binding sites and their interactions with glycosaminoglycans in glycoside hydrolase family 79 Endo-{beta}-D-Glucuronidase (Heparanase). *Glycobiology*, 22, 35–55.

Gilson, M. K. and Honig, B. 1988. Calculation of the total electrostatic energy of a macromolecular system: Solvation energies, binding energies, and conformational analysis. *Proteins*, 4, 7–18.

Glennon, T. M. and Merz, K. M. 1997. A carbohydrate force field for and its application to the study of saccharide to surface adsorption. *Theochem J Mol Struct*, 395–396, 157–171.

Gohlke, H. and Case, D. A. 2004. Converging free energy estimates: MM-PB(GB)SA studies on the protein–protein complex Ras–Raf. *J Comput Chem*, 25, 238–250.

Goodford, P. J. 1985. A computational procedure for determining energetically favorable binding sites on biologically important macromolecules. *J Med Chem*, 28, 849–857.

Grootenhuis, P. D. J. and Vanboeckel, C. A. A. 1991. Constructing a molecular-model of the interaction between anti-thrombin III and a potent heparin analog. *J Am Chem Soc*, 113, 2743–2747.

Guieysse, D., Cortés, J., Puech-Guenot, S. et al. 2008. A structure-controlled investigation of lipase enantioselectivity by a path-planning approach. *Chembiochem*, 9, 1308–1317.

Guvench, O., Greene, S. N., Kamath, G. et al. 2008. Additive empirical force field for hexopyranose monosaccharides. *J Comput Chem*, 29, 2543–2564.

Guvench, O., Hatcher, E. R., Venable, R. M., Pastor, R. W., and Mackerell, A. D. 2009. CHARMM additive all-atom force field for glycosidic linkages between hexopyranoses. *J Chem Theory Comput*, 5, 2353–2370.

Handel, T. M., Johnson, Z., Crown, S. E. et al. 2005. Regulation of protein function by glycosaminoglycans—As exemplified by chemokines. *Annu Rev Biochem*, 74, 385–410.

Hansen, H. S. and Hünenberger, P. H. 2011. A reoptimized GROMOS force field for hexopyranose-based carbohydrates accounting for the relative free energies of ring conformers, anomers, epimers, hydroxymethyl rotamers, and glycosidic linkage conformers. *J Comput Chem*, 32(6), 998–1032.

Hashimoto, H. 2006. Recent structural studies of carbohydrate-binding modules. *Cell Mol Life Sci*, 63, 2954–2967.

Haspel, N., Moll, M., Baker, M., Chiu, W., and Kavraki, L. 2010. Tracing conformational changes in proteins. *BMC Struct Biol*, 10, S1.

Hatcher, E., Guvench, O., and Mackerell, A. D. 2009. CHARMM additive all-atom force field for acyclic polyalcohols, acyclic carbohydrates and inositol. *J Chem Theory Comput*, 5, 1315–1327.

Heissigerova, H., Breton, C., Moravcova, J., and Imberty, A. 2003. Molecular modeling of glycosyltransferases involved in the biosynthesis of blood group A, blood group B, Forssman, and iGb3 antigens and their interaction with substrates. *Glycobiology*, 13, 377–386.

Höltje, H.-D., Sippl, W., Rognan, D., and Folkers, G. 2008. *Molecular Modeling: Basic Principles and Applications*. Weinheim: Wiley-VCH.

Homans, S. W. 1990. A molecular mechanical force field for the conformational analysis of oligosaccharides: Comparison of theoretical and crystal structures of Man .alpha.1-3Man.beta.1-4GlcNAc. *Biochemistry*, 29, 9110–9118.

Hricovini, M., Guerrini, M., Bisio, A. et al. 2001. Conformation of heparin pentasaccharide bound to antithrombin III. *Biochem J*, 359, 265–272.

Imai, T., Boisset, C., Samejima, M., Igarashi, K., and Sugiyama, J. 1998. Unidirectional processive action of cellobiohydrolase Cel7A on *Valonia* cellulose microcrystals. *FEBS lett*, 432, 113–116.

Imberty, A., Hardman, K. D., Carver, J. P., and Pérez, S. 1991. Molecular modelling of protein-carbohydrate interactions. Docking of monosaccharides in the binding site of concanavalin A. *Glycobiology*, 1, 631–642.

Imberty, A., Mikros, E., Koca, J. et al. 1995. Computer simulation of histo-blood group oligosaccharides: Energy maps of all constituting disaccharides and potential energy surfaces of 14 ABH and Lewis carbohydrate antigens. *Glycoconj J*, 12, 331–349.

Imberty, A., Mollicone, R., Mikros, E. et al. 1996. How do antibodies and lectins recognize histo-blood group antigens? A 3D-QSAR study by comparative molecular field analysis (CoMFA). *Bioorg Med Chem*, 4, 1979–1988.

Imberty, A. and Pérez, S. 2000. Structure, conformation, and dynamics of bioactive oligosaccharides: Theoretical approaches and experimental validations. *Chem Rev*, 100, 4567–4588.

Jansma, A. L., Kirkpatrick, J. P., Hsu, A. R., Handel, T. M., and Nietlispach, D. 2010. NMR Analysis of the structure, dynamics, and unique oligomerization properties of the chemokine CCL27. *J Biol Chem*, 285, 14424–14437.

Jin, L., Abrahams, J. P., Skinner, R. et al. 1997. The anticoagulant activation of antithrombin by heparin. *Proc Natl Acad Sci USA*, 94, 14683–14688.

Johnson, D. J. D., Li, W., Adams, T. E., and Huntington, J. A. 2006. Antithrombin-S195A factor Xa-heparin structure reveals the allosteric mechanism of antithrombin activation. *EMBO J*, 25, 2029–2037.

Jorgensen, W. L., Maxwell, D. S., and Tirado-Rives, J. 1996. Development and testing of the OPLS all-atom force field on conformational energetics and properties of organic liquids. *J Am Chem Soc*, 118, 11225–11236.

Kadirvelraj, R., Gonzalez-Outeirino, J., Foley, B. L. et al. 2006. Understanding the bacterial polysaccharide antigenicity of *Streptococcus agalactiae* versus *Streptococcus pneumoniae*. *Proc Natl Acad Sci USA*, 103, 8149–8154.

Kim, C. U., Lew, W., Williams, M. A. et al. 1997. Influenza neuraminidase inhibitors possessing a novel hydrophobic interaction in the enzyme active site: Design, synthesis, and structural analysis of carbocyclic sialic acid analogues with potent anti-influenza activity. *J Am Chem Soc*, 119, 681–690.

Kirillova, S., Cortés, J., Stefaniu, A., and Siméon, T. 2008. An NMA-guided path planning approach for computing large-amplitude conformational changes in proteins. *Proteins*, 70, 131–143.

Kirschner, K. N. and Woods, R. J. 2001. Solvent interactions determine carbohydrate conformation. *Proc Natl Acad Sci USA*, 98, 10541–10545.

Kirschner, K. N., Yongye, A. B., Tschampel, S. M. et al. 2008. GLYCAM06: A generalizable biomolecular force field. Carbohydrates. *J Comput Chem*, 29, 622–655.

Kjellen, L. and Lindahl, U. 1991. Proteoglycans: Structures and interactions. *Ann Rev Biochem*, 60, 443–475.

Kollman, P. A., Massova, I., Reyes, C. et al. 2000. Calculating structures and free energies of complex molecules: Combining molecular mechanics and continuum models. *Acc Chem Res*, 33, 889–897.

Kony, D., Damm, W., Stoll, S., and Van Gunsteren, W. F. 2002. An improved OPLS-AA force field for carbohydrates. *J Comput Chem*, 23, 1416–1429.

Kozmon, S. and Tvaroska, I. 2006. Catalytic mechanism of glycosyltransferases: Hybrid quantum mechanical/molecular mechanical study of the inverting *N*-acetylglucosaminyltransferase I. *J Am Chem Soc*, 128, 16921–16927.

Kranjc, A. 2009. Predicting structural determinants and ligand poses in proteins involved in neurological diseases: Bioinformatics and molecular simulation studies. Doctoral Thesis, SISSA, The International School for Advanced Studies, Italy.

Kraulis, P. J., Clore, G. M., Nilges, M. et al. 1989. Determination of the three-dimensional solution structure of the C-terminal domain of cellobiohydrolase I from *Trichoderma reesei*. A study using nuclear magnetic resonance and hybrid distance geometry-dynamical simulated annealing. *Biochemistry*, 28, 7241–7257.

Krengel, U. and Imberty, A. 2007. Crystallography and lectin structure database. In *Lectins*, Carol, L. N. (ed) pp. 15–50. Amsterdam: Elsevier Science B. V.

Krupicka, M. and Tvaroska, I. 2009. Hybrid quantum mechanical/molecular mechanical investigation of the beta-1,4-galactosyltransferase-I mechanism. *J Phys Chem B*, 113, 11314–11319.

Lafaquière, V., Barbe, S., Puech-Guenot, S. et al. 2009. Control of lipase enantioselectivity by engineering the substrate binding site and access channel. *Chembiochem*, 10, 2760–2771.

Laine, R. A. 1994. Invited commentary: A calculation of all possible oligosaccharide isomers both branched and linear yields 1.05×10^{12} structures for a reducing hexasaccharide: The isomer barrier to development of single-method saccharide sequencing or synthesis systems. *Glycobiology*, 4, 759–767.

Laughrey, Z. R., Kiehna, S. E., Riemen, A. J., and Waters, M. L. 2008. Carbohydrate-pi interactions: What are they worth? *J Am Chem Soc*, 130, 14625–14633.

Lengauer, T. and Rarey, M. 1996. Computational methods for biomolecular docking. *Curr Opin Struct Biol*, 6, 402–406.

Li, W., Johnson, D. J. D., Esmon, C. T., and Huntington, J. A. 2004. Structure of the antithrombin-thrombin-heparin ternary complex reveals the antithrombotic mechanism of heparin. *Nat Struct Mol Biol*, 11, 857–862.

Liang, G., Schmidt, R. K., Yu, H. A., Cumming, D. A., and Brady, J. W. 1996. Free energy simulation studies of the binding specificity of mannose-binding protein. *J Phys Chem*, 100, 2528–2534.

Lii, J.-H. and Allinger, N. L. 1991. The MM3 force field for amides, polypeptides and proteins. *J Comput Chem*, 12, 186–199.

Lins, R. D. and Hünenberger, P. H. 2005. A new GROMOS force field for hexopyranose-based carbohydrates. *J Comput Chem*, 26, 1400–1412.

Lortat-Jacob, H., Grosdidier, A., and Imberty, A. 2002. Structural diversity of heparan sulfate binding domains in chemokines. *Proc Natl Acad Sci USA*, 99, 1229–1234.

MacKerell, A. D., Banavali, N., and Foloppe, N. 2000. Development and current status of the CHARMM force field for nucleic acids. *Biopolymers*, 56, 257–265.

MacKerell, A. D., Bashford, D., Bellott et al. 1998. All-atom empirical potential for molecular modeling and dynamics studies of proteins. *J Phys Chem B*, 102, 3586–3616.

Mackerell, A. D., Feig, M., and Brooks, C. L. 2004. Extending the treatment of backbone energetics in protein force fields: Limitations of gas-phase quantum mechanics in reproducing protein conformational distributions in molecular dynamics simulations. *J Comput Chem*, 25, 1400–1415.

McCullers, J. A., Hoffmann, E., Huber, V. C., and Nickerson, A. D. 2005. A single amino acid change in the C-terminal domain of the matrix protein M1 of influenza B virus confers mouse adaptation and virulence. *Virology*, 336, 318–326.

Momany, F. A. and Willett, J. L. 2000. Computational studies on carbohydrates: In vacuo studies using a revised AMBER force field, AMB99C, designed for [alpha]-(1—>4) linkages. *Carbohydr Res*, 326, 194–209.

Morris, G. M., Goodsell, D. S., Halliday, R. S. et al. 1998. Automated docking using a Lamarckian genetic algorithm and an empirical binding free energy function. *J Comput Chem*, 19, 1639–1662.

Muegge, I. and Martin, Y. C. 1999. A general and fast scoring function for protein-ligand interactions: A simplified potential approach. *J Med Chem*, 42, 791–804.

Murase, T., Zheng, R. B., Joe, M. et al. 2009. Structural insights into antibody recognition of mycobacterial polysaccharides. *J Mol Biol*, 392, 381–392.

Nurisso, A., Kozmon, S., and Imberty, A. 2008. Comparison of docking methods for carbohydrate binding in calcium-dependent lectins and prediction of the carbohydrate binding mode to sea cucumber lectin CEL-III. *Mol Simul*, 34, 469–479.

Parsiegla, G., Reverbel-Leroy, C., Tardif, C. et al. 2000. Crystal structures of the cellulase Cel48F in complex with inhibitors and substrates give insights into its processive action. *Biochemistry*, 39, 11238–11246.

Parthasarathy, S., Ravindra, G., Balaram, H., Balaram, P., and Murthy, M. R. 2002. Structure of the *Plasmodium falciparum* triosephosphate isomerase-phosphoglycolate complex in two crystal forms: Characterization of catalytic loop open and closed conformations in the ligand-bound state. *Biochemistry*, 41, 13178–13188.

Pathiaseril, A. and Woods, R. J. 2000. Relative energies of binding for antibody-carbohydrate-antigen complexes computed from free-energy simulations. *J Am Chem Soc*, 122, 331–338.

Pellegrini, L., Burke, D. F., von Delft, F., Mulloy, B., and Blundell, T. L. 2000. Crystal structure of fibroblast growth factor receptor ectodomain bound to ligand and heparin. *Nature*, 407, 1029–1034.

Pérez, S. 2007. Molecular modeling in glycoscience. In *Comprehensive Glycoscience*, Kamerling, J. P. (ed) V. 2, pp. 347–388. Oxford: Elsevier.

Pérez, S., Gautier, C., Imberty, A. et al. 2000. Oligosaccharide conformations by diffraction methods. In *Carbohydrates in Chemistry & Biology*, Ernst, B., Hart, G. and Sinay, P. (eds) pp. 969–1001. Weinheim: Wiley-VCH.

Pérez, S., Imberty, A., Engelsen, S. B. et al. 1998. A comparison and chemometric analysis of several molecular mechanics force fields and parameter sets applied to carbohydrates. *Carbohydr Res*, 314, 141–155.

Pérez, S. and Kouwijzer, M. 1999. Shapes and interactions of polysaccharide chains. In *Carbohydrates: Structures, Syntheses, and Dynamics*, Finch, P. (ed), pp. 258–293. Dordrecht: Kluwer Academic Publishers.

Pérez, S., Meyer, C. and Imberty, A. 1995. Practical tools for molecular modeling of complex carbohydrates and their interactions with proteins. *Mol Eng*, 5, 271–300.

Peters, T. and Pinto, B. M. 1996. Structure and dynamics of oligosaccharides: NMR and modeling studies. *Curr Opin Struct Biol*, 6, 710–720.

Petrova, P., Koca, J. and Imberty, A. 1999. Potential energy hypersurfaces of nucleotide sugars: Ab initio calculations, force-field parametrization, and exploration of the flexibility. *J Am Chem Soc*, 121, 5535–5547.

Poveda, A., Asensio, J. L., Espinosa, J. F. et al. 1997. Applications of nuclear magnetic resonance spectroscopy and molecular modeling to the study of protein-carbohydrate interactions. *J Mol Graph Model*, 15, 9–17, 53.

Raab, M. and Tvaroska, I. 2011. The binding properties of the H5N1 influenza virus neuraminidase as inferred from molecular modeling. *J Mol Model*, 17, 1445–1456.

Raghothama, S., Simpson, P. J., Szabo, L. et al. 2000. Solution structure of the CBM10 cellulose binding module from *Pseudomonas xylanase* A. *Biochemistry*, 39, 978–984.

Raman, R., Venkataraman, G., Ernst, S., Sasisekharan, V., and Sasisekharan, R. 2003. Structural specificity of heparin binding in the fibroblast growth factor family of proteins. *Proc Natl Acad Sci USA*, 100, 2357–2362.

Ramirez-Gualito, K., Alonso-Rios, R., Quiroz-Garcia, B. et al. 2009. Enthalpic nature of the CH/pi interaction involved in the recognition of carbohydrates by aromatic compounds, confirmed by a novel interplay of NMR, calorimetry, and theoretical calculations. *J Am Chem Soc*, 131, 18129–18138.

Ramsland, P. A., Farrugia, W., Bradford, T. M., Mark Hogarth, P., and Scott, A. M. 2004. Structural convergence of antibody binding of carbohydrate determinants in Lewis Y tumor antigens. *J Mol Biol*, 340, 809–818.

Rice, K. G., Pengguang, W., Brand, L., and Lee, Y. C. 1993. Experimental determination of oligosaccharide three-dimensional structure. *Curr Opin Struct Biol*, 3, 669–674.

Rossi, D. and Zlotnik, A. 2000. The biology of chemokines and their receptors. *Annu Rev Immunol*, 18, 217–243.

Sapay, N., Cabannes, E., Petitou, M., and Imberty, A. 2012. Molecular model of human heparanase with proposed binding mode of a heparan sulfate oligosaccharide and catalytic amino acids. *Biopolymers*, 97, 21–34.

Schirmer, T., Keller, T. A., Wang, Y. F., and Rosenbusch, J. P. 1995. Structural basis for sugar translocation through maltoporin channels at 3.1 A resolution. *Science*, 267, 512–514.

Schlessinger, J., Plotnikov, A. N., Ibrahimi, O. A. et al. 2000. Crystal structure of a ternary FGF-FGFR-heparin complex reveals a dual role for heparin in FGFR biniding and dimerization. *Mol Simul*, 6, 743–750.

Schuler, L. D., Daura, X., and van Gunsteren, W. F. 2001. An improved GROMOS96 force field for aliphatic hydrocarbons in the condensed phase. *J Comput Chem*, 22, 1205–1218.

Schuler, L. D. and Van Gunsteren, W. F. 2000. On the choice of dihedral angle potential energy functions for n-alkanes. *Mol Simul*, 25, 301–319.

Schulz, E.-C., Bergfeld, A., Muehlenhoff, M., and Ficner, R. 2011. Crystal structure of the polysialic acid specific O-acetyltransferase NeuO. *PLoS One*, 6, e17403.

Simmons, D. A. and Romanowska, E. 1987. Structure and biology of Shigella flexneri O antigens. *J Med Microbiol*, 23, 289–302.

Sitkoff, D., Sharp, K. A., and Honig, B. 1994. Accurate calculation of hydration free energies using macroscopic solvent models. *J Phys Chem*, 98, 1978–1988.

Skinner, R., Abrahams, J.-P., Whisstock, J. C. et al. 1997. The 2.6 Å structure of antithrombin indicates a conformational change at the heparin binding site. *J Mol Biol*, 266, 601–609.

Sorin, E. J. and Pande, V. S. 2005. Empirical force-field assessment: The interplay between backbone torsions and noncovalent term scaling. *J Comput Chem*, 26, 682–690.

Spiwok, V., Lipovova, P., Skalova, T. et al. 2004. Role of CH/pi interactions in substrate binding by *Escherichia coli* beta-galactosidase. *Carbohydr Res*, 339, 2275–2280.

Spiwok, V., Lipovová, P., Skálová, T. et al. 2005. Modelling of carbohydrate–aromatic interactions: Ab initio energetics and force field performance. *J Comput Aided Mol Des*, 19, 887–901.

Srinivasan, J., Cheatham, T. E., Cieplak, P., Kollman, P. A., and Case, D. A. 1998. Continuum solvent studies of the stability of DNA, RNA, and phosphoramidate–DNA Helices. *J Am Chem Soc*, 120, 9401–9409.

Sulak, O., Cioci, G., Delia, M. et al. 2010. A TNF-like trimeric lectin domain from *Burkholderia cenocepacia* with specificity for fucosylated human histo-blood group antigens. *Structure*, 18, 59–72.

Sulak, O., Lameignere, E., Wimmerova, M., and Imberty, A. 2009. Specificity and affinity studies in lectin/carbohydrate interactions. *Carbohydr Chem*, 357–372.

Tempel, W., Tschampel, S., and Woods, R. J. 2002. The xenograft antigen bound to *Griffonia simplicifolia* Lectin 1-B4. *J Biol Chem*, 277, 6615–6621.

Tessier, M. B., DeMarco, M. L., Yongye, A. B., and Woods, R. J. 2008. Extension of the GLYCAM06 biomolecular force field to lipids, lipid bilayers and glycolipids. *Mol Simul*, 34, 349–364.

Theillet, F. X., Frank, M., Normand, B. V. et al. 2011. Dynamic aspects of antibody: oligosaccharide complexes characterized by molecular dynamics simulations and STD-NMR. *Glycobiology*, 21, 1570–1579.

Tormo, J., Lamed, R., Chirino, A. J. et al. 1996. Crystal structure of a bacterial family-III cellulose-binding domain: A general mechanism for attachment to cellulose. *EMBO J*, 15, 5739–5751.

Tschampel, S. M. and Woods, R. J. 2003. Quantifying the role of water in protein-carbohydrate interactions. *J Phys Chem A*, 107, 9175–9181.

Tvaroska, I. and Pérez, S. 1986. Conformational-energy calculations for oligosaccharides: A comparison of methods and a strategy of calculation. *Carbohydr Res*, 149, 389–410.

van Roon, A. M., Pannu, N. S., de Vrind, J. P. et al. 2004. Structure of an anti-Lewis X Fab fragment in complex with its Lewis X antigen. *Structure*, 12, 1227–1236.

van Roon, A. M., Pannu, N. S., Hokke, C. H., Deelder, A. M., and Abrahams, J. P. 2003. Crystallization and preliminary x-ray analysis of an anti-LewisX Fab fragment with and without its Lewis X antigen. *Acta Crystallogr D Biol Crystallogr*, 59, 1306–1309.

Vandenbussche, S., Diaz, D., Fernandez-Alonso, M. C. et al. 2008. Aromatic-carbohydrate interactions: An NMR and computational study of model systems. *Chemistry*, 14, 7570–7578.

Varghese, J. N., Laver, W. G., and Colman, P. M. 1983. Structure of the influenza virus glycoprotein antigen neuraminidase at 2.9 Å resolution. *Nature*, 303, 35–40.

Varghese, J. N., McKimm-Breschkin, J. L., Caldwell, J. B., Kortt, A. A., and Colman, P. M. 1992. The structure of the complex between influenza virus neuraminidase and sialic acid, the viral receptor. *Proteins*, 14, 327–332.

von Itzstein, M., Wu, W. Y., Kok, G. B. et al. 1993. Rational design of potent sialidase-based inhibitors of influenza virus replication. *Nature*, 363, 418–423.

Vulliez-Le Normand, B., Saul, F. A., Phalipon, A. et al. 2008. Structures of synthetic O-antigen fragments from serotype 2a *Shigella flexneri* in complex with a protective monoclonal antibody. *Proc Natl Acad Sci USA*, 105, 9976–9981.

Walker, R. C. 2003. The development of a QM/MM based linear response method and its application to proteins. Doctoral Thesis, Imperial College London, UK.

Woods, R. J. 1996. The application of molecular modeling techniques to the determination of oligosaccharide solution conformations. In *Reviews in Computational Chemistry*, Kenny B. Lipkowitz, K. B. and Donald B. Boyd, D. B. (eds), V. 9, pp. 129–165. New York: Wiley-VCH.

Woods, R. J. and Chappelle, R. 2000. Restrained electrostatic potential atomic partial charges for condensed-phase simulations of carbohydrates. *Theochem-J Mol Struct*, 527, 149–156.

Woods, R. J., Dwek, R. A., Edge, C. J., and Fraser-Reid, B. 1995. Molecular mechanical and molecular dynamic simulations of glycoproteins and oligosaccharides. 1. GLYCAM_93 parameter development. *J Phys Chem*, 99, 3832–3846.

Woods, R. J., Khalil, M., Pell, W., Moffat, S. H., and Smith, V. H. 1990. Derivation of net atomic charges from molecular electrostatic potentials. *J Comput Chem*, 11, 297–310.

Woods, R. J. and Tessier, M. B. 2010. Computational glycoscience: Characterizing the spatial and temporal properties of glycans and glycan-protein complexes. *Curr Opin Struct Biol*, 20, 575–583.

Xu, G.-Y., Ong, E., Gilkes, N. R. et al. 1995. Solution structure of a cellulose-binding domain from *Cellulomonas fimi* by nuclear magnetic resonance spectroscopy. *Biochemistry*, 34, 6993–7009.

Yang, M., Zhou, L., Zuo, Z., Tang, X., Liu, J., and Ma, X. 2008. Structure-based virtual screening for glycosyltransferase 51. *Mol Simul*, 34, 849–856.

Yui, T., Shiiba, H., Tsutsumi, Y. et al. 2010. Systematic docking study of the carbohydrate binding module protein of Cel7A with the cellulose Ialpha crystal model. *J Phys Chem B*, 114, 49–58.

5 Docking of Carbohydrates into Protein Binding Sites

Mark Agostino, Paul A. Ramsland,
and Elizabeth Yuriev

CONTENTS

5.1 INTRODUCTION

Structures of protein–carbohydrate complexes can be investigated experimentally by x-ray crystallography or nuclear magnetic resonance (NMR) (Chapters 1 and 2). However, the crystallization of these complexes is often complicated by the inherent flexibility of carbohydrates, whereas NMR is more appropriate for the determination of conformation of bound carbohydrates rather than the nature of the interactions occurring with a protein. Computational molecular modeling techniques offer attractive alternatives for the study of protein–carbohydrate interactions. These techniques have been widely used in conformational studies of carbohydrates, as well as in the investigation of carbohydrate–protein recognition. Notably, molecular docking can provide insight into protein–ligand interactions in systems that are difficult to study experimentally.

In general, molecular modeling methods require some structural knowledge of the ligand and the receptor of interest. Carbohydrate ligands are typically constructed using molecular modeling software or directly sourced from structural databases. Receptor structures are typically obtained from x-ray crystallography and NMR, and those that are unavailable can be generated by homology modeling, threading, and de novo methods.

Docking is a computational technique that places a small molecule (ligand) in the binding site of its macromolecular target (receptor) and estimates its binding affinity. Generally, automated in silico docking involves ligand (and, ideally, receptor) sampling and scoring. Sampling itself entails conformational and orientational sampling of the ligand within the constraints of the receptor binding site. A scoring function selects the best ligand pose (i.e., ligand conformation, orientation, and translation) and also rank orders ligands, if a ligand database is docked. To be considered successful, docking must accurately predict (relative to experimentally available data) either or both ligand structure (pose prediction) and its binding propensity (affinity prediction). For pose prediction, a program's ability to produce a ligand pose within a reasonable root-mean-square deviation (RMSD) to the experimental structure (commonly, ≤ 2 Å) as its top or highly ranked pose is usually considered a docking success. Available docking methods differ mainly in ligand placement in the active site, exploration of conformational space, and scoring or binding affinity estimation (Bottegoni et al. 2009; Corbeil and Moitessier 2009; Friesner et al. 2004; Jain 2007; Lang et al. 2009; Morris et al. 2009; Trott and Olson 2010; Verdonk et al. 2003).

In most popular docking programs, the ligand is treated as flexible, whereas the protein conformation is often kept rigid. This relies on the lock-and-key hypothesis for protein–ligand binding. However, it is now widely recognized that ligand binding is not a static event but a dynamic process, in which both the ligand and the protein may undergo conformational changes. Properly accounting for receptor flexibility is much more computationally expensive than doing so for ligand flexibility (B-Rao et al. 2009). Several programs and procedures are able to account for receptor flexibility by "soft docking," which emulates some plasticity of the receptor by allowing an overlap between the protein and the ligand; by allowing the implementation of explicit side-chain flexibility; and/or by ensemble docking, where docking is performed into multiple rigid receptor conformations (Cerqueira et al. 2009; Davis and Baker 2009; Morris et al. 2009; Sherman et al. 2006).

It has been long recognized in the docking field that various docking programs and scoring functions perform differently for different targets (Warren et al. 2006). Similarly, varying docking performance has been observed for different ligand types (Yuriev et al. 2011). A majority of scoring functions currently in use have been designed for protein interactions with ligands, which are mostly small organic molecules. However, docking studies for ligands other than small organic molecules, such as carbohydrates, offer novel ways to investigate mechanisms of ligand interactions with proteins. Therefore, novel docking programs and, more importantly, scoring functions need to be developed for these systems. Alternatively, existing programs and functions need to be evaluated for these systems before they can be used reliably.

Compared to protein interactions with small organic molecules, there are some unique features in protein–carbohydrate interactions that make carbohydrate docking particularly challenging. One must consider the extreme flexibility of carbohydrates and the presence of a large number of hydroxyl groups. The latter often leads to the formation of extensive hydrogen bonding networks, significantly contributing to the overall binding. A distinctive feature of protein–carbohydrate recognition is the interaction between aromatic side chains of the protein and C–H bonds of the carbohydrate's hydrophobic faces, which results in the formation of crucial CH–π contacts (Laughrey et al. 2008). As a result of these peculiarities, widely used docking programs and scoring functions, which account differently for these types of interactions, may not perform as well for protein–carbohydrate complexes.

Carbohydrate-specific scoring functions (empirical free energy function [Hill and Reilly 2008b; Laederach and Reilly 2003] and SLICK [Kerzmann et al. 2006]) and docking algorithms (DARWIN [Taylor and Burnett 2000] and BALLDock/SLICK [Kerzmann et al. 2008]) aim to address the specific issues associated with carbohydrate–protein docking. Despite the development of these specialized methods, it is always important to evaluate a range of docking approaches (where suitable validation cases exist) to determine which performs best for a given protein–ligand system. Popular docking software is routinely used to investigate carbohydrate–protein recognition because it is often shown to perform well at this task, for example, see the study by Agostino et al. (2009a). Molecular docking results are often considered in conjunction with molecular dynamics (MD) simulation or NMR-based approaches.

5.2 CARBOHYDRATE DOCKING COMPARISON AND VALIDATION

In order to demonstrate that docking can be confidently used for protein–carbohydrate complexes, a number of studies have been carried out to compare docking programs for their specific use in this area of research. Targets studied include enzymes (Alexacou et al. 2008), lectins (Adam et al. 2008; Agostino et al. 2011; Nurisso et al. 2008; Siebert et al. 2009), and antibodies (Agostino et al. 2009a). In one recent study, different types of protein targets were used to dock glycosaminoglycans (GAGs) into interleukin-8 (IL-8) by a range of docking programs (Samsonov et al. 2011). In most studies described in this chapter, where individual programs have been used for carbohydrate docking, validation against relevant crystal structures or an assessment of a docked binding mode in conjunction with other biochemical or biophysical data has been performed (see Tables 5.1 through 5.4).

5.2.1 EVALUATION OF CARBOHYDRATE DOCKING INTO ENZYMES

Glide and GOLD have been compared for docking of glucosyltriazolylacetamide to glycogen phosphorylase b (Alexacou et al. 2008). Both programs performed well in predicting the experimentally determined binding poses for the ligand, with Glide, in particular, producing impressive results. These researchers also used Glide Extra Precision (Glide XP) to dock a series of glucose-based spiroisoxazolines into glycogen phosphorylase b (Benltifa et al. 2009). The docked structures were found to be in excellent agreement with experimental results, both in terms of docked pose/crystal structure RMSDs and docked score/experimental free binding energy correlations.

5.2.2 Evaluation of Carbohydrate Docking into Lectins

The performance of AutoDock, DOCK, and Glide was evaluated for lectin–carbohydrate systems using seven crystal structures (Nurisso et al. 2008). It was found that Glide marginally outperformed the other programs in terms of pose RMSD to the crystal structure. However, AutoDock was found to most accurately reproduce calcium-binding geometry.

AutoDock and DOCK were also compared with respect to their performance in docking monosaccharide ligands to the calcium-dependent bacterial lectin PA-IIL (from *Pseudomonas aeruginosa*) and its in silico mutants (Adam et al. 2008). The comparison was carried out to check whether the programs could be used to predict changes in lectin affinity and specificity caused by minor mutations of the binding-site residues, critical to host recognition by the bacterial lectin. Both programs were evaluated in terms of both pose prediction and binding-energy prediction accuracy. This was achieved by comparing the docked structures and scoring function-based estimated binding energies with the experimental data. It was found that AutoDock outperformed DOCK on both counts.

Siebert et al. (2009) used AutoDock, FlexX, and Glide, in conjunction with NMR methods, to determine the binding mode of sialic acid-terminating carbohydrates in complex with the lectin SHL-2 from the Chinese bird-hunting spider. It was found that the poses obtained using Glide agreed excellently with the NMR data.

More recently, we used 15 high-resolution (≤ 2.0 Å) human-derived carbohydrate–lectin complexes to evaluate the pose prediction performance of Glide, GOLD, AutoDock, and DOCK (Agostino et al. 2011). Although the four evaluated programs were generally able to identify reasonable carbohydrate poses in each test case, the top ranked pose was seldom found to be the best pose obtained by any of the programs. The program DOCK correctly ranked the best pose in 6 out of the 15 cases, closely followed by Glide. We suggest that this ranking failure in the majority of cases is most likely due to a scoring problem. However, for Glide and GOLD, the best poses were normally found within the top 10 ranked poses. The program AutoDock generally failed to accurately identify the correct pose. This kind of performance discrepancy among Glide, GOLD, and AutoDock was previously observed in both antibody–carbohydrate (Agostino et al. 2009a) and lectin–carbohydrate (Nurisso et al. 2008) docking. In terms of pose prediction accuracy, regardless of scoring/ranking, GOLD was clearly the best program with an average RMSD for the best poses of 1.4 Å, compared to 2.2–2.7 Å for the other three programs. This docking study highlighted that considering the top ranked docking pose alone is not suitable for studying lectin–carbohydrate interactions and alternative and/or complementary approaches (discussed in Section 5.3.4) are required for this purpose.

5.2.3 Evaluation of Carbohydrate Docking into Antibodies

We compared several docking programs (Glide, AutoDock, GOLD, and FlexX) in antibody–carbohydrate cognate and cross-docking simulations and evaluated the performance of Glide and AutoDock in rigid versus flexible protein docking (Agostino et al. 2009a). A total of 11 high-resolution antibody–carbohydrate crystal structures were selected, each having short carbohydrates (2–4 residues in length) with a similar level of flexibility. In this study, all antibody binding sites belonged to a cavity class (according to Webster and Rees [1995]). We also used the scheme of Lee et al. (2006) to further characterize the binding sites based on familiar geological features. Several binding-site types (valleys, craters, and canyons) were chosen.

Our results demonstrated that Glide was the most accurate program for rigidly docking carbohydrates to antibodies. The programs GOLD and AutoDock had several problems and FlexX performed poorly. The induced-fit docking (IFD) protocol (Sherman et al. 2006) implemented in the Schrödinger package for flexible receptor docking combines Glide with the protein structure refinement module Prime (Jacobson et al. 2004). The IFD produced results with a level of accuracy comparable to that of cognate rigid receptor docking. The accuracy of the IFD protocol in both cognate and cross-docking is promising for the study of the dynamic nature of protein–carbohydrate recognition. Our findings are in agreement with the experiences of others. Although a perfect docking program does not exist currently, Glide ranks consistently high in docking studies (Englebienne et al. 2007; Kellenberger et al. 2004; Nurisso et al. 2008) with some studies listing its treatment of flexible ligands as an important factor (Vaque et al. 2008). More specifically, Glide seems to be a popular choice for carbohydrate docking, and this choice is supported by a range of studies in which carbohydrate docking into diverse proteins has been structurally validated (Benltifa et al. 2009; Kolomiets et al. 2009; Koppisetty et al. 2010).

Specifically for carbohydrate docking to antibodies, we found that the diverse antibody binding sites and carbohydrate ligand lengths present challenges for accurate docking (Agostino et al. 2009a). Large, flexible noncarbohydrate residues and crater-shaped binding sites, accommodating large carbohydrate ligands, were found to be the most limiting features. The best results were produced for antibodies with canyon-shaped binding sites, which offered good results for both small and large carbohydrate ligands.

5.2.4 EVALUATION OF GAG DOCKING

Due to their negatively charged nature, GAGs require accurate consideration of electrostatic and water-mediated interactions. This requirement adds another dimension of the complexity of generating protein–GAG complexes by docking. Until recently, docking GAGs represented a very impractical task, and modeling of their complexes with proteins was usually done in conjunction with other methods (Costa et al. 2010; Gandhi and Mancera 2009; Rogers et al. 2011; Sapay et al. 2011). To address this challenge, Samsonov and coworkers (2011) recently developed a method for de novo placement of explicit water molecules in order to improve docking outcomes. They used a dataset of 11 protein complexes and tested the docking performance of four programs (AutoDock 3, eHiTs, MOE, and FlexX 3.1) with and without explicit water. Test cases were chosen to be high-resolution structures (not greater than 2.2 Å) and to contain ligands with no more than four saccharide monomers. The performance of the four programs was found to be in the order of FlexX < MOE < eHiTs < AutoDock 3 with all four programs producing better poses when coupled with explicit water molecules. This finding is very important given that most docking programs, including the four tested ones, already include solvation effects implicitly in their scoring functions. Therefore, it was significant to note that the addition of explicit water molecules to the binding sites could further improve docking outcomes.

The program AutoDock 3 performed significantly better than the other three programs in the validation experiments. Specifically, it produced top poses with average 1.94 and 1.60 Å RMSDs for the binding sites without and with explicit water molecules, respectively. Yet it failed to correctly reproduce binding-site residues experimentally known (by site-directed mutagenesis) to be important for binding IL-8 to heparin/heparan sulfate disaccharides. The program FlexX performed significantly worse than the other three programs when tested with FlexX type 3 water molecules, but its performance was improved almost to the level of that of eHiTs when combined with crystallographic or GRID-generated water molecules. The failure of FlexX to dock carbohydrates well is similar to our own experience with FlexX and carbohydrate–antibody complexes (discussed in Section 5.2.3). The authors have suggested that FlexX's anchor-based ligand placement algorithm may be a reason for its unsuitability for GAG docking. Unfortunately, other programs commonly used for docking carbohydrates, such as Glide, GOLD, and Surflex, were not evaluated in this study.

5.3 MODELING CARBOHYDRATE–PROTEIN RECOGNITION BY MOLECULAR DOCKING

Sections 5.3.1 through 5.3.4 describe docking studies of carbohydrate recognition by enzymes, lectins, and antibodies. Proteins, recognizing and binding to carbohydrates, are not limited to these groups and docking methods have indeed been used to investigate their interactions with carbohydrates. For example, binding of GAGs to cellular growth factors has been studied by docking (the study by Sapay et al. [2011] and references to studies therein). However, in many cases docking alone is not sufficient to investigate the interaction and additional methods, such as MD (Costa et al. 2010; Sapay et al. 2011), free energy calculations (Gandhi and Mancera 2009), or microarray analysis (Rogers et al. 2011), are used in conjunction with docking. The studies described Sections 5.3.1 through 5.3.4 present the cases where outcomes of docking per se have been used to get an insight into protein–carbohydrate recognition.

In many earlier studies, carbohydrate docking was carried out by manually positioning the ligand pose based on the crystal structure of a related complex and then optimizing the new complex (e.g., see study by Meyer et al. [2005]). This approach is based on the assumption that the new complex is analogous to the template and is therefore biased toward input structures. Such manual docking cannot be validated and is not amenable to the prediction of truly novel complexes. Thus, we are not reviewing such studies here.

5.3.1 MODELING CARBOHYDRATE–GLYCOSIDASE RECOGNITION

Carbohydrate recognition by glycosidases and glycosidase-associated carbohydrate-binding modules (CBMs) has been extensively studied by automated molecular docking (Table 5.1). The majority of studies focus on elucidating the binding modes and catalytic mechanisms of carbohydrate cleavage by glycosidases. Selected examples are discussed in more detail here.

TABLE 5.1

Recently Published Molecular Docking Studies in Carbohydrate–Glycosidase Recognition

Ligand	Target	Research Aim	Programs	Validation[a,b]	Reference
α-Glucosides	Rhizopus oryzae glucoamylase	Complex structure elucidation	AutoDock 3.05	*(NMR, CSP)	(Liu et al. 2007)
	Solanum tuberosum α-amylase	Determine effect of ligand phosphorylation on complex stability	DOCK 6, Glide 4.5	+	(Dudkiewicz et al. 2008)
α-Glycosides	Glycoside hydrolase family 6 enzymes	Complex structure elucidation and complex stability	AutoDock 3.06	+	(Mertz et al. 2007)
	Glycoside hydrolase family 1 enzymes	Complex structure elucidation and complex stability	AutoDock 3.06	+	(Hill and Reilly 2008a)
	Fusarium oxysporum Cel7B	Determine force exerted by enzyme on substrate	AutoDock 3.06	+	(Mulakala and Reilly 2005)
	Cel9A cellulase CBM	Determine role of CBM in cellulase catalysis	DOCK 5	*(MD)	(Oliveira et al. 2009)
α/β-Glucosides	Glycogen phosphorylase	Complex structure elucidation; analog design	Glide 4.0, GOLD 3.1.1	+	(Alexacou et al. 2008)
			Glide 4.5 (XP and QPLD)	\(Alexacou et al. 2008)	(Benltifa et al. 2009)
			Glide 5.0 (XP and QPLD)	\(Alexacou et al. 2008)	(Tsirkone et al. 2010)
β-Glucosides	Candida albicans β1,3-exoglucanase	Understanding of catalytic mechanism	GOLD 3.1[c]	*(MD)	(Moura–Tamames et al. 2009)
Gluco-configured tetrahydroimidazopyridines	Thermotoga maritima β-glucosidase	Complex structure elucidation; analog design	Glide 5.0	–	(Li et al. 2011)
α1,2-Mannobioside	ERManI	Determination of likely ligand transition state	AutoDock[d]	+	(Mulakala et al. 2006)

(Continued)

TABLE 5.1 (*Continued*)
Recently Published Molecular Docking Studies in Carbohydrate–Glycosidase Recognition

Ligand	Target	Research Aim	Programs	Validation[a,b]	Reference
		Determination of cleaved product exit mechanism and return to ground-state conformation	AutoDock[d]	\ (Mulakala et al. 2006)	(Mulakala et al. 2007)
		Determine catalytic proton donor	AutoDock 3.0	+	(Cantu et al. 2008)
		Determine binding modes of conformationally constrained mimics	AutoDock[d]	+	(Sivapriya et al. 2007)
Rhamnogalacturonan-I	Rhamnogalacturonase	Complex structure elucidation	AutoDock 3.0	*(MD)	(Choi et al. 2004)
Xylotriose	*Pseudoaltermonas haloplanktis* TAH3a β1,4-xylanase	Determination of likely ligand transition state	AutoDock 3.0.5	+	(De Vos et al. 2006)
Sucrose	*Arabidopsis* invertase mutant	Complex structure elucidation	FlexX 2	+	(Matrai et al. 2008)
Galactomannan	*Saccharomyces cerevisiae* α-galactosidase	Complex structure elucidation	AutoDock 4.2	*	(Fernandez-Leiro et al. 2010)
Tri-*N*-acetylchitotriose	OsChia1b	Complex structure elucidation	AutoDock 4	−	(Kezuka et al. 2010)

[a] Guide to symbols: Docking method validated against relevant structures (+), no method validation performed or cited (−), followed earlier validated method (\), binding mode predicted in conjunction with other biochemical or biophysical data relating to interactions (*).

[b] Technique abbreviations: NMR—nuclear magnetic resonance; CSP—chemical shift perturbations; MD—molecular dynamics.

[c] Covalent docking approach used.

[d] Program version not specified.

Amylases are used industrially to catalyze the formation of glucose from starch and were among the earliest enzymes to be studied by molecular docking (see the study by Laederach and Reilly [2005] and references to earlier studies listed therein). Liu and coworkers (2007) determined the solution structure of *Rhizopus oryzae* glucoamylase CBM in complex with β-cyclodextrin and various maltooligosaccharides, using a combination of high-field NMR and docking simulations in AutoDock. Dudkiewicz et al. (2008) used a combination of DOCK 6 and Glide 4.5 to calculate the binding energies of phosphorylated maltooligosaccharides in complex with *Solanum tuberosum* α-amylase. It was found that complex stability increased with phosphorylation and chain elongation.

Cellulases are being engineered to make the generation of glucose from cellulose more economically viable. The Reilly group (Mertz et al. 2007) used AutoDock 3.06 to investigate substrate binding by glycoside hydrolase family 6 enzymes from *Hypocrea jecorina* and *Thermobifida fusca*. It was determined that β-glucose binds most tightly to the enzyme subsite approximately two carbohydrate residues away from the glycosidic cleavage point (i.e., the -2 subsite). A subsequent study by the Reilly group (Hill and Reilly 2008a) investigated the specificity of a variety of glycoside hydrolase family 1 enzymes for β-glucosides using AutoDock 3.06. A series of eight active-site residues were identified by the docking studies as being largely responsible for carbohydrate specificity. Oliveira et al. (2009) used a combination of docking in DOCK 5 and MD simulation to investigate cellulose recognition by the CBM of the Cel9A enzyme from *T. fusca*. The docking results suggested that cellulose chains could bind to the enzyme active site via the CBM, thus explaining why the removal of the CBM results in a reduced turnover of cellulose by the enzyme.

Sugar derivatives are of interest as therapeutic agents, particularly in the treatment of diabetes. Alexacou et al. (2008) used a combination of crystallography, molecular docking, and MD simulation to investigate the interaction of glucosyltriazolylacetamide with glycogen phosphorylase. The programs GOLD and Glide were compared for their ability to reproduce the crystallographic binding mode, and it was found that Glide performed best at this task both with and without waters bound to the active site. In a subsequent study (Benltifa et al. 2009), Glide was used to determine the binding modes of glucose-based spiroisoxazolines in complex with glycogen phosphorylase. For this purpose, Glide XP mode and quantum mechanics-polarized ligand docking (QPLD) were utilized. It was found that the methods performed comparably. The Glide XP and QPLD protocols were also used to investigate the binding of fluorinated glucosides to glycogen phosphorylase (Tsirkone et al. 2010). Li et al. (2011) synthesized derivatives of gluco-configured tetrahydroimidazopyridines, which were proposed as mimics of the β-glucosidase transition state. The program Glide was used to dock the derivatives, and the structures determined were used to justify the observed activities.

Although flexible receptor docking techniques are generally not used for investigating carbohydrate–glycosidase recognition, several studies have used automated docking followed by MD to simulate receptor flexibility (Choi et al. 2004; Moura-Tamames et al. 2009; Oliveira et al. 2009). Covalent docking was used in a recent study to elucidate the enzymatic mechanism of β-glucosidase (Moura-Tamames et al. 2009). Flexible receptor docking and covalent docking methods are relatively new innovations; thus, the literature describing their use in studying carbohydrate–enzyme recognition is currently sparse.

5.3.2 MODELING ENZYMATIC TRANSGLYCOSYLATION

Glycosyltransferases catalyze the extension of glycan chains (transglycosylation) by coupling an activated donor substrate to a specific acceptor substrate. They are emerging therapeutic targets for the treatment of glycosylation disorders and bacterial infections; they are also valuable in organ transplant engineering, where their presence or absence can affect the presentation of specific carbohydrate epitopes. Docking studies on the mechanisms of glycosyltransferases are challenging due to methodological limitations in handling multiple ligands and the creation of new covalent bonds. Although these limitations have recently been addressed (Li and Li 2010; Morris et al. 2009), such methods are yet to be broadly applied to investigate glycosyltransferase mechanisms. Recently published studies in transglycosylation are described in Table 5.2.

Rao and Tvaroška (2001) developed a homology model of bovine α1,3-galactosyltransferase to which a donor substrate, uridine diphosphate (UDP)-galactose, was docked. Homologues of this enzyme are responsible for the production of the major carbohydrate xenoantigen in pigs, which forms an immunological barrier to xenotransplantation. The model was also used to study the binding of an N-acetyllactosamine-derived inhibitor. The Tvaroška group (Kozmon and Tvaroška 2006) also investigated the reaction catalyzed by GnT-1, an enzyme that transfers N-acetylglucosamine to the nonreducing end mannose residue on high-mannose type glycans. The crystal structure of the catalytic fragment of rabbit GnT-1 was used as a starting point to which the missing acceptor substrate was docked using Glide. Mixed quantum mechanical (QM) density functional theory (DFT)/molecular mechanical (MM) simulations were then used to predict the likely transition state (Kozmon and Tvaroška 2006). The predicted transition state was subsequently used to determine the likely structures of transition-state analogs again using a combination of docking in Glide and QM(DFT)/MM simulations (Tvaroška et al. 2008).

Receptor flexibility is rarely considered in molecular docking studies of glycosyltransferases. Takaoka and coworkers (2010) carried out a study to evaluate the conjugation capacity of mutants of UDP-glucuronosyltransferase 1A1 (UGT1A1). The donor substrate, UDP-glucuronic acid, was first docked to the mutants using AutoDock 4. Active-site movements as a result of donor binding were modeled using the induced-fit method of MOE 2009.10. The program AutoDock 4 was then used to dock the acceptor substrate, bilirubin, to each mutant enzyme; 100 bilirubin orientations were collected. The number of bilirubin orientations obtained, in which the hydroxyl group of bilirubin pointed toward the donor substrate, was tallied for each mutant enzyme. The tallies obtained were in excellent agreement with the biochemically determined conjugation capacity of each of the mutants, thus indicating the utility of the molecular modeling approach.

In glycosidases, the transglycosylation reaction is weakly competitive with cleavage because the product of transglycosylation is typically a cleavage substrate.

TABLE 5.2

Recently Published Molecular Docking Studies in Enzymatic Transglycosylation

Donor	Acceptor	Target	Research Aim	Programs	Validation[a,b]	Reference
UDP-galactose	N-acetyllactosamine[c]	Bovine α1,3-galactosyltransferase	Donor and inhibitor complex structure elucidation	AutoDock[d]	−	(Rao and Tvaroška 2001)
UDP-N-acetylglucosamine	High-mannose glycans	GnT-1	Determine binding mode of transition-state analog[e]	Glide	*(QM/DFT)	(Tvaroška et al. 2008)
UDP-glucuronic acid	Bilirubin	UGT1A1	Determine donor and acceptor binding modes, incorporating protein flexibility	AutoDock 4, MOE 2009.10[f]	−	(Takaoka et al. 2010)
UDP-glucose	Flavonoids, isoflavonoids	Medicago truncatula UGT85H2	Determine specificity of enzyme for acceptor substrates	GOLD[d]	+	(Li et al. 2007)
GDP-mannose[g]	Phosphatidyl-myo-inositol	PimA	Determine binding mode of acceptor substrate	ICM v3.6	*(SDM, XRC)	(Guerin et al. 2009)
Xyloglucan[c]	XLLG xyloglucan oligosaccharide	HvXETs 3,4, and 6	Determine binding mode of acceptor substrate	AutoDock 4.1	*(XRC)	(Vaaje-Kolstad et al. 2010)
Galactose/glucose	Galactose/glucose	β-glycosidases[h]	Determine molecular basis of transglycosylation stereoselectivity	GOLD[i,i]	+	(Bras et al. 2009)

[a] Guide to symbols: Docking method validated against relevant structures (+), no method validation performed or cited (−), binding mode predicted in conjunction with other techniques (*). [b] Technique abbreviations: QM—quantum mechanics; DFT—density functional theory; SDM—site-directed mutagenesis; XRC—comparison of docked binding mode to closely related x-ray crystal structure complexes. [c] Binding mode not docked and unavailable from experimental structures. [d] Version not specified. [e] Donor and acceptor binding not explicitly studied. [f] The IFD mode used. [g] Binding mode not docked but sourced from experimental structures. [h] Multiple protein systems studied with combinations of donors and acceptors listed. [i] Covalent docking mode used.

However, in some glycosidases, the competing reaction proceeds efficiently. Such glycosidases are used as starting points to engineer efficient synthetic tools for the enzymatic production of oligosaccharides. The transglycosylation reaction was investigated using the recently developed covalent docking method in GOLD (Bras et al. 2009). Docking calculations against high-resolution x-ray crystal structures of apo and substrate-bound forms of a series of bacterial glycosidases were used to predict the regioselectivity of the enzymes for specific transglycosylation products. The glycosyl enzyme intermediate was generated using covalent docking, which was followed by noncovalent docking of the acceptor substrate and MD simulation. As demonstrated by the method validation in the study, this procedure is likely to be generally useful for studying transglycosylation in silico.

5.3.3 MODELING CARBOHYDRATE–LECTIN RECOGNITION

A wide variety of plant and mammalian lectins have been studied using molecular docking (for a review of earlier studies, see the study by Neumann et al. [2004]). Accurate determination of carbohydrate–lectin complexes remains a nontrivial matter due to the shallow and multichambered binding sites of many lectins. Recent docking studies in carbohydrate–lectin recognition are summarized in Table 5.3. Selected examples are discussed here.

One of the most comprehensive studies on the validation of carbohydrate–lectin docking was performed by the Imberty group (Nurisso et al. 2008). Their study compared the abilities of AutoDock, DOCK, and Glide to reproduce the carbohydrate-binding modes of seven calcium-dependent lectins. It was found that Glide performed best at this task, in terms of pose RMSD to the crystal structure. However, AutoDock was found to most accurately reproduce calcium-binding geometry. Therefore, both Glide and AutoDock were subsequently used to predict the binding modes of galactose and N-acetylgalactosamine in complex with the sea cucumber lectin CEL-III. A subsequent study by the Imberty group (Blanchard et al. 2008) used AutoDock followed by MD simulation to investigate the recognition of Galα(1-4)Gal-terminating glycosphingolipids by $P.$ $aeruginosa$ lectin I (PA-IL). The docking method was validated by docking Galα(1-3)Gal into the PA-IL binding site and comparing the result to that obtained crystallographically. The docking study was later extended to investigate several α-galactosyl disaccharides (Nurisso et al. 2010).

The DC-SIGN is a C-type (calcium dependent) lectin that binds to Lewis x and high-mannose carbohydrate structures. Pathogens that present these carbohydrates on their surface utilize DC-SIGN to promote infection, making it a potentially useful target for anti-infective agents. Reina et al. (2007) generated a mannobioside mimic of Manα(1-2)Man, and its complex with DC-SIGN was determined using QPLD. The procedure gave rise to several poses with distorted φ (phi) angles (i.e., non-exo-anomeric conformations). In a subsequent study (Reina et al. 2008), FlexiDock was used to dock a series of mannosyl trisaccharides to DC-SIGN. The study highlighted the importance of multiple binding modes in carbohydrate complexes with DC-SIGN. Obermajer et al. (2011) synthesized a series of mannobioside mimics with varying amide substituents on the second residue. The compounds were docked to DC-SIGN using FlexX, which was demonstrated to perform well

TABLE 5.3

Recently Published Molecular Docking Studies in Carbohydrate–Lectin Recognition

Ligand	Target	Research Aim	Program(s)	Validation[a,b]	Reference
Galactose/N-acetylgalactosamine	CEL-III	Complex structure elucidation	AutoDock 3.0.5, DOCK 6.1, Glide[c]	+	(Nurisso et al. 2008)
α-Galactobiosides	P. aeruginosa lectin I	Investigate dynamic behavior of galactoside recognition	AutoDock 3.0	\(Nurisso et al. 2008)	(Blanchard et al. 2008; Nurisso et al. 2010)
Monosaccharides	P. aeruginosa lectin II	Complex structure elucidation	AutoDock 3.0, DOCK 6.1	+	(Adam et al. 2008)
α-Mannoside mimics	DC-SIGN	Complex structure elucidation	Glide QPLD[c]	+	(Reina et al. 2007)
		Structure-based design of mannoside mimics	FlexX 3.1.2	+	(Obermajer et al. 2011)
α-Mannosides	DC-SIGN	Complex structure elucidation	FlexiDock[c]	−	(Reina et al. 2008)
	Cymbosema roseum lectin CRLI	Complex structure elucidation	MolDock[c]	−	(Rocha et al. 2011)
	Canavalia brasiliensis lectin	Complex structure elucidation	MolDock[c]	*(XRC)	(Bezerra et al. 2011)
	FimH	Structure-based design of mannosides	FlexX 1.10.0	+	(Sperling et al. 2006)
		Determine spacer length for targeting distant carbohydrate-binding sites	FlexX[c]	\(Sperling et al. 2006)	(Lindhorst et al. 2010)
		Determine if mannosyl squarates are covalent inhibitors of FimH	FlexX[c]	\(Sperling et al. 2006)	(Grabosch et al. 2011)

(Continued)

TABLE 5.3 (Continued)
Recently Published Molecular Docking Studies in Carbohydrate–Lectin Recognition

Ligand	Target	Research aim	Program(s)	Validation[a,b]	Reference
α-Fucosyl glycopeptides	P. aeruginosa LecB	Complex structure elucidation	Glide[c]	+	(Kolomiets et al. 2009)
β-Galactosides	CG-14 and CG-16	Determine structural basis of fine ligand specificity	HADDOCK v1.3	*(XRC)	(Wu et al. 2007)
	Human galectin-3	Determine structural basis of fine ligand specificity	HADDOCK v1.3	\(Wu et al. 2007)	(Krzeminski et al. 2011)
	Jacalin	Determine structure–activity relationship	GOLD 3.2	+	(Kumar et al. 2010)
Glucosyl trisaccharides	Human surfactant protein D	Determine structural basis of observed linkage specificity	AutoDock 3.06	*(XRC)	(Allen et al. 2001)
Glucose/N-acetylglucosamine	Human surfactant protein D	Determine structural basis of Glc vs. GlcNAc recognition	AutoDock 3.06	*(XRC)	(Allen et al. 2004)
Ganglioside GQ1bα fragments	Myelin-associated glycoprotein	Complex structure elucidation	AutoDock 3.0	*(NMR, trNOE)	(Bhunia et al. 2008)
Neuraminic acid	GNA-related lectins	Identify lectins suitable as anti-influenza drugs, on the basis of binding energy	DOCK 6.3	*(MD)	(Xu et al. 2012)
Sialyllactoses	SHL-1	Complex structure elucidation	AutoDock 4.0, FlexX[c], Glide[c]	*(NMR, CIDNP)	(Siebert et al. 2009)
Manno-configured septanosides	Concanavalin A	Determine structure–energy relationships	Glide 5.0	*(QM/MM)	(Duff et al. 2011)

Chitooligosaccharides	Pumpkin phloem exudate lectin	Determine length of optimally fitting oligosaccharide	FlexiDock	*(MD)	(Narahari et al. 2011)
A range of ligands	A range of human lectins (innate immunity)	Docking and site-mapping validation	Glide 5.6, AutoDock 4.2, GOLD 4.1.1, DOCK 6.4	+	(Agostino et al. 2011)
Selenium derivatives of histo-blood group ABH antigens	Eight lectins	Effect of selenium on the complex structure	Glide[c]	*(XRC)	(Strino et al. 2010)

[a] Guide to symbols: Docking method validated against relevant structures (+), no method validation performed or cited (−), followed earlier validated method (√), binding mode predicted in conjunction with other biochemical or biophysical data relating to interactions (*).

[b] Technique abbreviations: XRC—comparison of docked binding mode to closely related x-ray crystal structure complexes; NMR—nuclear magnetic resonance; trNOE—transferred nuclear Overhauser effect; MD—molecular dynamics; CIDNP—chemically induced nuclear polarization; QM/MM—mixed quantum mechanics/molecular mechanics.

[c] Version not specified.

at reproducing the binding mode of Man$_4$. The docking model highlighted the importance of hydrophobic substituents in generating mannoside mimics.

The bacterial adhesion protein FimH is known to bind to α-mannosides. The Lindhorst group investigated agents that mimic α-mannosides for binding to FimH. Mannoside derivatives were docked into the carbohydrate recognition domain of the bacterial adhesin FimH using FlexX (Sperling et al. 2006). The structures obtained were used to optimize ligand interactions with FimH made by the pendant groups attached to mannose. It has recently been suggested that FimH contains multiple carbohydrate-binding sites. In a subsequent study (Lindhorst et al. 2010), molecular modeling was used to optimize the spacer length of a bivalent glycopeptide, which targeted two putative carbohydrate-binding sites on FimH. However, it was found that the developed glycopeptide did not offer any significant improvements in activity. Most recently, squaric acid mannosides were evaluated for their ability to act as covalent inhibitors of FimH (Grabosch et al. 2011). The assay results, in combination with docking, indicated that these compounds did not act as covalent inhibitors.

Strino and coworkers (2010) investigated the effect of substituting glycosyl oxygen atoms with selenium in the derivatives of histo-blood group ABH antigens on their binding to anti-ABH lectins. Specifically, Glide XP was used to dock eight test cases of selenoglycosides as lectin ligands, which were chosen on the basis of availability of experimental data for unmodified glycosides. The OPLS2005 force field was used because it is parameterized for selenium and is superior to OPLS2001 for modeling disaccharides. This docking protocol was evaluated by comparing the docked selenoglycosides to their unmodified counterparts and docking scores to experimental binding energies. In some cases, where docking failed to predict the bound poses, a semimanual approach was implemented, whereby the crystallized ligands were used as starting points for modification. It was found that the selenoglycosides were accommodated at the same site as the natural counterpart and, in most cases, the orientation of the key residues was conserved. Where the conformations and/or orientations of key and remaining residues differed from the natural ligands, they were commonly compensated by more favorable interactions with proteins. An interesting example of such an accommodation was the case of PA-IL, in which the terminal α-galactose of most ligands docked similarly to the crystal structure, whereas other carbohydrate residues assumed different orientations. This reflects the ability of this lectin to bind a variety of carbohydrates with a terminal α-galactose. This mechanism of specificity for a terminal anchor residue, combined with a diverse tolerance for the rest of the ligand, is similar to the mechanism proposed by us for the recognition of carbohydrate xenoantigens by anti-Gal antibodies (Agostino et al. 2010) (described in more detail in Section 5.3.4 and in Chapter 9). This study (Strino et al. 2010) indicated that selenoglycosides are biologically active as lectin ligands with a potential for improved affinity. Therefore, they can serve as nonhydrolyzable mimics of histo-blood group ABH determinants and prove therapeutically useful.

Molecular docking and site mapping were applied to investigate carbohydrate recognition by lectins involved in innate immunity (Agostino et al. 2011). High-resolution structures of human lectins cocrystallized with carbohydrates were used to validate the techniques. Four popular molecular docking programs (Glide, GOLD, AutoDock, and DOCK) were evaluated for their ability to reproduce the crystal bound conformation of the carbohydrate in each case. It was found that GOLD generated the most accurate binding modes; however, these generally could not be accurately ranked by the scoring function. This highlighted the need for alternative scoring functions when considering carbohydrate–lectin interactions. The site-mapping technique (discussed in more detail in Section 5.3.4), which considers multiple alternative binding modes, was able to identify key lectin residues involved in carbohydrate recognition with improved accuracy over the top pose obtained from molecular docking. In conjunction with ligand-based mapping approaches (Agostino et al. 2010), the site-mapping technique may be useful for identifying likely carbohydrate-binding modes when studying carbohydrate–lectin recognition.

5.3.4 MODELING CARBOHYDRATE–ANTIBODY RECOGNITION

In comparison with the large number of docking studies carried out on carbohydrate–enzyme and carbohydrate–lectin recognition, there are relatively few published docking studies on carbohydrate–antibody recognition. The currently published studies are summarized in Table 5.4. Selected examples are discussed here.

Oomen et al. (1991) reported one of the earliest attempts to model carbohydrate recognition by an antibody. The mAb YsT9.1 is specific for a repeating internal motif of the *Brucella* A-antigen, which consists of α1,2-linked 4-formamido-4,6-dideoxy-α-D-mannose (RhaNFo). Models of the antigen were prepared using the guided evolutionary simulated annealing (GESA) algorithm and manually docked to a homology model of YsT9.1 using Quanta. Miller et al. (1998) reported the first use of automated docking to model carbohydrate–antibody recognition. The interaction of Rhaα(1-2)Galα(1-OMe), a derivative of the *Shigella dysenteriae* O-antigen, with mAb 3707 E9 was studied using the ICM program. The model highlighted the importance of tryptophan residues in binding the O-antigen. Möller et al. (2002) generated structures of mucin-derived glycopeptides using a combination of NMR and MD. The generated structures were docked into mAb SM3 using FlexiDock and the poses subjected to further MD simulation. Krengel et al. (2004) attempted to determine the x-ray crystal structure complex of mAb 14F7 with *N*-glycolyl GM3; however, only the unliganded Fab structure could be obtained. Its complex with *N*-glycolyl GM3 was modeled using a modified version of DOCK. Subsequent crystallographic studies on another anti-*N*-glycolyl GM3 antibody, chimeric P3, yielded only an unliganded Fab structure into which the carbohydrate was docked using AutoDock (Talavera et al. 2009). Since part of the HCDR3 loop was not well resolved in the crystallographic data, several representative conformations of this loop were generated for docking.

Antibody–carbohydrate docking, in conjunction with antibody–peptide docking, has been used to investigate peptide mimicry of carbohydrate antigens (Clement et al. 2006). AutoDock was used to dock *Shigella flexneri* 5a O-specific polysaccharide (O-SP) ligands into the homology model of IgA I3. Because of the large size of the studied ligands, starting structures were based on those known experimentally, and the conformational flexibility in the ligands was restricted, whereby only hydroxyl and *N*-acetyl groups were treated as flexible. One antibody–carbohydrate structure was found to be compatible with experimental data, including saturation transfer

TABLE 5.4

Published Molecular Docking Studies in Carbohydrate–Antibody Recognition

Ligand	Target	Research Aim	Programs	Validation[a,b]	Reference
Mimics of *Shigella flexneri* 5a O-SP	IgA I3	Peptide–carbohydrate mimicry	AutoDock 3	–	(Clement et al. 2006)
Galα(1-3)Gal	8.17	Peptide–carbohydrate mimicry	DOCK 4.0	+	(Yuriev et al. 2008)
Galα(1-3)Gal	8.17 and 15.101	Complex structure elucidation	DOCK 4.0	\(Yuriev et al. 2008)	(Milland et al. 2007)
α-Mannose/α-glucose	1H7, 2D10	Determine role of antibody flexibility in molecular mimicry	AutoDock 3.05	*(MD)	(Krishnan et al. 2008)
Galα(1-2)[Abeα(1-3)]Manα	Se155-4	Docking method validation	DARWIN[c]	+	(Taylor and Burnett 2000)
GD3 ganglioside	ME36.1	Docking method validation	DARWIN[c]	+	(Taylor and Burnett 2000)
RhaNFo	YsT9.1	Complex structure elucidation	Quanta[c,d]	–	(Oomen et al. 1991)
Rhaα(1-2)Galα(1-OMe)	3707 E9	Complex structure elucidation	ICM[c]	–	(Miller et al. 1998)
PDT(O-αGalNAc)RP	SM3	Complex structure elucidation	FlexiDock[c]	*(NMR, trNOE, MD)	(Möller et al. 2002)
N-glycolyl GM3	14F7	Complex structure elucidation	DOCKdyna[c]	*(SDM)	(Krengel et al. 2004)
	chP3	Complex structure elucidation	AutoDock 4.0	*(SDM)	(Talavera et al. 2009)
Neu2en5Ac	SIC172	Investigate enzyme properties of antibody	Affinity 97.2	*(SDM)	(Kamei et al. 2001)
Cardiac glycosides	1B3	Complex structure elucidation	GOLD 2.0	+	(Paula et al. 2005)
Streptococcus capsular polysaccharides	1B1	Determine structure basis of antibody selectivity for specific polysaccharides	AutoDock 3.05	*(MD)	(Kadirvelraj et al. 2006)
Trimeric Lewis x	54–5C10-A	Complex structure elucidation	Molegro Virtual Docker[c]	–	(de Geus et al. 2009)

Di-, tri-, and tetrasaccharides	11 different antibodies	Docking method validation	Glide 4.5,[e] GOLD 3.1.1, FlexX 1.2, AutoDock 4.0[e]	+	(Agostino et al. 2009a)
αGal-terminating saccharides	A panel of six anti-Gal monoclonal antibodies	Site-mapping method validation	Glide 4.5	√(Agostino et al. 2009a)	(Agostino et al. 2009b)
αGal-terminating saccharides	A panel of six anti-Gal monoclonal antibodies	Complex structure elucidation	Glide 4.5	√(Agostino et al. 2009a)	(Agostino et al. 2010)

[a] Guide to symbols: Docking method validated against relevant structures (+), followed earlier validated method (√), no method validation performed or cited (−), binding mode predicted in conjunction with other biochemical or biophysical data relating to interactions (*).

[b] Technique abbreviations: MD—molecular dynamics; NMR—nuclear magnetic resonance; trNOE—transferred nuclear Overhauser effect; SDM—site-directed mutagenesis.

[c] Version not specified.

[d] Manual docking procedure used.

[e] Flexible receptor docking mode was used.

difference (STD) NMR, and provided insight into the molecular mechanism of mimicry of complex polysaccharides by peptides.

Automated docking has been used to explore the potential antibody-binding modes of carbohydrates containing terminal Galα(1-3)Gal epitopes (the αGal epitope), which is known to be the main immunological barrier to xenotransplantation (see Chapter 9 for more details). Initially, docking into the binding sites of anti-αGal monoclonal antibodies (mAb) 8.17 and 15.101 was performed on the αGal disaccharide and αGal-terminating trisaccharides (Milland et al. 2007; Yuriev et al. 2008) using DOCK. The binding modes of the top ranked carbohydrate poses were analyzed and it was found that end-on insertion was the preferred mode. The terminal galactose residue entered the deepest part of the cavity and interacted with a series of conserved residues at the floor of the cavity as previously predicted by Ramsland et al. (2003). In particular, aromatic residues lining the floor of the binding site provided numerous van der Waals interactions to the hydrophobic face of the terminal galactose residue and also participated in hydrogen bonding of the carbohydrate ligands. Unlike lectins, the large size of the antibody binding site was found to be able to accommodate Galα(1-3)Galβ(1-4)GlcNAc/Glc trisaccharides by end-on insertion with complementarity-determining region (CDR) residues participating in interactions with all three monosaccharide units.

Following these initial studies, a site-mapping technique was developed to allow a more thorough exploration of antibody–carbohydrate recognition (Agostino et al. 2009b). It was validated against a series of high-resolution antibody–carbohydrate crystal structures. The technique involves the use of molecular docking to generate a series of antibody–carbohydrate complexes followed by an analysis of the hydrogen bonding and van der Waals interactions occurring in each complex, which are mapped (based on their relative contribution) to the antibody binding site surface (Figure 5.1). Thus, contributions to recognition of a variety of potential carbohydrate binding modes are taken into consideration to compensate for differences in antibody binding-site shapes and flexibility of the carbohydrates. When applied to αGal carbohydrate–antibody interactions, site mapping indicates that there is a significant overlap of the antibody regions engaging xenoantigens across a panel of six monoclonal antibodies (Agostino et al. 2009b). The consensus recognition site for αGal carbohydrates comprised nine amino acid positions in five out of the six antibody CDRs of light (L) and heavy (H) chain variable (V) domains. When an additional six amino acids were included (forming the extended consensus site), more than 80% of all potential αGal carbohydrate interactions were accounted for across the panel of monoclonal antibodies.

The site-mapping technique was further augmented by combining both receptor-based and ligand-based in silico mapping (Agostino et al. 2010). Using both mapping approaches and taking into consideration the conformational preferences of carbohydrate ligands, likely binding modes for carbohydrate xenoantigens in complex with anti-αGal antibodies were determined. Interaction-based filters using the antibody site maps and the carbohydrate maps were applied to select the carbohydrate poses exhibiting the most preferred binding characteristics. Principally, the protocol exploits the tendency of certain carbohydrate linkages to cluster at specific

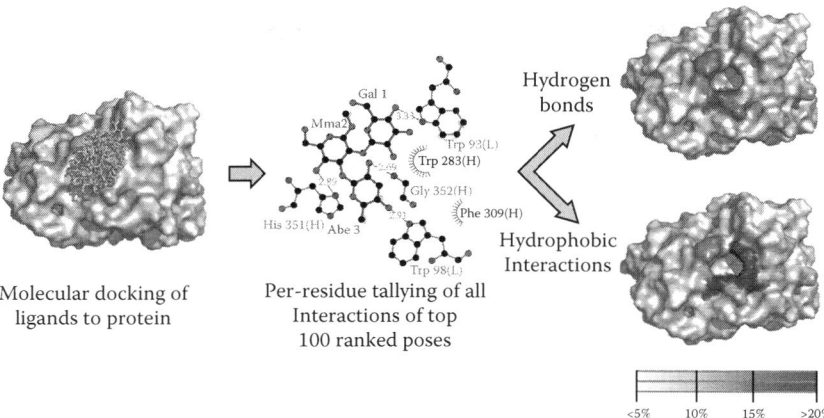

FIGURE 5.1 **(See color insert.)** Illustration of the site-mapping process. Left: A series of poses for the trisaccharide in complex with the SE155-4 antibody (PDB ID: 1MFD) is generated by molecular docking. Middle: Hydrogen bonding and van der Waals interactions are analyzed by LIGPLOT (Wallace et al. 1995) for each pose and tallied. Right: Hydrogen-bonding (red) and van der Waals interaction (blue) maps are obtained. The contribution of each residue to overall binding is represented as a percentage of the total number of interactions observed between the antibody and the carbohydrate ligand. The color intensity is based on the percentage contribution and indicates the importance of that residue for binding.

conformational minima; specifically, restricted conformations of the Galβ(1-4) Glc/GlcNAc linkage were observed, and this linkage was used as a filter to select preferred binding modes of αGal-terminating trisaccharide (or longer) carbohydrates. Furthermore, the most populated conformational minimum of this linkage was in excellent agreement with experimentally observed conformations in similar carbohydrates. In contrast, conformation of the terminal Galα(1-3)Gal linkage varied depending on the antibody binding-site topography, and it is possible that some of the studied antibodies recognize more than one Galα(1-3)Gal conformation. Carbohydrate interaction maps revealed that, across the panel of antibodies, the predominant interactions occurred with the terminal Galα(1-3)Gal disaccharide, but additional interactions occured with the third saccharide unit, which was either GlcNAc or Glc in the different xenoantigens examined. The binding modes obtained indicated that each antibody uses distinct mechanisms for recognizing the target antigens.

Molecular modeling has also been used to investigate the recognition of αGal epitopes by a human xenoreactive antibody, IGHV3-11 (Kearns-Jonker et al. 2007). This antibody was induced in human patients mounting an immune response to a bioartificial liver. Trisaccharide and pentasaccharide binding modes were determined, and it was found that a similar set of residues to those identified in mouse antibodies is important for xenoantigen recognition. The effect of multiple binding modes in recognition was not considered. This study suggests that the αGal epitope is recognized similarly by mouse and human antibodies and, thus, mouse antibodies represent a good model for the human immune response to the αGal epitope.

The limited number of carbohydrate–antibody docking studies published reflects both the limited number of suitable test systems for validation (i.e., high-resolution carbohydrate–antibody crystal structure complexes) and the difficulties associated with modeling carbohydrate–antibody recognition.

5.4   CONCLUSIONS

Computational modeling of protein–carbohydrate complexes faces multiple challenges associated with extreme ligand flexibility, formation of extensive hydrogen bonding networks (often water mediated), shortage of specialized scoring functions, and paucity of experimental structural data available for validation. However, protein–carbohydrate docking has come of age and, as evidenced by the studies published in recent years, has started producing reliable and insightful results. The reliability of such results is supported by structural validations and the correlation of predicted binding affinities with experimentally measured ones. The insights achieved in the field have illuminated such areas of research as catalytic mechanisms of carbohydrate cleavage and extension by enzymes, carbohydrate binding by lectins implicated in infection and cancer, and xenoantigen recognition, to name just a few. With respect to specific software, Glide and AutoDock appear to be programs of choice for carbohydrate docking; they are closely followed by GOLD.

But challenges still remain. Further research is needed to properly account for receptor flexibility in carbohydrate docking. Very few studies have implemented flexible receptor docking so far, and in cases where it was used for validation it was shown to be less successful than rigid docking (e.g., see the study by Agostino et al. [2009a]).

Another challenge is the ability of docking programs to correctly score docking poses for cases of small ligands in large or poorly defined binding sites, as demonstrated by the failure of AutoDock to do so in the case of IL-8–disaccharide complexes (Samsonov et al. 2011) or a range of programs to appropriately score lectin–carbohydrate complexes (Agostino et al. 2011). We have demonstrated that, in the absence of a perfect scoring function suitable for this task, our mapping approach (binding-site mapping [Agostino et al. 2009b], epitope mapping, and conformational clustering [Agostino et al. 2010]) can be utilized to determine likely carbohydrate-binding modes (Agostino et al. 2010).

For a better account of the role of water in carbohydrate–protein docking, the issues of displayable versus bridging water molecules still need to be resolved. Samsonov et al. (2011) have suggested that an on-the-fly sampling step may be included into the calculation to account for the solvent. Alternatively, the recently developed WaterMap module within the Schrödinger package, which predicts likely water binding sites prior to ligand docking (Abel et al. 2008), may be able to fulfill this task, and it can be used in conjunction with Glide.

ACKNOWLEDGMENTS

The research for this chapter was supported by a small grant from the Faculty of Pharmacy and Pharmaceutical Sciences, Monash University, Melbourne, Victoria, Australia, to Elizabeth Yuriev. Mark Agostino is a recipient of the Monash University Postgraduate Publication Award (PPA). Paul A. Ramsland is a Sir Zelman Cowen senior research fellow (Sir Zelman Cowen Fellowship Fund, Burnet Institute, Melbourne, Victoria, Australia). The authors gratefully acknowledge the contribution to this work by the Victorian Operational Infrastructure Support Program received by the Burnet Institute.

REFERENCES

Abel, R., Young, T., Farid, R., Berne, B. J., and Friesner, R. A. 2008. Role of the active-site solvent in the thermodynamics of factor Xa ligand binding. *J Am Chem Soc*, 130, 2817–2831.

Adam, J., Kriz, Z., Prokop, M., Wimmerova, M., and Koca, J. 2008. In silico mutagenesis and docking studies of *Pseudomonas aeruginosa* PA-IIL lectin—predicting binding modes and energies. *J Chem Inf Model*, 48, 2234–2242.

Agostino, M., Jene, C., Boyle, T., Ramsland, P. A., and Yuriev, E. 2009a. Molecular docking of carbohydrate ligands to antibodies: Structural validation against crystal structures. *J Chem Inf Model*, 49, 2749–2760.

Agostino, M., Sandrin, M. S., Thompson, P. E., Yuriev, E., and Ramsland, P. 2009b. In silico analysis of antibody–carbohydrate interactions and its application to xenoreactive antibodies. *Mol Immunol*, 47, 233–246.

Agostino, M., Sandrin, M. S., Thompson, P. E., Yuriev, E., and Ramsland, P. A. 2010. Identification of preferred carbohydrate binding modes in xenoreactive antibodies by combining conformational filters and binding site maps. *Glycobiology*, 20, 724–735.

Agostino, M., Yuriev, E., and Ramsland, P. 2011. A computational approach for exploring carbohydrate recognition by lectins in innate immunity. *Front Immun*, 2, 23.

Alexacou, K. M., Hayes, J. M., Tiraidis, C. et al. 2008. Crystallographic and computational studies on 4-phenyl-*N*-(beta-D-glucopyranosyl)-1H-1,2,3-triazole-1-acetamide, an inhibitor of glycogen phosphorylase: Comparison with alpha-D-glucose, *N*-acetyl-beta-D-glucopyranosylamine and *N*-benzoyl-*N'*-beta-D-glucopyranosyl urea binding. *Proteins*, 71, 1307–1323.

Allen, M. J., Laederach, A., Reilly, P. J., and Mason, R. J. 2001. Polysaccharide recognition by surfactant protein D: Novel interactions of a C-type lectin with nonterminal glucosyl residues. *Biochemistry*, 40, 7789–7798.

Allen, M. J., Laederach, A., Reilly, P. J., Mason, R. J., and Voelker, D. R. 2004. Arg343 in human surfactant protein D governs discrimination between glucose and *N*-acetylglucosamine ligands. *Glycobiology*, 14, 693–700.

Benltifa, M., Hayes, J. M., Vidal, S. et al. 2009. Glucose-based spiro-isoxazolines: A new family of potent glycogen phosphorylase inhibitors. *Bioorg Med Chem*, 17, 7368–7380.

Bezerra, E. H., Rocha, B. A., Nagano, C. S. et al. 2011. Structural analysis of ConBr reveals molecular correlation between the carbohydrate recognition domain and endothelial NO synthase activation. *Biochem Biophys Res Commun*, 408, 566–570.

Bhunia, A., Schwardt, O., Gathje, H. et al. 2008. Consistent bioactive conformation of the Neu5Acalpha(2—>3)Gal epitope upon lectin binding. *Chem Bio Chem*, 9, 2941–2945.

Blanchard, B., Nurisso, A., Hollville, E. et al. 2008. Structural basis of the preferential binding for globo-series glycosphingolipids displayed by *Pseudomonas aeruginosa* lectin I. *J Mol Biol*, 383, 837–853.

Bottegoni, G., Kufareva, I., Totrov, M., and Abagyan, R. 2009. Four-dimensional docking: A fast and accurate account of discrete receptor flexibility in ligand docking. *J Med Chem*, 52, 397–406.

B-Rao, C., Subramanian, J., and Sharma, S. D. 2009. Managing protein flexibility in docking and its applications. *Drug Discov Today*, 14, 394–400.

Bras, N. F., Fernandes, P. A., and Ramos, M. J. 2009. Docking and molecular dynamics studies on the stereoselectivity in the enzymatic synthesis of carbohydrates. *Theor Chem Account*, 122, 283–296.

Cantu, D., Nerinckx, W., and Reilly, P. J. 2008. Theory and computation show that Asp463 is the catalytic proton donor in human endoplasmic reticulum alpha-(1—>2)-mannosidase I. *Carbohydr Res*, 343, 2235–2242.

Cerqueira, N. M., Bras, N. F., Fernandes, P. A., and Ramos, M. J. 2009. MADAMM: A multistaged docking with an automated molecular modeling protocol. *Proteins*, 74, 192–206.

Choi, J. K., Lee, B. H., Chae, C. H., and Shin, W. 2004. Computer modeling of the rhamnogalacturonase-"hairy" pectin complex. *Proteins*, 55, 22–33.

Clement, M.-J., Fortune, A., Phalipon, A. et al. 2006. Towards a better understanding of the basis of molecular mimicry of polysaccharide antigens by peptides: The example of *Shigella flexneri* 5a. *J Biol Chem*, 281, 2317–2332.

Corbeil, C. R. and Moitessier, N. 2009. Docking ligands into flexible and solvated macromolecules. 3. Impact of input ligand conformation, protein flexibility, and water molecules on the accuracy of docking Programs. *J Chem Inf Model*, 49, 997–1009.

Costa, M. G., Batista, P. R., Shida, C. S. et al. 2010. How does heparin prevent the pH inactivation of cathepsin B? Allosteric mechanism elucidated by docking and molecular dynamics. *BMC Genomics*, 11 Suppl 5, S5.

Davis, I. W. and Baker, D. 2009. Rosetta ligand docking with full ligand and receptor flexibility. *J Mol Biol*, 385, 381–392.

de Geus, D. C., van Roon, A. M., Thomassen, E. A. J. et al. 2009. Characterization of a diagnostic Fab fragment binding trimeric Lewis X. *Proteins*, 76, 439–447.

De Vos, D., Collins, T., Nerinckx, W. et al. 2006. Oligosaccharide binding in family 8 glycosidases: Crystal structures of active-site mutants of the beta-1,4-xylanase pXyl from Pseudoaltermonas haloplanktis TAH3a in complex with substrate and product. *Biochemistry*, 45, 4797–4807.

Dudkiewicz, M., Siminska, J., Pawlowski, K., and Orzechowski, S. 2008. Bioinformatics analysis of oligosaccharide phosphorylation effect on the stabilization of the beta-amylase ligand complex. *J Carbohyd Chem*, 27, 479–495.

Duff, M. R., Jr., Fyvie, W. S., Markad, S. D. et al. 2011. Computational and experimental investigations of mono-septanoside binding by Concanavalin A: Correlation of ligand stereochemistry to enthalpies of binding. *Org Biomol Chem*, 9, 154–164.

Englebienne, P., Fiaux, H., Kuntz, D. A. et al. 2007. Evaluation of docking programs for predicting binding of Golgi alpha-mannosidase II inhibitors: A comparison with crystallography. *Proteins*, 69, 160–176.

Fernandez-Leiro, R., Pereira-Rodriguez, A., Cerdan, M. E., Becerra, M., and Sanz-Aparicio, J. 2010. Structural analysis of *Saccharomyces cerevisiae* alpha-galactosidase and its complexes with natural substrates reveals new insights into substrate specificity of GH27 glycosidases. *J Biol Chem*, 285, 28020–28033.

Friesner, R. A., Banks, J. L., Murphy, R. B. et al. 2004. Glide: A new approach for rapid, accurate docking and scoring. 1. Method and assessment of docking accuracy. *J Med Chem*, 47, 1739–1749.

Gandhi, N. S. and Mancera, R. L. 2009. Free energy calculations of glycosaminoglycan–protein interactions. *Glycobiology*, 19, 1103–1115.

Grabosch, C., Hartmann, M., Schmidt-Lassen, J., and Lindhorst, T. K. 2011. Squaric acid monoamide mannosides as ligands for the bacterial lectin FimH: Covalent inhibition or not? *ChemBioChem*, 12, 1066–1074.

Guerin, M. E., Schaeffer, F., Chaffotte, A. et al. 2009. Substrate-induced conformational changes in the essential peripheral membrane-associated mannosyltransferase PimA from mycobacteria: Implications for catalysis. *J Biol Chem*, 284, 21613–21625.

Hill, A. D. and Reilly, P. J. 2008a. Computational analysis of glycoside hydrolase family 1 specificities. *Biopolymers*, 89, 1021–1031.

Hill, A. D. and Reilly, P. J. 2008b. A Gibbs free energy correlation for automated docking of carbohydrates. *J Comput Chem*, 29, 1131–1141.

Jacobson, M. P., Pincus, D. L., Rapp, C. S. et al. 2004. A hierarchical approach to all-atom protein loop prediction. *Proteins*, 55, 351–367.

Jain, A. N. 2007. Surflex-Dock 2.1: Robust performance from ligand energetic modeling, ring flexibility, and knowledge-based search. *J Comput-Aided Mol Des*, 21, 281–306.

Kadirvelraj, R., Gonzalez-Outeirino, J., Foley, B. L. et al. 2006. Understanding the bacterial polysaccharide antigenicity of *Streptococcus agalactiae* versus *Streptococcus pneumoniae*. *Proc Natl Acad Sci USA*, 103, 8149–8154.

Kamei, H., Shimazaki, K., and Nishi, Y. 2001. Computational 3-D modeling and site-directed mutation of an antibody that binds Neu2en5Ac, a transition state analogue of a sialic acid. *Proteins*, 45, 285–296.

Kearns-Jonker, M., Barteneva, N., Mencel, R. et al. 2007. Use of molecular modeling and site-directed mutagenesis to define the structural basis for the immune response to carbohydrate xenoantigens. *BMC Immunol*, 8, 3.

Kellenberger, E., Rodrigo, J., Muller, P., and Rognan, D. 2004. Comparative evaluation of eight docking tools for docking and virtual screening accuracy. *Proteins*, 57, 225–242.

Kerzmann, A., Fuhrmann, J., Kohlbacher, O., and Neumann, D. 2008. BALLDock/SLICK: A new method for protein–carbohydrate docking. *J Chem Inf Model*, 48, 1616–1625.

Kerzmann, A., Neumann, D., and Kohlbacher, O. 2006. SLICK—scoring and energy functions for protein–carbohydrate interactions. *J Chem Inf Model*, 46, 1635–1642.

Kezuka, Y., Kojima, M., Mizuno, R. et al. 2010. Structure of full-length class I chitinase from rice revealed by x-ray crystallography and small-angle x-ray scattering. *Proteins*, 78, 2295–2305.

Kolomiets, E., Swiderska, M. A., Kadam, R. U. et al. 2009. Glycopeptide dendrimers with high affinity for the fucose-binding lectin LecB from *Pseudomonas aeruginosa*. *ChemMedChem*, 4, 562–569.

Koppisetty, C. A., Nasir, W., Strino, F. et al. 2010. Computational studies on the interaction of ABO-active saccharides with the norovirus VA387 capsid protein can explain experimental binding data. *J Comput-Aided Mol Des*, 24, 423–431.

Kozmon, S. and Tvaroška, I. 2006. Catalytic mechanism of glycosyltransferases: Hybrid quantum mechanical/molecular mechanical study of the inverting *N*-acetylglucosaminyltransferase I. *J Am Chem Soc*, 128, 16921–16927.

Krengel, U., Olsson, L. L., Martinez, C. et al. 2004. Structure and molecular interactions of a unique antitumor antibody specific for *N*-glycolyl GM3. *J Biol Chem*, 279, 5597–5603.

Krishnan, L., Sahni, G., Kaur, K. J., and Salunke, D.M. 2008. Role of antibody paratope conformational flexibility in the manifestation of molecular mimicry. *Biophys J*, 94, 1367–1376.

Krzeminski, M., Singh, T., Andre, S. et al. 2011. Human galectin-3 (Mac-2 antigen): Defining molecular switches of affinity to natural glycoproteins, structural and dynamic aspects of glycan binding by flexible ligand docking and putative regulatory sequences in the proximal promoter region. *Biochem Biophys Acta*, 1810, 150–161.

Kumar, A., Ramanujam, B., Singhal, N. K., Mitra, A., and Rao, C. P. 2010. Interaction of aromatic imino glycoconjugates with jacalin: Experimental and computational docking studies. *Carbohydr Res*, 345, 2491–2498.

Laederach, A. and Reilly, P. J. 2003. Specific empirical free energy function for automated docking of carbohydrates to proteins. *J Comput Chem*, 24, 1748–1757.

Laederach, A. and Reilly, P. J. 2005. Modeling protein recognition of carbohydrates. *Proteins*, 60, 591–597.

Lang, P. T., Brozell, S. R., Mukherjee, S. et al. 2009. DOCK 6: Combining techniques to model RNA-small molecule complexes. *RNA*, 15, 1219–1230.

Laughrey, Z. R., Kiehna, S. E., Riemen, A. J., and Waters, M. L. 2008. Carbohydrate-pi interactions: What are they worth? *J Am Chem Soc*, 130, 14625–14633.

Lee, M., Lloyd, P., Zhang, X. et al. 2006. Shapes of antibody binding sites: Qualitative and quantitative analyses based on a geomorphic classification scheme. *J Org Chem*, 71, 5082–5092.

Li, H. and Li, C. 2010. Multiple ligand simultaneous docking: Orchestrated dancing of ligands in binding sites of protein. *J Comput Chem*, 31, 2014–2022.

Li, L., Modolo, L. V., Escamilla-Trevino, L. L. et al. 2007. Crystal structure of *Medicago truncatula* UGT85H2—insights into the structural basis of a multifunctional (iso)flavonoid glycosyltransferase. *J Mol Biol*, 370, 951–963.

Li, T., Guo, L., Zhang, Y. et al. 2011. Structure-activity relationships in a series of C2-substituted gluco-configured tetrahydroimidazopyridines as beta-glucosidase inhibitors. *Bioorg Med Chem*, 19, 2136–2144.

Lindhorst, T. K., Bruegge, K., Fuchs, A., and Sperling, O. 2010. A bivalent glycopeptide to target two putative carbohydrate binding sites on FimH. *Beilstein J Org Chem*, 6, 801–809.

Liu, Y. N., Lai, Y. T., Chou, W. I., Chang, M. D., and Lyu, P. C. 2007. Solution structure of family 21 carbohydrate-binding module from *Rhizopus oryzae* glucoamylase. *Biochem J*, 403, 21–30.

Matrai, J., Lammens, W., Jonckheer, A. et al. 2008. An alternate sucrose binding mode in the E203Q Arabidopsis invertase mutant: An x-ray crystallography and docking study. *Proteins*, 71, 552–564.

Mertz, B., Hill, A. D., Mulakala, C., and Reilly, P. J. 2007. Automated docking to explore subsite binding by glycoside hydrolase family 6 cellobiohydrolases and endoglucanases. *Biopolymers*, 87, 249–260.

Meyer, S., van Liempt, E., Imberty, A. et al. 2005. DC-SIGN mediates binding of dendritic cells to authentic pseudo-LewisY glycolipids of Schistosoma mansoni cercariae, the first parasite-specific ligand of DC-SIGN. *J Biol Chem*, 280, 37349–37359.

Milland, J., Yuriev, E., Xing, P. X. et al. 2007. Carbohydrate residues downstream of the terminal Galα(1,3)Gal epitope modulate the specificity of xenoreactive antibodies. *Immunol Cell Biol*, 85, 623–632.

Miller, C. E., Mulard, L. A., Padlan, E. A., and Glaudemans, C. P. 1998. Binding of modified fragments of the Shigella dysenteriae type 1 O-specific polysaccharide to monoclonal IgM 3707 E9 and docking of the immunodeterminant to its modeled Fv. *Carbohydr Res*, 309, 219–226.

Möller, H., Serttas, N., Paulsen, H., Burchell, J. M., and Taylor-Papadimitriou, J. 2002. NMR-based determination of the binding epitope and conformational analysis of MUC-1 glycopeptides and peptides bound to the breast cancer-selective monoclonal antibody SM3. *Eur J Biochem*, 269, 1444–1455.

Morris, G. M., Huey, R., Lindstrom, W. et al. 2009. AutoDock4 and AutoDockTools4: Automated docking with selective receptor flexibility. *J Comput Chem*, 30, 2785–2791.

Moura-Tamames, S. A., Ramos, M. J., and Fernandes, P. A. 2009. Modelling beta-1,3-exoglucanase-saccharide interactions: Structure of the enzyme-substrate complex and enzyme binding to the cell wall. *J Mol Graph Model*, 27, 908–920.

Mulakala, C., Nerinckx, W., and Reilly, P. J. 2006. Docking studies on glycoside hydrolase Family 47 endoplasmic reticulum alpha-(1—>2)-mannosidase I to elucidate the pathway to the substrate transition state. *Carbohydr Res*, 341, 2233–2245.

Mulakala, C., Nerinckx, W., and Reilly, P. J. 2007. The fate of beta-D-mannopyranose after its formation by endoplasmic reticulum alpha-(1—>2)-mannosidase I catalysis. *Carbohydr Res*, 342, 163–169.

Mulakala, C. and Reilly, P. J. 2005. Force calculations in automated docking: Enzyme-substrate interactions in *Fusarium oxysporum* Cel7B. *Proteins*, 61, 590–596.

Narahari, A., Singla, H., Nareddy, P. K. et al. 2011. Isothermal titration calorimetric and computational studies on the binding of chitooligosaccharides to pumpkin (*Cucurbita maxima*) phloem exudate lectin. *J Phys Chem B*, 115, 4110–4117.

Neumann, D., Lehr, C. M., Lenhof, H. P., and Kohlbacher, O. 2004. Computational modeling of the sugar-lectin interaction. *Adv Drug Deliv Rev*, 56, 437–457.

Nurisso, A., Blanchard, B., Audfray, A. et al. 2010. Role of water molecules in structure and energetics of *Pseudomonas aeruginosa* lectin I interacting with disaccharides. *J Biol Chem*, 285, 20316–20327.

Nurisso, A., Kozmon, S., and Imberty, A. 2008. Comparison of docking methods for carbohydrate binding in calcium-dependent lectins and prediction of the carbohydrate binding mode to sea cucumber lectin CEL-III. *Mol Simul*, 34, 469–479.

Obermajer, N., Sattin, S., Colombo, C. et al. 2011. Design, synthesis and activity evaluation of mannose-based DC-SIGN antagonists. *Mol Divers*, 15, 347–360.

Oliveira, O. V., Freitas, L. C., Straatsma, T. P., and Lins, R. D. 2009. Interaction between the CBM of Cel9A from *Thermobifida fusca* and cellulose fibers. *J Mol Recognit*, 22, 38–45.

Oomen, R. P., Young, N. M., and Bundle, D. R. 1991. Molecular modeling of antibody-antigen complexes between the *Brucella abortus* O-chain polysaccharide and a specific monoclonal antibody. *Protein Eng*, 4, 427–433.

Paula, S., Monson, N., and Ball, W. J., Jr. 2005. Molecular modeling of cardiac glycoside binding by the human sequence monoclonal antibody 1B3. *Proteins*, 60, 382–391.

Ramsland, P. A., Farrugia, W., Yuriev, E., Edmundson, A. B., and Sandrin, M. S. 2003. Evidence for structurally conserved recognition of the major carbohydrate xenoantigen by natural antibodies. *Cell Mol Biol*, 49, 307–317.

Rao, M. and Tvaroška, I. 2001. Structure of bovine alpha-1,3-galactosyltransferase and its complexes with UDP and DPGal inferred from molecular modeling. *Proteins*, 44, 428–434.

Reina, J. J., Diaz, I., Nieto, P. M. et al. 2008. Docking, synthesis, and NMR studies of mannosyl trisaccharide ligands for DC-SIGN lectin. *Org Biomol Chem*, 6, 2743–2754.

Reina, J. J., Sattin, S., Invernizzi, D. et al. 2007. 1,2-Mannobioside mimic: Synthesis, DC-SIGN interaction by NMR and docking, and antiviral activity. *ChemMedChem*, 2, 1030–1036.

Rocha, B. A., Delatorre, P., Oliveira, T. M. et al. 2011. Structural basis for both pro- and anti-inflammatory response induced by mannose-specific legume lectin from *Cymbosema roseum*. *Biochimie*, 93, 806–816.

Rogers, C. J., Clark, P. M., Tully, S. E. et al. 2011. Elucidating glycosaminoglycan–protein–protein interactions using carbohydrate microarray and computational approaches. *Proc Natl Acad Sci USA*, 108, 9747–9752.

Samsonov, S. A., Teyra, J., and Pisabarro, M. T. 2011. Docking glycosaminoglycans to proteins: Analysis of solvent inclusion. *J Comput-Aided Mol Des*, 25, 477–489.

Sapay, N., Cabannes, E., Petitou, M., and Imberty, A. 2011. Molecular modeling of the interaction between heparan sulfate and cellular growth factors: Bringing pieces together. *Glycobiology*, 21, 1181–1193.

Sherman, W., Day, T., Jacobson, M. P., Friesner, R. A., and Farid, R. 2006. Novel procedure for modeling ligand/receptor induced fit effects. *J Med Chem*, 49, 534–553.

Siebert, H. C., Lu, S. Y., Wechselberger, R. et al. 2009. A lectin from the Chinese bird-hunting spider binds sialic acids. *Carbohydr Res*, 344, 1515–1525.

Sivapriya, K., Hariharaputran, S., Suhas, V. L., Chandra, N., and Chandrasekaran, S. 2007. Conformationally locked thiosugars as potent alpha-mannosidase inhibitors: Synthesis, biochemical and docking studies. *Bioorg Med Chem*, 15, 5659–5665.

Sperling, O., Fuchs, A., and Lindhorst, T. K. 2006. Evaluation of the carbohydrate recognition domain of the bacterial adhesin FimH: Design, synthesis and binding properties of mannoside ligands. *Org Biomol Chem*, 4, 3913–3922.

Strino, F., Lii, J. H., Koppisetty, C. A., Nyholm, P. G., and Gabius, H. J. 2010. Selenoglycosides in silico: Ab initio-derived reparameterization of MM4, conformational analysis using histo-blood group ABH antigens and lectin docking as indication for potential of bioactivity. *J Comput-Aided Mol Des*, 24, 1009–1021.

Takaoka, Y., Ohta, M., Takeuchi, A. et al. 2010. Ligand orientation governs conjugation capacity of UDP-glucuronosyltransferase 1A1. *J Biochem*, 148, 25–28.

Talavera, A., Eriksson, A., Okvist, M. et al. 2009. Crystal structure of an anti-ganglioside antibody, and modelling of the functional mimicry of its NeuGc-GM3 antigen by an anti-idiotypic antibody. *Mol Immunol*, 46, 3466–3475.

Taylor, J. S. and Burnett, R. M. 2000. DARWIN: A program for docking flexible molecules. *Proteins*, 41, 173–191.

Trott, O. and Olson, A. J. 2010. AutoDock Vina: Improving the speed and accuracy of docking with a new scoring function, efficient optimization, and multithreading. *J Comput Chem*, 31, 455–461.

Tsirkone, V. G., Tsoukala, E., Lamprakis, C. et al. 2010. 1-(3-Deoxy-3-fluoro-beta-D-glucopyranosyl) pyrimidine derivatives as inhibitors of glycogen phosphorylase b: Kinetic, crystallographic and modelling studies. *Bioorg Med Chem*, 18, 3413–3425.

Tvaroška, I., Sihelnikova, L., and Kozmon, S. 2008. DFT and docking study of potential transition state analogue inhibitors of glycosyltransferases. *Collect Czech Chem Commun*, 73, 591–607.

Vaaje-Kolstad, G., Farkas, V., Fincher, G. B., and Hrmova, M. 2010. Barley xyloglucan xyloglucosyl transferases bind xyloglucan-derived oligosaccharides in their acceptor-binding regions in multiple conformational states. *Arch Biochem Biophys*, 496, 61–68.

Vaque, M., Ardrevol, A., Blade, C. et al. 2008. Protein–ligand docking: A review of recent advances and future perspectives. *Curr Pharm Anal*, 4, 1–19.

Verdonk, M. L., Cole, J. C., Hartshorn, M. J., Murray, C. W., and Taylor, R. D. 2003. Improved protein–ligand docking using GOLD. *Proteins*, 52, 609–623.

Wallace, A. C., Laskowski, R. A., and Thornton, J. M. 1995. LIGPLOT: A program to generate schematic diagrams of protein–ligand interactions. *Protein Eng*, 8, 127–134.

Warren, G. L., Andrews, C. W., Capelli, A. M. et al. 2006. A critical assessment of docking programs and scoring functions. *J Med Chem*, 49, 5912–5931.

Webster, D. M. and Rees, A. R. 1995. Molecular modeling of antibody-combining sites. *Methods Mol Biol*, 51, 17–49.

Wu, A.M., Singh, T., Liu, J. H. et al. 2007. Activity-structure correlations in divergent lectin evolution: Fine specificity of chicken galectin CG-14 and computational analysis of flexible ligand docking for CG-14 and the closely related CG-16. *Glycobiology*, 17, 165–184.

Xu, H., Li, C., He, X. et al. 2012. Molecular modeling, docking and dynamics simulations of GNA-related lectins for potential prevention of influenza virus (H1N1). *J Mol Model*, 18, 27–37.

Yuriev, E., Agostino, M., and Ramsland, P. A. 2011. Challenges and advances in computational docking: 2009 in review. *J Mol Recognit*, 24, 149–164.

Yuriev, E., Sandrin, M. S., and Ramsland, P. 2008. Antibody-ligand docking: Insights into peptide-carbohydrate mimicry. *Mol Simulat*, 34, 461–468.

Section III

*Alternative Approaches
for Yielding Detailed
Structural Information*

6 Mass Spectrometry for Glycomics Analysis of *N*- and *O*-Linked Glycoproteins

Daniel Kolarich and Nicolle H. Packer

CONTENTS

6.1 INTRODUCTION

Mass spectrometry (MS) has become the most popular technique in the analysis of complex biomolecules such as glycoproteins and carbohydrates. Standardized methods for the assessment of the protein primary structure using proteolytically derived peptides have been successfully implemented in high-throughput proteomics workflows based on tandem MS fragmentation data (Mann et al. 2001). The preceding sequencing of the genome contributed significantly to the ability of tandem MS to identify proteins and the success of current standard proteomics workflows.

In contrast to peptides, oligosaccharides do not follow template-driven biosynthesis and are rarely assembled in a linear way, making their structural determination from tandem MS spectra more challenging (Varki et al. 2009). The isomeric monosaccharide building blocks of oligosaccharides present an additional challenge as they are not distinguishable by their mass alone (Dell 1990), and monosaccharide analysis by itself does not provide any information on the structures of the resulting oligosaccharides (Lindberg and Lonngren 1978; Townsend 1995). Fortunately for the analyst, nature does not use all the theoretically possible repertoire of monosaccharides to synthesize eukaryotic protein-bound oligosaccharides but uses a rather limited collection (Figure 6.1) (Varki et al. 2009). In addition, the pathways of oligosaccharide biosynthesis, especially for protein-bound glycans, are highly conserved, and this knowledge assists the assignment of structures from MS and/or tandem MS data. Although the current MS technology offers capabilities only imagined just two decades ago, the crucial key to successful analysis still remains the initial preparation of a sample. Enrichment/isolation of target compounds and simultaneous depletion of contaminating substances, while concomitantly minimizing the introduction of artifacts, are the essential steps that must be optimized for successful mass spectrometric analysis.

	Residue mass		Monosaccharide	Symbol[a]		
	Mono-isotopic	Average			Color	Black/white
Hexose	162.0528	162.1424	Glucose	Blue circle	●	
			Galactose	Yellow circle	◐	
			Mannose	Green circle	●	
N-acetyl hexosamine	203.0794	203.1950	N-acetyl glucosamine	Blue square	■	
			N-acetyl galactosamine	Yellow square	▨	
Desoxyhexose	146.0579	146.1430	Fucose	Red triangle	▲	
Neuraminic acid	291.0954	291.2579	N-acetyl neuraminic acid	Purple diamond	◆	
	307.0903	307.2573	N-glycolyl neuraminic acid[b]	Light blue diamond	◇	
Phosphate	79.9663	79.9799		P	P	
Sulfate	79.9568	80.0642		S	S	
Acetate	42.0106	42.0373				

FIGURE 6.1 Monosaccharide building blocks and common modifications found on mammalian protein-bound N- and O-glycans: [a]Symbols used are as suggested by the database CFG (www.functionalglycomics.org). [b]N-glycolyl neuraminic acid is not a building block of human oligosaccharides.

6.2   GLYCOMICS: GENERAL CONSIDERATIONS FOR SAMPLE PREPARATION

The term "glycome" describes the complete repertoire of glycans and glycoconjugates that cells produce under specified conditions of time, space, and environment (Varki et al. 2009). The term "glycomics," in general, refers to studies that profile the glycome. This chapter focuses on the analysis of the part of the glycome that is found attached to proteins.

Most glycomics approaches for the structural determination of protein-bound N- and/or O-linked glycans include steps for the enrichment of glycoproteins followed by the release of the attached sugars for characterization. Glycoproteins frequently found in secreted fluids such as saliva, tears, and plasma are usually highly soluble in aqueous buffers and can be comparably easily enriched under mild- and non-denaturing conditions. In contrast, membrane-bound or membrane-associated glycoproteins require, in the majority of cases, the addition of detergents such as sodium dodecyl sulfate (SDS) or Triton X-114. Detergents increase solubility in buffers and reduce losses during sample handling; they have been used for membrane proteomics and glycomics analyses (Lee, Chick et al. 2010; Lee et al. 2009; Lee, Nakano et al. 2010). For example, ultracentrifugation enrichment of membranes followed by Triton X-114 phase partitioning enriched for a glycoprotein subfraction with a significant reduction of "contaminating," higher abundant nonglycosylated proteins (Lee, Chick et al. 2010).

Electrophoretic techniques such as SDS–polyacrylamide gel electrophoresis (SDS-PAGE) or two-dimensional (2D)-PAGE offer a relatively straightforward and quick means of protein separation. Gel separation is still a major step in proteomics and glycomics sample preparation of individual proteins, and numerous protocols have been developed using various in-gel digestion methods for further processing of a protein sample (Kolarich and Altmann 2000; Kuster et al. 1998; Rendic et al. 2007). In addition, gel-separated proteins can be electroblotted onto membrane carriers such as polyvinylidene fluoride (PVDF), which have been successfully used to isolate and characterize the N- and O-glycoprofiles of individual glycoproteins (Christiansen et al. 2010; Deshpande et al. 2010; Karlsson, Schulz et al. 2004; Karlsson and Thomsson 2009; Thomsson et al. 2005; Wilson et al. 2002). Immobilization of the protein on PVDF offers the advantage that the analyst is able to sequentially release and isolate N- and O-glycans, respectively, from the very same protein spot or band, thereby reducing the amount of sample required for glycomic analysis of individual proteins. This approach can also provide the basis for general glycome investigations of a complex mixture of glycoproteins, for example, from tissue or body fluids, by immobilization of the entire protein fraction onto the membrane followed by subsequent enzymatic N-glycan release by peptide:N-glycanase F (PNGase F) and chemical O-glycan release using reductive β-elimination (Jensen et al. 2012; Wilson et al. 2008). It should be noted that gel or membrane staining by extended exposure to high acidic conditions can result in the loss of sialic acid and/or fucose residues as well as other acid-labile molecules such as sulfates and phosphates (Honda et al. 2003; Woodward et al. 1987) and, hence, it should be used sparingly. Fixation of the gel after isoelectric focusing (IEF) or SDS-PAGE using the highly acidic conditions of 11.5% trichloroacetic acid resulted in a partial loss of sialic acids (Kolarich, Weber et al. 2006).

Another widely used sample preparation method for global glycomics analyses of a tissue or fluid involves early-stage proteolytic digestion of extracted proteins. This approach mostly eliminates the solubility issues of hydrophobic glycoproteins by converting them into smaller pieces (Ashida et al. 2008; Jang-Lee et al. 2006). Subsequently, peptides and glycopeptides can be further separated by various approaches before glycan release and isolation. Depending on the favored choice of mass spectrometric analysis, native glycans can be analyzed directly by matrix-associated laser desorption ionization (MALDI) MS/MS and direct infusion electrospray ionization (ESI)-MS (Ashida et al. 2008) or separated prior to detection using liquid chromatography (LC)-ESI-MS/MS (Harris et al. 2010). Alternatively, glycans can be derivatized by reductive amination with a variety of fluorescent dyes (often in combination with LC separation) (Pabst et al. 2009; Royle et al. 2006) or be permethylated prior to analysis (Ashline et al. 2007; Azadi and Heiss 2009; Jang-Lee et al. 2006).

In this chapter, we focus on various aspects, advantages, and limitations of the most frequently used approaches for the characterization of *N*- and *O*-glycans by mass spectrometric techniques.

6.3 RELEASING GLYCANS FROM PROTEINS

Separation of glycans and proteins into distinct fractions has certain benefits when more detailed structural analyses of the respective compounds is required. Whereas enzymatic approaches usually allow for analysis of both, the glycan and peptide/protein moieties after glycan release and sample cleanup, chemical methods are usually more optimised to maintain either one or the other component of a glycoprotein. This section summarises the most commonly applied approaches for releasing glycans from the respective protein carriers.

6.3.1 RELEASING *N*-GLYCANS FROM PROTEINS

The *N*-glycans from mammalian glycoproteins are most commonly released enzymatically (Figure 6.2). The PNGase F from *Flavobacterium meningosepticum* (Plummer et al. 1984) is a rugged deamidase that removes most mammalian *N*-linked oligosaccharides from peptides/proteins while keeping both the proteins and the oligosaccharides intact. The hydrophilic *N*-glycans can be separated from the hydrophobic protein/peptides by reversed-phase chromatography and both can be recovered for further analysis (Kolarich, Weber et al. 2006). Another advantage of the enzymatic glycan cleavage is the fact that the asparagine amino acid carrying the glycan is deamidated in the reaction to form aspartic acid and the reaction results in a +1 Da mass increase to the previously glycosylated peptide (Plummer et al. 1984) (Figure 6.2b). This feature allows high-throughput identification of previously glycosylated peptides in proteomics (Zhang and Aebersold 2006).

The PNGase F has been successfully used to release mammalian *N*-glycans from both intact proteins and proteolytically produced peptides with the addition of a low

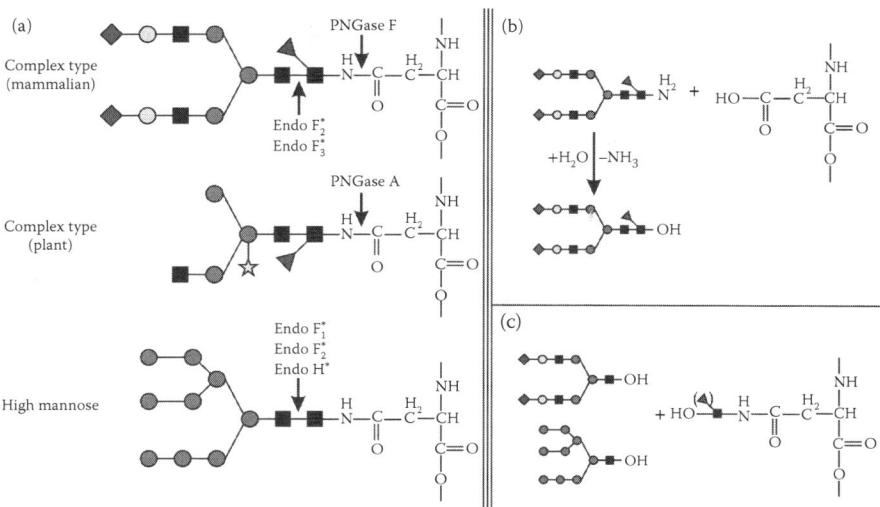

FIGURE 6.2 Examples for complex mammalian, plant, and high-mannose *N*-glycan structures. (a) Arrows indicate the place of action of different endoglycosidases commonly used for protein deglycosylation. Endoglycosidase H (Endo H) essentially acts on high-mannose-type structures; the presence of core α1,6-fucose increases the activity of endoglycosidases F_2 and F_3. (b) The PNGase F and PNGase A cleave between the amine linked to the reducing-end monosaccharide of the glycan and asparagine, converting it to aspartic acid. The amine initially remains on the glycan and is subsequently converted to the free reducing terminus autocatalytically. The *N*-glycans carrying core α1,3-fucose, for example, in plants and insects, require the use of PNGase A for release from the peptide backbone. (c) In contrast, Endo H and F_{1-3} and other similar enzymes conserve the innermost GlcNAc of the *N*-glycan (including core fucose, if present) on the peptide backbone.

concentration of a detergent, for example, SDS, CHAPS, Nonidet P40, and octylglycoside, significantly increasing the deglycosylation efficiency (Nuck et al. 1990). Although PNGase F has its optimal activity in the basic range of pH values 7–9, it has been reported to retain its activity at pH values 5–7 (Plummer et al. 1984) and the digestion conditions do not seem to have any negative effects on acid-labile glycan residues such as sialic acid. Glycans on proteins from sources such as insects, plants, and parasites, which carry the nonmammalian core α1,3-linked fucose residue, inhibit PNGase F action (Altmann et al. 1995) (Figure 6.2a). In these cases, PNGase A, an enzyme purified from almonds, can be used (Ashida et al. 2008; Kolarich, Altmann et al. 2006; Kolarich et al. 2005; Lehr et al. 2007; Paschinger et al. 2008; Rendic et al. 2007). However, PNGase A does not show significant deglycosidase activity on intact glycoproteins but requires proteolytic digestion of the protein prior to *N*-glycan release (Altmann et al. 1995).

Endoglycosidases F_1, F_2, F_3, and H (Tarentino and Plummer 1994) can also be used to cleave the majority of glycan structures from the protein (Figure 6.2). In contrast to PNGase F/A, Endoglycosidases F_1, F_2, F_3, and H cleave between the two GlcNAc moieties of the chitobiose core leaving the innermost GlcNAc on the glycoprotein. These enzymes are less sensitive to protein conformation compared to PNGase F and are suitable for removing *N*-glycans from intact glycoproteins and determining sites of prior glycosylation. However, they have distinct specificities regarding the outer arm structures and, depending on the structures present, a combination of different endoglycosidases is required to achieve complete removal of the *N*-glycans (Tarentino and Plummer 1994).

Hydrazinolysis is a commonly used chemical method for releasing *N*-glycans (Natsuka and Hase 1998; Natsuka et al. 2011). The use of different temperatures allows the sequential release of *O*-glycans (60°C; see Section 6.3.2) followed by that of *N*-glycans (95°C). It is generally considered to be effective in removing all types of *N*-glycans, but in contrast to the enzymatic approach the protein is destroyed under the releasing conditions. In general, anhydrous hydrazine is used, but hydrazine monohydrate is also reported to provide similar results (Nakakita et al. 2007). Following hydrazinolysis, the amino sugars in the released glycans need to be reacetylated.

6.3.2 RELEASING *O*-GLYCANS FROM PROTEINS

Global release of complete *O*-glycan oligosaccharides cannot be achieved by enzymatic methods. The available *O*-glycosidase removes only the core 1 *O*-glycan [Galβ(1-3)GalNAcα-Ser/Thr], and the majority of *O*-glycans need to be sequentially truncated by a cocktail of exoglycosidases before this single remaining disaccharide structure is released from the protein (Figure 6.3). Single *O*-linked GalNAc also

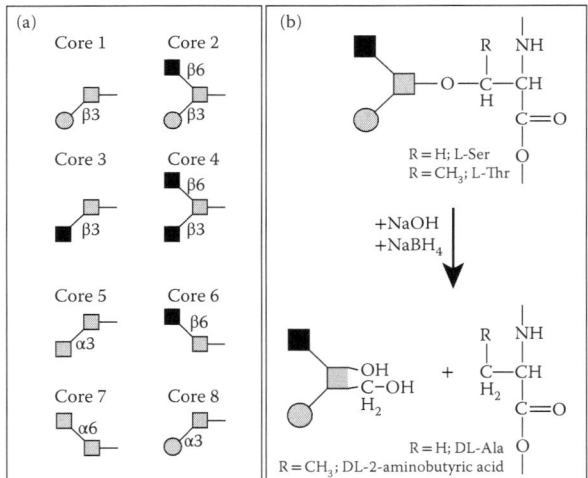

FIGURE 6.3 Figure showing *O*-glycan structures. (a) Eight defined core *O*-glycan structures are known. In human beings, cores 1–4 are most commonly found. The only known *O*-glycanase can just act on a core 1–type disaccharide; thus, global *O*-glycomics requires chemical release procedures. (b) Reductive β-elimination releases *O*-glycans from the protein backbone quantitatively. Peeling reactions known to start from the reducing terminus under basic conditions are prevented by the subsequent reduction of the reducing-end terminus. Abbreviations: L-Ser = L-Serine; L-Thr = L-Threonine; DL-Ala = D or L-Alanine.

cannot be removed by this enzyme (Ashida et al. 2008). This approach may give information on O-linked site glycosylation but does not allow characterization of heterogeneity of the attached O-glycan structures at that site.

Thus, chemical approaches are most frequently used to quantitatively release O-glycans from glycoproteins. The commonly used reductive β-elimination chemistry releases O-glycans under basic conditions (e.g., 50 mM of KOH) and elevated temperatures (50°C) in the presence of a reducing agent (sodium borohydrate) (Jensen et al. 2010) (Figure 6.3). The addition of the reducing agent produces a reduced saccharide on the terminus of the O-glycan immediately after release and avoids the destructive "peeling" of the reducing end under the required alkaline conditions. Excess salt and borohydride need to be removed prior to mass spectrometric analysis (Karlsson and Packer 2002).

If the reducing terminus is required to be kept intact for further labeling reactions with fluorescent dyes, hydrazinolysis at 60°C is the best option (Merry et al. 2002). As described for N-glycans in Section 6.3.1, deacetylation of amino sugars occurs. The method of β-elimination using nonreducing conditions has been shown to provide less reliable and reproducible results compared to the aforementioned approaches and is, thus, in the opinion of the authors, not recommended for structural analysis of protein O-glycans (Wada et al. 2010).

6.4 DERIVATIZATION OF OLIGOSACCHARIDES FOR MS DETECTION

Mass spectrometric analysis of N- and O-glycans can be performed without any further modification (see Section 6.4.1); nevertheless, various types of reported chemical derivatization approaches can provide glycans with different properties, which can be advantageous or might be required for further downstream analyses. However, it needs to be considered that chemical modification can result in sample losses and the introduction of artifacts.

6.4.1 Labeling of the Reducing End

Labeling of the reducing end of the released glycans with fluorescent or ultraviolet (UV)-absorbing dyes, enabling sensitive detection by optical methods, has been a long-standing analytical method, especially in combination with high-performance LC (HPLC) (Hase et al. 1979a, 1979b). Many compounds for labeling oligosaccharides by reductive amination have been described and most are compatible with mass spectrometric detection (Anumula 1994; Bigge et al. 1995; Charlwood et al. 2000; Harvey 2000b; Morelle et al. 2005; Pabst et al. 2009). This derivatization can change the properties of glycans such that they have beneficial analytical effects: Glycans can be made more hydrophobic, a charge can be introduced, sensitive detection by fluorescence can be achieved, and ionization can be enhanced (Harvey 2000a; Klein et al. 1998; Seveno et al. 2008). However, reductive amination requires an excess of the labeling reagent for the reaction to occur, which needs to be removed before further analysis (Pabst et al. 2009). Incomplete derivatization, modifications to the glycan structure, and sample losses may also occur.

A recent comprehensive study comparing the mass spectrometric features of 15 frequently used oligosaccharide-labeling compounds showed minor differences between native and labeled oligosaccharides under ESI (positive and negative modes). The most sensitive label (2-aminobenzoic acid ethyl ester for ESI+ MS and procaine [4-aminobenzoic acid 2-diethylaminoethyl ester] for MALDI+ time-of-flight [TOF] MS) resulted in about twice the ion intensity of the underivatized glycans (Pabst et al. 2009). Similarly, the spot preparation technique and selection of "hot spots" had a greater influence on signal intensity than the derivatization labels in MALDI TOF positive ion mode analysis. On the other hand, negative ion mode MALDI TOF analysis using 2,4,6-trihydroxyacetophenone (THAP) as a matrix showed that derivatization of oligosaccharides with labels containing a single acidic group resulted in significantly better ionization compared to those with labels containing three acidic groups (Pabst et al. 2009).

6.4.2 PERMETHYLATION OF OLIGOSACCHARIDES

Permethylation of glycans results in the conversion of all free hydroxyl groups on an oligosaccharide to stable methyl ethers (Powell and Harvey 1996). This modification turns a hydrophilic glycan into a hydrophobic molecule and neutralizes the negative charge of sialic acid; this explains its ionization behavior (Harvey 2005d; Zaia 2004). This derivatization allows the simultaneous relative quantitation of neutral and acidic oligosaccharides by MALDI TOF MS since the signal intensities of uncharged glycans can be correlated with their amount in a sample (Harvey 2005d). In addition, the sialic acids are stabilized for MALDI TOF analysis by this step and do not show the in-source and postsource fragmentation often detected for underivatized glycans in MALDI MS (Zaia 2004). For data interpretation, it needs to be considered that permethylation results in overall mass increase for every glycan structure (Dell 1990). Permethylation is also a classical approach allowing detailed determination of glycosidic linkages of the component monosaccharide residues by gas chromatography (GC)-MS (Price 2008).

6.5 MS IN GLYCOMICS ANALYSIS

Mass Spectrometry (MS) has emerged as one of the major technologies for the detection and characterisation of biomolecules. Current Proteomics and Glycomics workflows would not have been possible without the tremendous advances in MS-technologies made in recent decades. This section provides a broad overview on different approaches for solving glycomic challenges.

6.5.1 LC-ESI ANALYSIS OF UNDERIVATIZED GLYCANS

Mass spectrometry allows direct and sensitive (generally, femtomol–picomol range) analysis of underivatized glycans (Dreisewerd et al. 2006; Harvey 2005d, 2005e; Kolarich and Altmann 2000; Kuster et al. 1998). Analysis of released glycans without further derivatization reduces the number of sample preparation steps; however, simultaneous detection of both neutral and acidic glycans requires some consideration to be taken into account regarding the analysis methods.

Neutral and acidic glycans can be ionized well under online LC-ESI-MS conditions (Huang and Riggin 2000; Pabst and Altmann 2008). The sample complexity can be reduced by introducing a chromatographic separation step prior to mass spectrometric detection. Reversed-phase chromatographic matrices are generally not suited for the separation of underivatized glycans as the hydrophilic glycans have insufficient interaction with the separation matrix. Porous graphitized carbon (PGC) LC-MS proves to be the most robust and widespread method used in the analysis of underivatized glycans by LC-ESI-MS/MS, and its applications, benefits, and limitations have been excellently reviewed by Ruhaak et al. (2009). Many research groups are using this approach for the global glycomic analysis of a tissue, a fluid, or specific proteins (Deshpande et al. 2010; Karlsson and Packer 2002; Karlsson, Schulz et al. 2004; Karlsson et al. 2005; Karlsson, Wilson et al. 2004; Ninonuevo et al. 2005, 2008; Pabst et al. 2007; Ruhaak et al. 2009; Tao et al. 2008, 2009; Wilson et al. 2008; Wilson et al. 2002). Except for monosaccharides, *N*- and *O*-glycans are generally well retained on PGC, which is able to chromatographically separate isobaric glycan structures differing only in linkage. This allows the acquisition of separate MS/MS spectra for individual isobaric glycan structures, thereby facilitating detailed structural determination (Figure 6.4). Reduction of the reducing end of *N*- and *O*-linked oligosaccharides is recommended for PGC analysis, since α- and β-anomers of the same structural isomer of a reducing sugar can be separated on the carbon stationary phase resulting in two chromatographic peaks for an identical structure (Ruhaak et al. 2009). Released *O*-glycans are already present in the reduced form if reductive β-elimination release of the glycans is used, and simple procedures can be applied to reduce enzymatically released *N*-glycans or free glycans with sodium borohydride prior to analysis (Jensen et al. 2012; Ninonuevo et al. 2005, 2008). Reduction provides the additional advantage of introducing a +2 Da mass tag on the reducing-end sugar, which facilitates interpretation of MS/MS spectra by unambiguous identification of this glycan terminus.

In the mass spectrometric analysis of *N*- and *O*-glycans, negative ionization is often used. Although negative ionization MS is often considered less sensitive than its positive ion counterpart, negatively charged acidic glycans as well as neutral glycans can be detected simultaneously using this technique down to low femtomol sensitivities (Harvey 2005a, 2005b, 2005c; Karlsson, Schulz et al. 2004; Karlsson, Wilson et al. 2004; Peter-Katalinic 1994; Zaia 2004). Under the conditions used for negative-mode MS, negatively charged glycans usually have higher signal intensities compared with similar amounts of neutral oligosaccharides. Nevertheless, both neutral and acidic oligosaccharides ionize well in negative ion mode and relative quantitation can be reliably obtained using normalization factors (Olson et al. 2005). Another advantage of negative ionization PGC LC-ESI-MS analysis is the facilitated detection of sulfated and phosphorylated glycan structures (Karlsson and Thomsson 2009; Thomsson et al. 2005).

An alternative approach for LC-ESI-MS analysis of released, underivatized, protein-bound glycans using homemade Amide-80 nanocolumns for normal-phase nanoscale separation of oligosaccharides has been described (Wuhrer et al. 2004). This approach uses oligosaccharide hydroxyl groups for polar interactions with the stationary phase under hydrophobic solvent conditions.

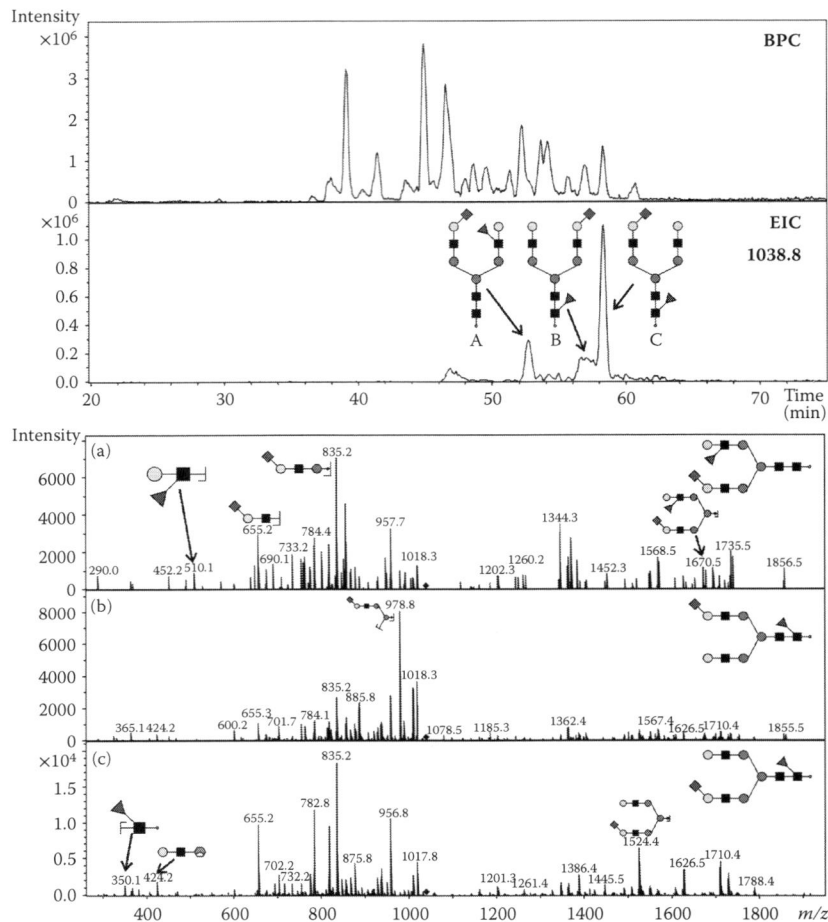

FIGURE 6.4   Example for the LC-MS/MS analysis of released *N*-glycans by PGC chromatography. The base peak chromatogram (BPC) analysis of the *N*-glycans released from sIgA
provides an overview of the entity of structures present; the extracted ion chromatograms
(EICs) deliver elution information of different isobaric structures separated by PGC chromatography. By chromatographic separation, separate MS/MS data can be acquired for isobaric
glycan structures just differing in particular monosaccharide linkages. This enhances detailed
structural characterization of isobaric glycoforms A, B, and C. Bottom panels (a), (b), and (c)
exemplify the respective different MS/MS spectra of structures A, B and C, respectively,
acquired for the doubly negatively charged signal of *m/z* value 1038.9 corresponding to the
composition of Hex5HexNAc4NeuAcFuc, which is present in three distinct isoforms in the
analyzed sample. Abbreviations: *m/z* = mass over charge ratio.

6.5.2   Off-Line Analysis of Underivatized Glycans

Off-line direct-injection nano-ESI-MS analysis has also been reported for the detailed
mass spectrometric elucidation of protein-bound oligosaccharides (Sagi et al. 2002;
Vakhrushev et al. 2006). Very often, a series of various off-line purification steps

such as reversed-phase cleanups and gel filtration or weak anion exchange, to name a few, precedes this type of MS analysis. This approach has provided useful data on the fragmentation patterns of milk oligosaccharide structures under various ionization conditions (Chai et al. 2001, 2002, 2006; Kogelberg et al. 2004; Pfenninger et al. 2002a, 2002b; Seymour et al. 2006). Interestingly, more diagnostic cross-ring fragmentation cleavages have been identified to be produced from negatively charged ions under low collision induced dissociation (CID) conditions compared to positive-mode fragmentation (Chai et al. 2001; Harvey et al. 2008). Negative-mode ESI ion trap multistage fragmentation (MS^n) has been used to evaluate isomeric mixtures of milk oligosaccharides with specific fragmentation patterns (Pfenninger et al. 2002a, 2002b). Specific MS/MS fragmentation masses were also used to determine the branching pattern of other oligosaccharides (Chai et al. 2001, 2002, 2006, 2005; Kogelberg et al. 2004). The definition of general fragmentation rules for oligosaccharides is very useful for detailed structural assignments, although fragmentation patterns can vary with ionization conditions and between instruments.

6.5.3 ESI-MS^N OF PERMETHYLATED GLYCANS

The combination of off-line injection of permethylated glycans and MS^n analysis of defined fragments has been described to accomplish detailed structural assignment of oligosaccharides (Ashline et al. 2005, 2007; Lapadula et al. 2005; Zhang et al. 2005). In contrast to LC-ESI-MS methods using online separation, the total glycan pool is analyzed simultaneously when the sample is injected off-line, making it more challenging to distinguish between isobaric structures, especially if there is a considerable concentration difference between structures present in the mixture. This was partially overcome by Costello et al. (2007) by using PGC for online LC separation prior to ESI-MS detection. Separation of isobaric oligosaccharides could be achieved in their permethylated state, although for the price of significantly increased chromatographic base peak width compared to underivatized glycans.

6.5.4 GLYCOMICS USING MALDI IONIZATION

The MALDI MS is an often-used, contamination-tolerant, and robust method for the analysis of oligosaccharides. In this approach, oligosaccharides can be analyzed directly without any prior modification (Finke et al. 1999; Kolarich and Altmann 2000; Kuster et al. 1998; Pfenninger et al. 1999; Rendic et al. 2007; Stahl et al. 1994) or more commonly as their permethylated derivatives (Jang-Lee et al. 2006; North et al. 2012).

In contrast to positive- and negative-ion ESI, where singly and multiply protonated $[M + xH]^{x+}$ or deprotonated $[M - yH]^{y-}$ ions, respectively, are the major ions produced during the ionization, singly charged metal adduct ions are the primary ion species observed in positive-ion MALDI MS analysis of underivatized oligosaccharides (Harvey 2005e) (Figure 6.5). Sodiated ions $[M + Na]^+$ are generally the most abundant, but this can be shifted toward different cationic ions by doping the matrix

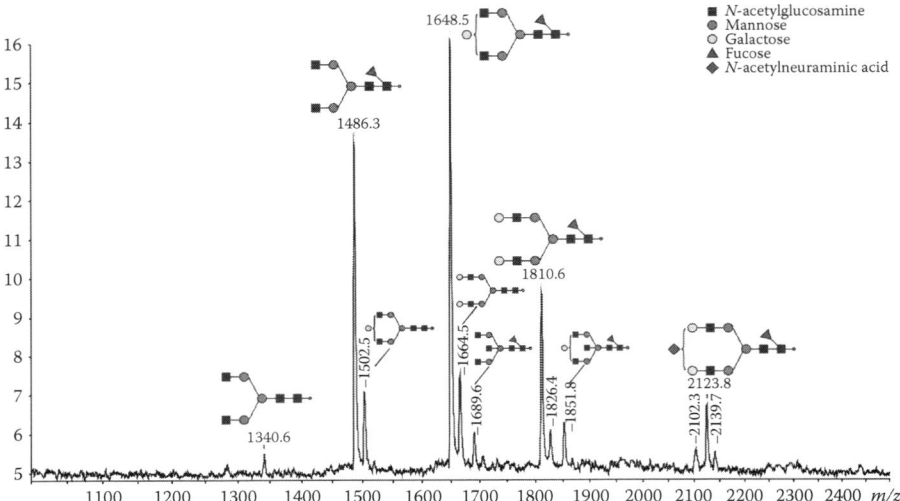

FIGURE 6.5　The MALDI TOF MS spectrum of unmodified *N*-glycans released and isolated from human IgG. Different isobaric glycans cannot be distinguished separately; nevertheless, the speed of analysis is significantly increased. The signals correspond to sodiated [M + Na]⁺ ions; the spectrum was recorded in linear mode to enable detection of the sialylated *N*-glycans, which otherwise would not be detectable if the spectra were recorded in reflector mode (see also Section 6.5.4).

with the appropriate salt (Harvey 2005e). However, sialic acid–containing molecules usually produce a mixture of ions ([M – H]⁻, [M + Na]⁺, [M – nH + (n + 1)Na]⁺) due to salt formation. These molecules are more susceptible to in-source and postsource fragmentation with the concomitant loss of sialic acid (Harvey 2005e; Zaia 2004). In the use of reflectron TOF instruments, postsource fragmentation is reported to produce metastable peaks (Harvey 2005e) and permethylation is usually the method of choice for the analysis of sialylated glycans by MALDI TOF MS (Zaia 2004). For more details on the theory of MALDI MS in the context of oligosaccharides, readers can refer to the excellent reviews by Harvey and Zaia (Harvey 1999, 2003, 2005e; Zaia 2004).

Another factor significantly influencing the analysis success of MALDI MS for oligosaccharides is use of the appropriate matrix. A large number of different MALDI matrices have been published for oligosaccharide analysis; however, 2,5-dihydroxybenzoic acid (DHB) remains the most commonly used matrix (Harvey 2005d). For the analysis of sialylated oligosaccharides, matrices such as THAP have been shown to give best results when mixed with ammonium citrate (Papac et al. 1996). It is noted that 2,6-azathiothymine (ATT) has successfully been used to analyze both sialylated and neutral oligosaccharides by MALDI MS. Being a nonacidic matrix, ATT offers the additional advantage that "on-target" exoglycosidase treatments can be applied for determining specific linkages (Geyer et al. 1999). Pfenninger et al. (1999) compared different matrices and conditions to optimize the conditions for MALDI MS analysis of milk oligosaccharides. They identified DHB, 3-aminoquinoline, ATT, and 5-chloromercaptobenzothiazole to be the most suitable matrices for neutral oligosaccharides (Pfenninger et al. 1999). Additives such as NaCl improved ionization with DHB especially for neutral oligosaccharides, whereas diammonium hydrogen citrate (DAHC) as an additive in ATT was their preferred matrix for acidic oligosaccharides (Pfenninger et al. 1999). Tzeng et al. (2009) described the use of alkali hydroxide–doped matrices for structural characterization of neutral underivatized oligosaccharides by MALDI TOF MS. In this study, the partial alkaline degradation that occurred on laser desorption/ionization facilitated the structural characterization of oligosaccharides by postsource decay TOF MS.

The majority of glycomic studies by MALDI TOF MS, however, use permethylated oligosaccharides as the stabilization of neuraminic acid and charge neutralization (see Section 6.4.2) facilitates the relative quantitative comparison of neutral and previously negatively charged glycans (Azadi and Heiss 2009; Jang-Lee et al. 2006). Recently, this approach was applied to the glycomic characterization of the hyaluronic acid receptor for endocytosis (HARE) (Figure 6.6) (Harris et al. 2010) and of human mast cells, eosinophils, and basophils, showing that the different

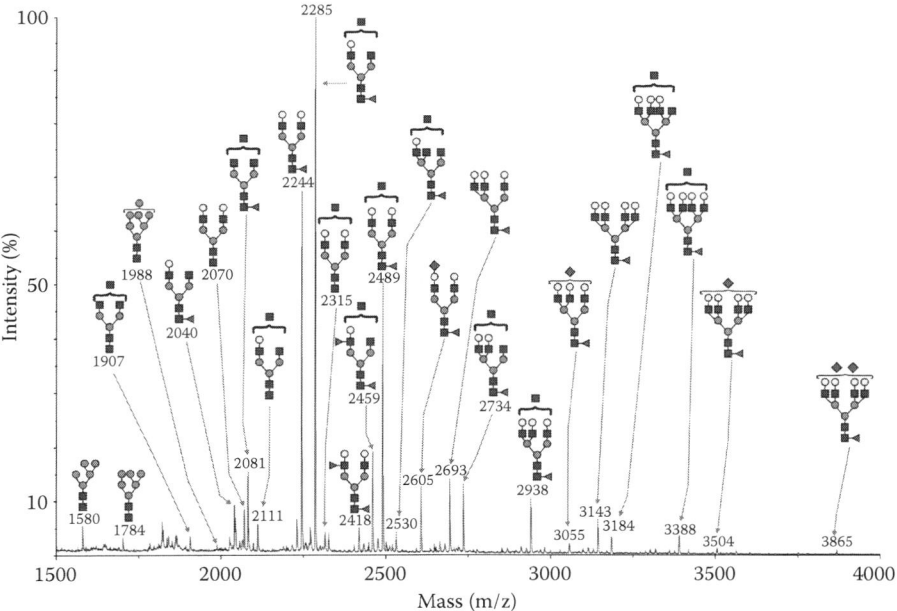

FIGURE 6.6 Typical example for a glycomic MALDI TOF MS analysis of permethylated *N*-glycans from HARE. The spectrum shows the glycomic analysis of s190-HARE by MALDI TOF analysis of permethylated *N*-glycans. All molecular ions are [M + Na]⁺, and nominal masses of the ¹²C isotope are shown. The putative structures presented are based on composition, tandem MS analysis, and knowledge of biosynthetic pathways. (Harris, E. N. et al. *N*-Glycans on the link domain of human HARE/Stabilin-2 are needed for hyaluronan binding to purified ecto-domain, but not for cellular endocytosis of hyaluronan, *Glycobiology*, 20, 991–1001, 2010. By permission of Oxford University Press and the Society of Glycobiology.)

cell types differ significantly in their repertoire of glycoprotein glycan structures (North et al. 2012). Since glycoproteins and their glycans are often the first line of interaction between cells because of their location on the cell surface, results similar to the aforementioned ones are important to study diseases that are related to these cell types, such as allergy and inflammation. The glycome of different mouse and human tissue glycoproteins has been studied by this approach and made publicly available through the efforts of the Consortium for Functional Glycomics (www .functionalglycomics.org) (Haslam et al. 2006).

Relative quantitation between different sets of samples can also be achieved by permethylation of two samples using C12 and C13-labeled methyl iodide, respectively (Alvarez-Manilla et al. 2007). The two samples are mixed prior to analysis, which allows quantitative comparison.

It is noted that TOF detectors are frequently used for MALDI ionization sources; however, combination Fourier transform ion cyclotron resonance (FT-ICR) MS or quadrupole ion trap (QIT) TOF can also be used (Kronewitter et al. 2010; Takemori et al. 2006). These techniques provide the specific advantages of ultrahigh-resolution or multistage tandem MS to MALDI-produced ions.

6.6 GLYCOMICS DATA STANDARDIZATION AND REPORTING

The increasing speeds at which data can be and is generated from current mass spectrometric glycomics approaches require appropriate means of data handling, processing, storing, and maintenance. In contrast to proteomics where such standards have already been acknowledged in the Minimum Information about a Proteomics Experiment (MIAPE) project (Binz et al. 2008; Taylor et al. 2007, 2008), similar efforts are currently only starting to develop in glycomics. The Minimum Information Required for a Glycomics Experiment (MIRAGE) project (glycomics .ccrc.uga.edu/MIRAGE) is working to set the urgently required standards in this area. Different glycosylation analysis methods provide different sets of data that are often manually evaluated by scientists for the determination of glycan structures, and standardization projects are facing the challenge of defining reliable and strong data reporting guidelines suitable for the majority of glycomics approaches.

6.7 PUBLICALLY AVAILABLE BIOINFORMATIC
TOOLS FOR GLYCOMICS

In contrast to the collection of well-known genome and proteome databases, the number of glycan bioinformatic databases is still comparably sparse. Five publicly available carbohydrate databases (CFG, Eurocarb DB, Glycosciences.de, GlycosuiteDB, and KEGG GLYCAN) containing sets of data on glycan structures have been created, which are currently at various stages of development and funding (Aoki-Kinoshita 2008). The recently initiated UnicarbDB glycan structural MS database (UnicarbDB; www.unicarb-db.org) and the planned integration of resources into a glycomics knowledge base (Unicarb KB; www.unicarbkb.org) will hopefully support glycomics research by providing a high-quality annotated platform for data exchange and a database of high-quality datasets (Campbell et al. 2011; Hayes et al. 2011).

The work of glycobiology MS researchers is also supported by a few freely available tools that facilitate structure determination by mass spectrometric analyses. The GlycoWorkBench (www.glycoworkbench.org) is one such freely available software tool that combines the drawing of glycan structures with their predicted fragmentation by MS (Ceroni et al. 2007, 2008). The easy-to-use graphical user interface can be quickly understood by even nonexperts and helps to reduce the time needed for the interpretation of glycan MS spectra. It is hoped that in time the development of informatics tools for glycomics will reach the development level of tools available to genomics and proteomics analysts and will help to realize the discovery potential of this area of biological research.

6.8 CONCLUDING REMARKS

It is noted that MS is without doubt the most powerful approach for glycomic studies of glycoproteins, especially if it is combined with selective and appropriate off-line and online sample preparation. The soft ionization techniques of ESI and MALDI and the tremendous mass spectrometric innovations made in the last two decades have contributed significantly to the fact that scientists can now gain novel and unprecedented insights into biomolecules. Encouraging developments covering the technical, informatic, and biological aspects of glycomics can be expected to increase in the near and medium-term future, which makes it exciting to be a scientist in this ever-inspiring field that provides one with new surprises every day.

REFERENCES

Altmann, F., Schweiszer, S., and Weber, C. 1995. Kinetic comparison of peptide: N-glycosidases F and A reveals several differences in substrate specificity. *Glycoconjugate J*, 12, 84–93.

Alvarez-Manilla, G., Warren, N. L., Abney, T. et al. 2007. Tools for glycomics: Relative quantitation of glycans by isotopic permethylation using 13CH3I. *Glycobiology*, 17, 677–687.

Anumula, K. R. 1994. Quantitative determination of monosaccharides in glycoproteins by high-performance liquid chromatography with highly sensitive fluorescence detection. *Anal Biochem*, 220, 275–283.

Aoki-Kinoshita, K. F. 2008. An introduction to bioinformatics for glycomics research. *PLoS Comput Biol*, 4, e1000075.

Ashida, H., Maki, R., Ozawa, H. et al. 2008. Characterization of two different endo-alpha-N-acetylgalactosaminidases from probiotic and pathogenic enterobacteria, Bifidobacterium longum and clostridium perfringens. *Glycobiology*, 18, 727–734.

Ashline, D. J., Lapadula, A. J., Liu, Y. H. et al. 2007. Carbohydrate structural isomers analyzed by sequential mass spectrometry. *Anal Chem*, 79, 3830–3842.

Ashline, D., Singh, S., Hanneman, A., and Reinhold, V. 2005. Congruent strategies for carbohydrate sequencing. 1. Mining structural details by MSn. *Anal Chem*, 77, 6250–6262.

Azadi, P. and Heiss, C. 2009. Mass spectrometry of N-linked glycans. *Methods Mol Biol*, 534, 37–51.

Bigge, J. C., Patel, T. P., Bruce, J. A. et al. 1995. Nonselective and efficient fluorescent labeling of glycans using 2-amino benzamide and anthranilic acid. *Anal Biochem*, 230, 229–238.

Binz, P. A., Barkovich, R., Beavis, R. C. et al. 2008. Guidelines for reporting the use of mass spectrometry informatics in proteomics. *Nat Biotechnol*, 26, 862.

Campbell, M. P., Hayes, C. A., Karlsson, N. G. et al. 2011. UniCarbKB: Putting the pieces together for glycomics research. *Proteomics*, 11, 4117–4121.

Ceroni, A., Dell, A., and Haslam, S. M. 2007. The GlycanBuilder: A fast, intuitive and flexible software tool for building and displaying glycan structures. *Source Code Biol Med*, 2, 3.

Ceroni, A., Maass, K., Geyer, H. et al. 2008. GlycoWorkbench: A tool for the computer-assisted annotation of mass spectra of glycans. *J Proteome Res*, 7, 1650–1659.

Chai, W., Piskarev, V., and Lawson, A. M. 2001. Negative-ion electrospray mass spectrometry of neutral underivatized oligosaccharides. *Anal Chem*, 73, 651–657.

Chai, W., Piskarev, V., and Lawson, A. M. 2002. Branching pattern and sequence analysis of underivatized oligosaccharides by combined MS/MS of singly and doubly charged molecular ions in negative-ion electrospray mass spectrometry. *J Am Soc Mass Spectrom*, 13, 670–679.

Chai, W., Piskarev, V. E., Mulloy, B. et al. 2006. Analysis of chain and blood group type and branching pattern of sialylated oligosaccharides by negative ion electrospray tandem mass spectrometry. *Anal Chem*, 78, 1581–1592.

Chai, W., Piskarev, V. E., Zhang, Y., Lawson, A. M., and Kogelberg, H. 2005. Structural determination of novel lacto-N-decaose and its monofucosylated analogue from human milk by electrospray tandem mass spectrometry and 1H NMR spectroscopy. *Arch Biochem Biophys*, 434, 116–127.

Charlwood, J., Skehel, J. M., and Camilleri, P. 2000. Analysis of N-linked oligosaccharides released from glycoproteins separated by two-dimensional gel electrophoresis. *Anal Biochem*, 284, 49–59.

Christiansen, M. N., Kolarich, D., Nevalainen, H., Packer, N. H., and Jensen, P. H. 2010. Challenges of determining O-glycopeptide heterogeneity: A fungal glucanase model system. *Anal Chem*, 82, 3500–3509.

Costello, C. E., Contado-Miller, J. M., and Cipollo, J. F. 2007. A glycomics platform for the analysis of permethylated oligosaccharide alditols. *J Am Soc Mass Spectrom*, 18, 1799–1812.

Dell, A. 1990. Preparation and desorption mass spectrometry of permethyl and peracetyl derivatives of oligosaccharides. *Methods Enzymol*, 193, 647–660.

Deshpande, N., Jensen, P. H., Packer, N. H., and Kolarich, D. 2010. GlycoSpectrumScan: Fishing glycopeptides from MS spectra of protease digests of human colostrum sIgA. *J Proteome Res*, 9, 1063–1075.

Dreisewerd, K., Kolbl, S., Peter-Katalinic, J., Berkenkamp, S., and Pohlentz, G. 2006. Analysis of native milk oligosaccharides directly from thin-layer chromatography plates by matrix-assisted laser desorption/ionization orthogonal-time-of-flight mass spectrometry with a glycerol matrix. *J Am Soc Mass Spectrom*, 17, 139–150.

Finke, B., Stahl, B., Pfenninger, A. et al. 1999. Analysis of high-molecular-weight oligosaccharides from human milk by liquid chromatography and MALDI-MS. *Anal Chem*, 71, 3755–3762.

Geyer, H., Schmitt, S., Wuhrer, M., and Geyer, R. 1999. Structural analysis of glycoconjugates by on-target enzymatic digestion and MALDI-TOF-MS. *Anal Chem*, 71, 476–482.

Harris, E. N., Parry, S., Sutton-Smith, M. et al. 2010. N-Glycans on the link domain of human HARE/Stabilin-2 are needed for hyaluronan binding to purified ecto-domain, but not for cellular endocytosis of hyaluronan. *Glycobiology*, 20, 991–1001.

Harvey, D. J. 1999. Matrix-assisted laser desorption/ionization mass spectrometry of carbohydrates. *Mass Spectrom Rev*, 18, 349–450.

Harvey, D. J. 2000a. Electrospray mass spectrometry and fragmentation of N-linked carbohydrates derivatized at the reducing terminus. *J Am Soc Mass Spectrom*, 11, 900–915.

Harvey, D. J. 2000b. N-(2-diethylamino)ethyl-4-aminobenzamide derivative for high sensitivity mass spectrometric detection and structure determination of N-linked carbohydrates. *Rapid Commun Mass Spectrom*, 14, 862–871.

Harvey, D. J. 2003. Matrix-assisted laser desorption/ionization mass spectrometry of carbohydrates and glycoconjugates. *Int J Mass Spectrom*, 226, 1–35.

Harvey, D. J. 2005a. Fragmentation of negative ions from carbohydrates: Part 1. Use of nitrate and other anionic adducts for the production of negative ion electrospray spectra from N-linked carbohydrates. *J Am Soc Mass Spectrom*, 16, 622–630.

Harvey, D. J. 2005b. Fragmentation of negative ions from carbohydrates: Part 2. Fragmentation of high-mannose N-linked glycans. *J Am Soc Mass Spectrom*, 16, 631–646.

Harvey, D. J. 2005c. Fragmentation of negative ions from carbohydrates: Part 3. Fragmentation of hybrid and complex N-linked glycans. *J Am Soc Mass Spectrom*, 16, 647–659.

Harvey, D. J. 2005d. Proteomic analysis of glycosylation: Structural determination of N- and O-linked glycans by mass spectrometry. *Expert Rev Proteomics*, 2, 87–101.

Harvey, D. J. 2005e. Structural determination of N-linked glycans by matrix-assisted laser desorption/ionization and electrospray ionization mass spectrometry. *Proteomics*, 5, 1774–1786.

Harvey, D. J., Royle, L., Radcliffe, C. M., Rudd, P. M., and Dwek, R. A. 2008. Structural and quantitative analysis of N-linked glycans by matrix-assisted laser desorption ionization and negative ion nanospray mass spectrometry. *Anal Biochem*, 376, 44–60.

Hase, S., Ikenaka, T., and Matsushima, Y. 1979a. Analyses of oligosaccharides by tagging the reducing end with a fluorescent compound. I. Application to glycoproteins. *J Biochem*, 85, 989–994.

Hase, S., Ikenaka, T., and Matsushima, Y. 1979b. Analyses of oligosaccharides by tagging the reducing end with a fluorescent compound. II. Linkage point analyses. *J Biochem*, 85, 995–1002.

Haslam, S. M., North, S. J., and Dell, A. 2006. Mass spectrometric analysis of N- and O-glycosylation of tissues and cells. *Curr Opin Struct Biol*, 16, 584–591.

Hayes, C. A., Karlsson, N. G., Struwe, W. B. et al. 2011. UniCarb-DB: A database resource for glycomic discovery. *Bioinformatics*, 27, 1343–1344.

Honda, S., Suzuki, S., and Taga, A. 2003. Analysis of carbohydrates as 1-phenyl-3-methyl-5-pyrazolone derivatives by capillary/microchip electrophoresis and capillary electrochromatography. *J Pharm Biomed Anal*, 30, 1689–1714.

Huang, L. and Riggin, R. M. 2000. Analysis of nonderivatized neutral and sialylated oligosaccharides by electrospray mass spectrometry. *Anal Chem*, 72, 3539–3546.

Jang-Lee, J., North, S. J., Sutton-Smith, M. et al. 2006. Glycomic Profiling of Cells and Tissues by Mass Spectrometry: Fingerprinting and Sequencing Methodologies. *Methods Enzymol*, 415, 59–86.

Jensen, P. H., Karlsson, N. G., Kolarich, D., and Packer, N. H. 2012. Structural analysis of N- and O-glycans released from glycoproteins. *Nat Protoc* 7, 1299–1310. doi:10.1038/nprot.2012.063

Jensen, P. H., Kolarich, D., and Packer, N. H. 2010. Mucin-type O-glycosylation—putting the pieces together. *FEBS J*, 277, 81–94.

Karlsson, N. G. and Packer, N. H. 2002. Analysis of O-linked reducing oligosaccharides released by an in-line flow system. *Anal Biochem*, 305, 173–185.

Karlsson, N. G., Schulz, B. L., and Packer, N. H. 2004. Structural determination of neutral O-linked oligosaccharide alditols by negative ion LC-electrospray-MSn. *J Am Soc Mass Spectrom*, 15, 659–672.

Karlsson, N. G., Schulz, B. L., Packer, N. H., and Whitelock, J. M. 2005. Use of graphitised carbon negative ion LC-MS to analyse enzymatically digested glycosaminoglycans. *J Chromatogr B Analyt Technol Biomed Life Sci*, 824, 139–147.

Karlsson, N. G. and Thomsson, K. A. 2009. Salivary MUC7 is a major carrier of blood group I type O-linked oligosaccharides serving as the scaffold for sialyl Lewis x. *Glycobiology*, 19, 288–300.

Karlsson, N. G., Wilson, N. L., Wirth, H. J. et al. 2004. Negative ion graphitised carbon nano-liquid chromatography/mass spectrometry increases sensitivity for glycoprotein oligosaccharide analysis. *Rapid Commun Mass Spectrom*, 18, 2282–2292.

Klein, A., Lebreton, A., Lemoine, J. et al. 1998. Identification of urinary oligosaccharides by matrix-assisted laser desorption ionization time-of-flight mass spectrometry. *Clin Chem*, 44, 2422–2428.

Kogelberg, H., Piskarev, V. E., Zhang, Y., Lawson, A. M., and Chai, W. 2004. Determination by electrospray mass spectrometry and 1H-NMR spectroscopy of primary structures of variously fucosylated neutral oligosaccharides based on the iso-lacto-N-octaose core. *Eur J Biochem*, 271, 1172–1186.

Kolarich, D. and Altmann, F. 2000. N-Glycan analysis by matrix-assisted laser desorption/ionization mass spectrometry of electrophoretically separated nonmammalian proteins: Application to peanut allergen Ara h 1 and olive pollen allergen Ole e 1. *Anal Biochem*, 285, 64–75.

Kolarich, D., Altmann, F., and Sunderasan, E. 2006. Structural analysis of the glycoprotein allergen Hev b 4 from natural rubber latex by mass spectrometry. *Biochim Biophys Acta*, 1760, 715–720.

Kolarich, D., Leonard, R., Hemmer, W., and Altmann, F. 2005. The N-glycans of yellow jacket venom hyaluronidases and the protein sequence of its major isoform in Vespula vulgaris. *FEBS J*, 272, 5182–5190.

Kolarich, D., Weber, A., Turecek, P. L., Schwarz, H. P., and Altmann, F. 2006. Comprehensive glyco-proteomic analysis of human alpha1-antitrypsin and its charge isoforms. *Proteomics*, 6, 3369–3380.

Kronewitter, S. R., de Leoz, M. L., Peacock, K. S. et al. 2010. Human serum processing and analysis methods for rapid and reproducible N-glycan mass profiling. *J Proteome Res*, 9, 4952–4959.

Kuster, B., Hunter, A. P., Wheeler, S. F., Dwek, R. A., and Harvey, D. J. 1998. Structural determination of N-linked carbohydrates by matrix-assisted laser desorption/ionization-mass spectrometry following enzymatic release within sodium dodecyl sulphate-polyacrylamide electrophoresis gels: Application to species-specific glycosylation of alpha1-acid glycoprotein. *Electrophoresis*, 19, 1950–1959.

Lapadula, A. J., Hatcher, P. J., Hanneman, A. J. et al. 2005. Congruent strategies for carbohydrate sequencing. 3. OSCAR: An algorithm for assigning oligosaccharide topology from MSn data. *Anal Chem*, 77, 6271–6279.

Lee, A., Chick, J. M., Kolarich, D. et al. 2010. Liver membrane proteome glycosylation changes in mice bearing an extra-hepatic tumour. *Mol Cell Proteomics*, 2011, 10, M900538–MCP200. doi:10.1074/mcp.M900538-MCP200

Lee, A., Kolarich, D., Haynes, P. A. et al. 2009. Rat liver membrane glycoproteome: Enrichment by phase partitioning and glycoprotein capture. *J Proteome Res*, 8, 770–781.

Lee, A., Nakano, M., Hincapie, M. et al. 2010. The lectin riddle: Glycoproteins fractionated from complex mixtures have similar glycomic profiles. *OMICS*, 14, 487–499.

Lehr, T., Geyer, H., Maass, K., Doenhoff, M. J., and Geyer, R. 2007. Structural characterization of N-glycans from the freshwater snail Biomphalaria glabrata cross-reacting with Schistosoma mansoni glycoconjugates. *Glycobiology*, 17, 82–103.

Lindberg, B. and Lonngren, J. 1978. Methylation analysis of complex carbohydrates: General procedure and application for sequence analysis. *Methods Enzymol*, 50, 3–33.

Mann, M., Hendrickson, R. C. and Pandey, A. 2001. Analysis of proteins and proteomes by mass spectrometry. *Annu Rev Biochem*, 70, 437–473.

Merry, A. H., Neville, D. C., Royle, L. et al. 2002. Recovery of intact 2-aminobenzamide-labeled O-glycans released from glycoproteins by hydrazinolysis. *Anal Biochem*, 304, 91–99.

Morelle, W., Slomianny, M. C., Diemer, H. et al. 2005. Structural characterization of 2-aminobenzamide-derivatized oligosaccharides using a matrix-assisted laser desorption/ionization two-stage time-of-flight tandem mass spectrometer. *Rapid Commun Mass Spectrom*, 19, 2075–2084.

Nakakita, S., Sumiyoshi, W., Miyanishi, N., and Hirabayashi, J. 2007. A practical approach to N-glycan production by hydrazinolysis using hydrazine monohydrate. *Biochem Biophys Res Commun*, 362, 639–645.

Natsuka, S. and Hase, S. 1998. Analysis of N- and O-glycans by pyridylamination. *Methods Mol Biol*, 76, 101–113.

Natsuka, S., Hirohata, Y., Nakakita, S., Sumiyoshi, W., and Hase, S. 2011. Structural analysis of N-glycans of the planarian Dugesia japonica. *FEBS J*, 278, 452–460.

Ninonuevo, M., An, H., Yin, H. et al. 2005. Nanoliquid chromatography-mass spectrometry of oligosaccharides employing graphitized carbon chromatography on microchip with a high-accuracy mass analyzer. *Electrophoresis*, 26, 3641–3649.

Ninonuevo, M. R., Perkins, P. D., Francis, J. et al. 2008. Daily variations in oligosaccharides of human milk determined by microfluidic chips and mass spectrometry. *J Agric Food Chem*, 56, 618–626.

North, S. J., von Gunten, S., Antonopoulos, A. et al. 2012. Glycomic analysis of human mast cells, eosinophils and basophils. *Glycobiology*, 22, 12–22.

Nuck, R., Zimmermann, M., Sauvageot, D., Josi, D., and Reutter, W. 1990. Optimized deglycosylation of glycoproteins by peptide-N4-(N-acetyl-beta-glucosaminyl)-asparagine amidase from *Flavobacterium meningosepticum. Glycoconjugate J*, 7, 279–286.

Olson, F. J., Backstrom, M., Karlsson, H., Burchell, J., and Hansson, G. C. 2005. A MUC1 tandem repeat reporter protein produced in CHO-K1 cells has sialylated core 1 O-glycans and becomes more densely glycosylated if coexpressed with polypeptide-GalNAc-T4 transferase. *Glycobiology*, 15, 177–191.

Pabst, M. and Altmann, F. 2008. Influence of electrosorption, solvent, temperature, and ion polarity on the performance of LC-ESI-MS using graphitic carbon for acidic oligosaccharides. *Anal Chem*, 80, 7534–7542.

Pabst, M., Bondili, J. S., Stadlmann, J., Mach, L., and Altmann, F. 2007. Mass + retention time = structure: A strategy for the analysis of N-glycans by carbon LC-ESI-MS and its application to fibrin N-glycans. *Anal Chem*, 79, 5051–5057.

Pabst, M., Kolarich, D., Poltl, G. et al. 2009. Comparison of fluorescent labels for oligosaccharides and introduction of a new postlabeling purification method. *Anal Biochem*, 384, 263–273.

Papac, D. I., Wong, A., and Jones, A. J. 1996. Analysis of acidic oligosaccharides and glycopeptides by matrix-assisted laser desorption/ionization time-of-flight mass spectrometry. *Anal Chem*, 68, 3215–3223.

Paschinger, K., Gutternigg, M., Rendic, D., and Wilson, I. B. 2008. The N-glycosylation pattern of Caenorhabditis elegans. *Carbohyd Res*, 343, 2041–2049.

Peter-Katalinic, J. 1994. Analysis of glycoconjugates by fast atom bombardment mass spectrometry and related ms techniques. *Mass Spectrom Rev*, 13, 77–98.

Pfenninger, A., Karas, M., Finke, B., and Stahl, B. 2002a. Structural analysis of underivatized neutral human milk oligosaccharides in the negative ion mode by nano-electrospray MS(n) (part 1: methodology). *J Am Soc Mass Spectrom*, 13, 1331–1340.

Pfenninger, A., Karas, M., Finke, B., and Stahl, B. 2002b. Structural analysis of underivatized neutral human milk oligosaccharides in the negative ion mode by nano-electrospray MS(n) (part 2: application to isomeric mixtures). *J Am Soc Mass Spectrom*, 13, 1341–1348.

Pfenninger, A., Karas, M., Finke, B., Stahl, B., and Sawatzki, G. 1999. Matrix optimization for matrix-assisted laser desorption/ionization mass spectrometry of oligosaccharides from human milk. *J Mass Spectrom*, 34, 98–104.

Plummer, T. H., Jr., Elder, J. H., Alexander, S., Phelan, A. W., and Tarentino, A. L. 1984. Demonstration of peptide:N-glycosidase F activity in endo-beta-N-acetylglucosaminidase F preparations. *J Biol Chem*, 259, 10700–10704.

Powell, A. K. and Harvey, D. J. 1996. Stabilization of sialic acids in N-linked oligosaccharides and gangliosides for analysis by positive ion matrix-assisted laser desorption/ionization mass spectrometry. *Rapid Commun Mass Spectrom*, 10, 1027–1032.

Price, N. P. 2008. Permethylation linkage analysis techniques for residual carbohydrates. *Appl Biochem Biotechnol*, 148, 271–276.

Rendic, D., Wilson, I. B., Lubec, G. et al. 2007. Adaptation of the "in-gel release method" to N-glycome analysis of low-milligram amounts of material. *Electrophoresis*, 28, 4484–4492.

Royle, L., Dwek, R. A., and Rudd, P. M. 2006. Determining the structure of oligosaccharides N- and O-linked to glycoproteins. *Curr Protoc Protein Sci*, Chapter 12, Unit 12.6. doi:10.1002/0471140864.ps1206s43. http://onlinelibrary.wiley.com/doi/10.1002/0471140864.ps1206s43/abstract

Ruhaak, L. R., Deelder, A. M., and Wuhrer, M. 2009. Oligosaccharide analysis by graphitized carbon liquid chromatography-mass spectrometry. *Anal Bioanal Chem*, 394, 163–174.

Sagi, D., Peter-Katalinic, J., Conradt, H. S., and Nimtz, M. 2002. Sequencing of tri- and tetraantennary N-glycans containing sialic acid by negative mode ESI QTOF tandem MS. *J Am Soc Mass Spectrom*, 13, 1138–1148.

Seveno, M., Cabrera, G., Triguero, A. et al. 2008. Plant N-glycan profiling of minute amounts of material. *Anal Biochem*, 379, 66–72.

Seymour, J. L., Costello, C. E., and Zaia, J. 2006. The influence of sialylation on glycan negative ion dissociation and energetics. *J Am Soc Mass Spectrom*, 17, 844–854.

Stahl, B., Thurl, S., Zeng, J. et al. 1994. Oligosaccharides from human milk as revealed by matrix-assisted laser desorption/ionization mass spectrometry. *Anal Biochem*, 223, 218–226.

Takemori, N., Komori, N., and Matsumoto, H. 2006. Highly sensitive multistage mass spectrometry enables small-scale analysis of protein glycosylation from two-dimensional polyacrylamide gels. *Electrophoresis*, 27, 1394–1406.

Tao, N., DePeters, E. J., Freeman, S. et al. 2008. Bovine milk glycome. *J Dairy Sci*, 91, 3768–3778.

Tao, N., DePeters, E. J., German, J. B., Grimm, R., and Lebrilla, C. B. 2009. Variations in bovine milk oligosaccharides during early and middle lactation stages analyzed by high-performance liquid chromatography-chip/mass spectrometry. *J Dairy Sci*, 92, 2991–3001.

Tarentino, A. L. and Plummer, T. H., Jr. 1994. Enzymatic deglycosylation of asparagine-linked glycans: Purification, properties, and specificity of oligosaccharide-cleaving enzymes from Flavobacterium meningosepticum. *Methods Enzymol*, 230, 44–57.

Taylor, C. F., Binz, P. A., Aebersold, R. et al. 2008. Guidelines for reporting the use of mass spectrometry in proteomics. *Nat Biotechnol*, 26, 860–861.

Taylor, C. F., Paton, N. W., Lilley, K. S. et al. 2007. The minimum information about a proteomics experiment (MIAPE). *Nat Biotechnol*, 25, 887–893.

Thomsson, K. A., Schulz, B. L., Packer, N. H., and Karlsson, N. G. 2005. MUC5B glycosylation in human saliva reflects blood group and secretor status. *Glycobiology*, 15, 791–804.

Townsend, R. 1995. Analysis of glycoconjugates using high-pH anion exchange chromatography. In *Carbohydrate Analysis: High Performance Liquid Chromatography and Capillary Electrophoresis*, Rassi, Z.E. (ed) pp. 181–209. Amsterdam: Elsevier.

Tzeng, Y. K., Zhu, Z., and Chang, H. C. 2009. Alkali-hydroxide-doped matrices for structural characterization of neutral underivatized oligosaccharides by MALDI time-of-flight mass spectrometry. *J Mass Spectrom*, 44, 375–383.

Vakhrushev, S. Y., Mormann, M., and Peter-Katalinic, J. 2006. Identification of glycoconjugates in the urine of a patient with congenital disorder of glycosylation by high-resolution mass spectrometry. *Proteomics*, 6, 983–992.

Varki, A., Cummings, R. D., Esko, J. D. et al. 2009. *Essentials of Glycobiology*, 2nd ed. New York: Cold Spring Harbor Laboratory Press.

Wada, Y., Dell, A., Haslam, S. M. et al. 2010. Comparison of methods for profiling O-glycosylation: Human Proteome Organisation Human Disease Glycomics/Proteome Initiative multi-institutional study of IgA1. *Mol Cell Proteomics*, 9, 719–727.

Wilson, N. L., Robinson, L. J., Donnet, A. et al. 2008. Glycoproteomics of milk: Differences in sugar epitopes on human and bovine milk fat globule membranes. *J Proteome Res*, 7, 3687–3696.

Wilson, N. L., Schulz, B. L., Karlsson, N. G., and Packer, N. H. 2002. Sequential analysis of N- and O-linked glycosylation of 2D-PAGE separated glycoproteins. *J Proteome Res*, 1, 521–529.

Woodward, H. D., Ringler, N. J., Selvakumar, R. et al. 1987. Deglycosylation studies on tracheal mucin glycoproteins. *Biochemistry*, 26, 5315–5322.

Wuhrer, M., Koeleman, C. A., Deelder, A. M., and Hokke, C. H. 2004. Normal-phase nanoscale liquid chromatography-mass spectrometry of underivatized oligosaccharides at low-femtomole sensitivity. *Anal Chem*, 76, 833–838.

Zaia, J. 2004. Mass spectrometry of oligosaccharides. *Mass Spectrom Rev*, 23, 161–227.

Zhang, H. and Aebersold, R. 2006. Isolation of glycoproteins and identification of their N-linked glycosylation sites. *Methods Mol Biol*, 328, 177–185.

Zhang, H., Singh, S., and Reinhold, V. N. 2005. Congruent strategies for carbohydrate sequencing. 2. FragLib: An MSn spectral library. *Anal Chem*, 77, 6263–6270.

7 Glycan-Based Arrays for Determining Specificity of Carbohydrate-Binding Proteins

Xuezheng Song, David F. Smith,
and Richard D. Cummings

CONTENTS

7.1 INTRODUCTION

The functions of many glycans in biological systems involve their interactions with relevant carbohydrate (glycan)-binding proteins (CBPs or GBPs). Protein–glycan interactions are therefore extremely important and dependent on the structures of both proteins and glycans. Glycan microarrays (Feizi et al. 2003; Horlacher and Seeberger 2008; Paulson et al. 2006; Smith et al. 2010; Stevens et al. 2006), in which many glycan structures are presented simultaneously in a single microscope slide or chip for interrogation with labeled GBPs and microorganisms, have become very popular during the last decade largely due to the public availability of a large glycan array developed by the Consortium for Functional Glycomics (CFG). This platform of over 600 structurally defined glycan targets is a high-throughput tool to quickly explore the glycan-binding specificities of GBPs and microorganisms. This chapter serves as an overview of the development and application of this widely used technique.

7.2 PREPARATION OF GLYCAN MICROARRAYS

To construct a glycan microarray, glycan structures are immobilized either covalently or noncovalently onto a solid surface using various methods, as shown in
Figure 7.1. Typical noncovalent immobilization techniques are based on hydrophobic interactions, where plastic surfaces such as multi-well microtiter plates (Alvarez
and Blixt 2006), nitrocellulose membranes, or glass slides coated with nitrocellulose
are used as hydrophobic surfaces for immobilization of hydrophobic biomolecules.
Glycans by themselves, however, are not sufficiently hydrophobic to adhere strongly
to these surfaces for assay with GBPs. Therefore, free reducing glycans or periodate-
treated reduced glycans are commonly derivatized with a lipid tail and the resulting
"neoglycolipids" are successfully immobilized on the hydrophobic surfaces (Feizi
et al. 1994; Liu et al. 2007). The glycan microarrays generated from these glycan
derivatives have been successfully applied in the analysis of many GBPs.

    Covalent attachment of glycans onto solid surfaces can be built on various chemistries.
In most cases, glycans are synthesized to have a specific functional group at the reducing
end or free glycans are modified to have such a group installed in them. Accordingly, the
solid surface needs to be specifically activated so that it can react with the modified glycan structures and retain these structures covalently. For example, if a glycan structure is
installed with a primary amino group, both N-hydroxysuccinimide (NHS) ester-activated
and epoxy-activated glass surfaces can be used for the immobilization of these glycans
(Blixt et al. 2004). Similarly, other chemistries have been used, such as thiol–maleimide
chemistry (Houseman et al. 2003; Park et al. 2004; Park and Shin 2002; Ratner et al.
2004), disulfide formation (Bryan et al. 2004), and azide–alkyne click chemistry (Fazio
et al. 2002; Krishnamurthy et al. 2010). A method of directly immobilizing free reducing
glycans using hydrazide- or aminoxy-modified glass surfaces has also been developed

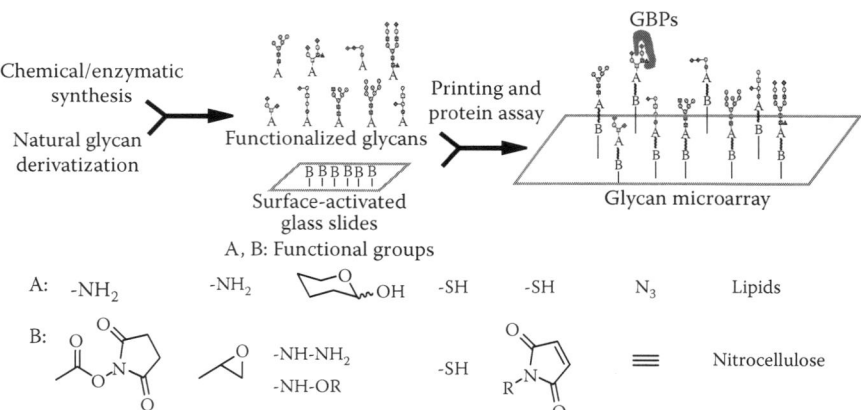

FIGURE 7.1 (**See color insert.**) The concept of glycan microarrays and the chemistry
involved in preparing glycan microarrays by covalent and noncovalent attachments. Glycans
are synthesized or derivatized so that a functional group A is presented usually at the reducing end. These functionalized glycans can be immobilized onto solid surfaces activated with
a corresponding functional group B covalently or noncovalently.

(Park et al. 2009). In principle, all these methods can be used to prepare a glycan micro-array, but each method has its unique advantages and drawbacks. In practice, it is more convenient to use activated glass surfaces that are commonly and commercially available as the substrate and focus on the chemistry of glycan modifications. The NHS-activated glass slides are commercially available from various vendors and are currently the most widely used surface chemistry, as developed by the CFG and used in thousands of analyses in publicly available databases. The reaction between primary amino group and NHS ester is well defined and used in the conjugation of two suitable biomolecules. It is therefore more widely used in the preparation of glycan microarrays.

7.3 GLYCAN MICROARRAYS FROM THE CFG AND THEIR APPLICATION IN DETERMINING GBP SPECIFICITIES

Although the chemistry of immobilization of glycan derivatives to activated glass surfaces is somewhat straightforward, the actual practice of preparing successful glycan microarrays is challenging. The success of a glycan microarray is critically dependent on the number, diversity, and relevance of the glycan library toward specific GBPs, antibodies, and microorganisms. It is, however, challenging to acquire a large and diversified glycan library, either through synthesis or by isolation from natural sources. The chemical and/or enzymatic synthesis of glycans is laborious, despite the development of many elegant synthetic methods, especially for complex glycan structures. The isolation of natural glycans is also hampered by their well-known heterogeneity. As a result, there are only a few glycan microarrays that can be used for general screening of GBPs, among which the glycan microarray from the CFG is the most widely used (Blixt et al. 2004) because it is publicly available and supported by the National Institutes of General Medical Sciences at no cost to investigators. This resource was established in 2004 and was gradually expanded into the current version comprising 611 glycan targets. In this microarray, most of the glycan structures were synthesized, either chemically or enzymatically. All the glycan structures possess a primary amino group that enables their direct immobilization onto NHS-activated glass slides. Due to the large and diversified glycan library, this glycan microarray has been used for determining binding specificities of many GBPs and microorganisms (Smith et al. 2010).

A typical glycan microarray experiment includes three steps: (1) Array printing, (2) assay with GBP, and (3) data analysis (as shown in Figure 7.2). To print glycan

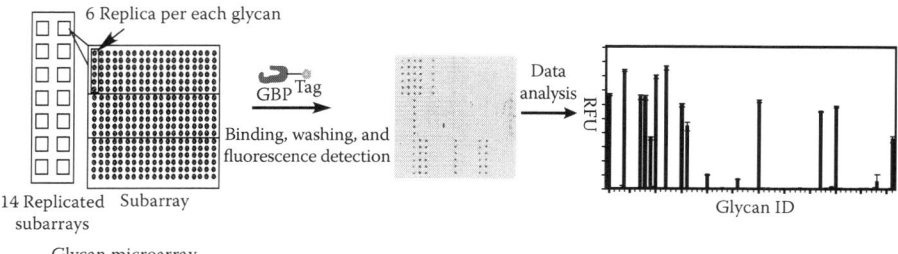

FIGURE 7.2   (See color insert.) Schematic of a glycan microarray experiment. A glycan microarray printed as 14 identical subarrays is used as an example.

microarrays, specialized instruments are required. Given the same glycan library and an appropriate substrate such as NHS-activated or epoxy-activated glass slides, the printing technologies can be categorized into contact printing and noncontact printing. In contact printing, printer pins dispense buffered glycan solutions onto the slides by physical contact with the slides. In noncontact printing, such as ink-jet and piezoelectric printing techniques, the printer head/pin does not contact the microarray slides. In piezoelectric printing, printer pins dispense buffered glycan solutions onto the slides using electronic pulses. Although contact printers can print many slides in a short time, they are usually less accurate than noncontact printers and require more quantity of sample for working. Piezoelectric printing, although generally slower than contact printing, is more accurate and conserves sample. It is more desirable to use noncontact printing in small batches when sample amounts are very limited, as in the case of many natural complex glycans. After the glycan solutions are spotted on surface-activated glass slides, which are then incubated, chemical reactions occur to immobilize the glycan structures onto the slides. A quenching process to destroy the remaining activated functional groups and a washing procedure to remove by-products and any unreacted glycan complete the preparation of the glycan microarray. After printing, the glycan microarray can be used immediately or stored desiccated at a controlled temperature. Most printed glycan arrays are stable for a year or more when stored desiccated at room temperature.

The binding assay of GBPs or microorganisms on a microarray is straightforward. Microarray slides are hydrated and equilibrated in the same buffer system in which the protein solution is prepared. The protein or microorganism is usually prelabeled either directly with fluorescent tags or indirectly with a tag such as biotin, which is used to introduce fluorescence through a secondary incubation process. For antibodies, many fluorescent secondary antibodies are commercially available making prelabeling unnecessary. The incubation of a protein solution with the microarray slides should be controlled by temperature and time according to the specific protein. If a fluorescent protein is used, no secondary incubation is needed and incubate in the dark to avoid any loss of fluorescence. If a nonfluorescent protein is used, secondary incubation is carried out to introduce fluorescence into the bound protein molecules. After washing, followed by a final washing with water to remove salt from the buffered solutions, excess water is removed from the microarray slides by centrifugation or using a gentle stream of nitrogen. After air drying, the slides are scanned by a microarray fluorescence scanner to generate a fluorescence image from which data are processed. After assay with protein, these slides can be stored only for a limited amount of time depending on the type of fluorescence carried by the protein. The fluorescence image of the array is processed by image-processing software, which is available from most scanner manufacturers. Each fluorescent spot represents a glycan structure that was specifically bound by the protein. According to the original microarray .GAL file, which is generated during the printing process and stores information on all the spots printed on the microarray, the fluorescence image is converted to a data file generally in the form of a Microsoft Excel file relating glycan to the fluorescent signal, which is generally expressed in tables and as a histogram. The histogram's x-axis represents the individual glycans and y-axis represents the binding strength in relative fluorescence units (RFUs). Microarray data are rich in information, and further sorting and comparison of data among different glycan structures and corresponding RFUs is often required for accurately determining the binding specificities of GBPs, including analysis of bound and related unbound structures.

7.4 GLYCAN MICROARRAY PREPARED FROM NATURAL GLYCOCONJUGATES

The 611 glycans on version 5.0 of the CFG glycan microarray evolved through several versions over several years as a result of chemical and enzymatic synthesis with confirmation of structure by mass spectrometry (MS) and/or nuclear magnetic resonance (NMR). It is well known that the synthesis of oligosaccharides is an extremely challenging and labor-intensive process, despite there being continuous development of new chemical and enzymatic approaches. The expansion of the glycan library that supports large glycan arrays such as the one maintained by the CFG becomes more difficult as more complex glycan structures are needed. On the other hand, nature represents the largest glycan library with potentially unlimited glycan structures. Natural glycans are also more biologically relevant. The challenge involved in dealing with natural glycans is the complexity of isolating them and the lack of structural analytical methods that can be applied to them. However, due to the quick evolution of chromatography and MS techniques, the isolation of a significant amount of structurally defined glycan structures is possible today.

To isolate natural glycans for a glycan microarray, several problems need to be addressed. Glycans are relatively invisible to most spectrometric detection, making their monitoring during chromatographic separation inconvenient. To overcome this problem, fluorescent tags are introduced to the reducing end, which render the glycans fluorescent. These fluorescent glycan derivatives can be easily monitored during a high-performance liquid chromatography (HPLC) separation. Another problem is that natural glycans do not possess a specific functional group that can be used to immobilize them onto reactive solid surfaces. We and other researchers have developed multiple bifunctional fluorescent linkers that can facilitate the separation of natural glycans and the immobilization process (de Boer et al. 2007; Song et al. 2009a, 2009b, 2009c). We have applied this approach to different classes of glycoconjugates, as described in Figure 7.3.

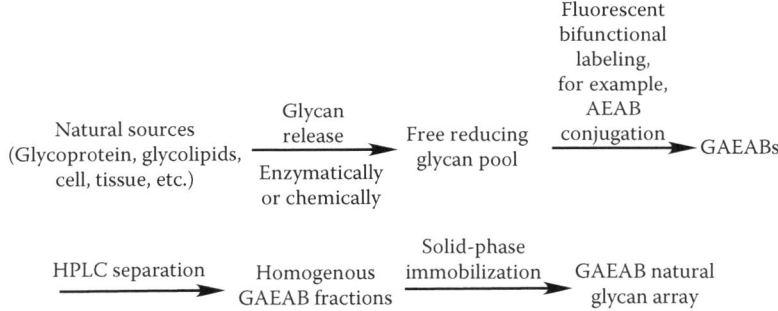

FIGURE 7.3 The natural glycan microarray approach. Natural glycans can be isolated and tagged for multidimensional HPLC separation. Isolated fractions can be printed as natural glycan microarrays for functional studies.

7.5 FREE GLYCANS RELEASED FROM N-GLYCANS

N-glycans are universally present in eukaryotic cells and known to play important roles in cell adhesion and signaling and protein quality control. Many of these functions are through their interactions with specific GBPs. Because of their relatively large size, the synthesis of N-glycans is a very labor-intensive process. On the other hand, various N-glycans can be specifically released by N-glycanase as free reducing glycans from common glycoproteins. To utilize such naturally occurring glycans for glycan microarrays, we have developed bifunctional fluorescent tags (Song et al. 2009c). These bifunctional fluorescent tags can be conjugated to free reducing glycans through reductive amination; these tags also possess another active amino group that is used for later immobilization of glycans onto reactive solid surfaces such as NHS-activated glass slides. Purified reducing glycans or a mixture of heterogeneous glycans can be tagged with these bifunctional fluorescent tags. Mixtures of labeled glycans can be purified or resolved into relatively homogeneous fractions after labeling by extensive chromatography, that is, two-dimensional (2D) chromatographic procedures. These separated fractions can be analyzed and defined using advanced MS and quantified based on fluorescence. The fluorescence is very important in not only monitoring the chromatography process but also quantifying glycans, especially when only microscale amounts of glycans can be isolated. The separated glycans can be adjusted to precise concentrations and directly immobilized onto microarray slides because of the existing active amino group, avoiding further functionalization. This approach minimizes the chemical treatment on a glycan pool to a single step and avoids undesired side reactions leading to possible structural degradation. The success of this approach is largely based on HPLC separation and MS analysis. The 2D HPLC based on different HPLC columns, including amino normal phase, Hypercarb, and C18 reverse phase, has been shown to successfully resolve the majority of fluorescent glycan conjugates. It is noted that MS and tandem MS analysis are also compatible with these glycan conjugates, which have increased sensitivity compared with underivatized glycans.

This approach has proved successful in determining binding specificities of certain GBPs. Using 2-amino-N-(2-aminoethyl)benzamide (AEAB) conjugates of glycan prepared and separated from commercially available glycoproteins, we have revealed the difference between binding specificities of galectin-1 and -3 (Song et al. 2009c). Other researchers, using 2-aminobenzoic acid and 2-aminobenzamide conjugates of glycans and by printing on epoxy-coated microarray slides, have successfully probed the specificities of several monoclonal antibodies (de Boer et al. 2007). The release of N-glycans by N-glycanase is easier to achieve than other classes of glycoconjugates; this approach can be generally applied to all free reducing glycans, including naturally occurring free glycans such as milk oligosaccharides and O-glycans released nonreductively.

7.6 GLYCOSPHINGOLIPIDS

Since glycosphingolipids (GSLs) are another important class of glycoconjugates playing important roles in cell adhesion and host–pathogen interactions, the incorporation of this class of glycans into glycan microarrays is of great interest. Although

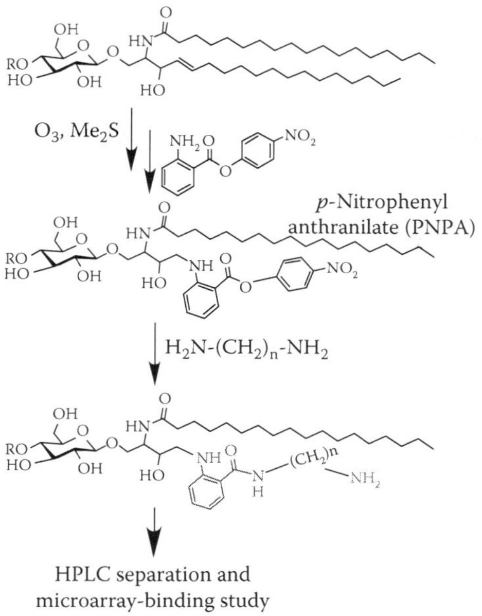

FIGURE 7.4 Derivatizations of GSLs: Ozone treatment of GSLs provides the aldehyde functional group. After reductive amination with the heterobifunctional linker PNPA and following diamine substitution, fluorescently and functionally tagged GSLs can be separated by HPLC for microarray preparation.

some of the GSL glycans have been synthesized, a systematic study of GSLs using glycan microarray requires the preparation of many more glycan structures from natural GSLs. Although endoglycoceramidases can be used to release free reducing glycans from GSLs, the specificities of these enzymes vary and their efficiency is generally low. Ozonolysis followed by base-catalyzed β-elimination releases the glycan portion; however, the glycans are usually reduced immediately to avoid base-induced degradation of the glycan at the reducing end, often referred to as "peeling" reactions. By using a commercially available chemical, such as p-nitrophenyl anthranilate (PNPA), as a bifunctional linker, we were able to fluorescently tag GSLs (Song et al. 2011), as shown in Figure 7.4. Similar to free glycan–AEAB conjugates, these derivatives have a primary alkylamine at the end of their hydrophobic tails, enabling direct solid-phase immobilization such as microarray printing. These derivatives can also be separated by 2D HPLC to nearly homogeneous fractions, which can be analyzed by MS and tandem MS.

7.7 OTHER CLASSES OF GLYCOCONJUGATES

The concept of building a tagged glycan library (TGL) based on natural glycans applies to other classes of glycoconjugates also. It is noted that O-glycans are another important class of glycans attached to glycoproteins. The release of O-glycans is more challenging than that of N-glycans as there are no O-glycanases available that can universally cleave O-glycans. The release of complex O-glycans relies on a chemical approach; it is most often base-catalyzed β-elimination. The released glycans under a basic condition are prone to peeling as mentioned in Section 7.6 for GSL-derived glycans and are often reduced to corresponding alditols by including $NaBH_4$ in the base elimination reaction. Nevertheless, there are nonreducing chemical methods for releasing O-glycans using hydrazine, ammonium hydroxide/ammonium carbonate, and so on. Once glycans with intact reducing ends are generated, they can be tagged, separated, and printed as described in Section 7.4 and Figure 7.3 for other microarrays.

Glycosaminoglycans (GAGs) are another class of glycans that pose a significant challenge in this field. The GAG chains are highly heterogeneous with respect to sulfate distribution and glucuronic/iduronic acid isomerization. Both chemical/enzymatic synthesis and separation of these glycans from natural mixtures are very difficult. In addition, MS analysis of GAGs is not trivial due to their many negatively charged sulfate groups. Nevertheless, GAG microarrays have been prepared from relatively small numbers of GAG structures either through synthesis or from natural sources (de Paz et al. 2006; Noti et al. 2006; Yamaguchi et al. 2006). Significant improvements have been made recently in the chromatographic separation and MS analysis of GAG oligosaccharides, and we anticipate that more comprehensive GAG microarrays will become available in the near future.

7.8 SHOTGUN GLYCOMICS

From the very beginning of the development of glycan microarrays, it has been assumed that structures of the glycans spotted on a microarray should be fully defined as the subsequent analyses of microarray data would generate a defined answer about the binding specificities of GBPs. Although this has been an extremely successful approach for defining GBP specificities, which provide important clues to the function of GBPs, this approach does not contribute directly to the discovery of novel protein–carbohydrate interactions. We recently developed a novel strategy for functional glycomics studies that we termed "shotgun glycomics" (Song et al. 2011), which is depicted in Figure 7.5. In this approach, natural glycans released from biological sources are fluorescently and functionally derivatized, as in glycan–AEAB conjugates and GSL–AOAB 2-amino-**N** -(2-aminooctyl)benzamide (AOAB) conjugates. These derivatives can be separated by multidimensional HPLC techniques to nearly homogeneous fractions and after a quick collection of MS data on all the fractions to obtain compositional data and partial sequencing, they are directly printed as microarray slides without complete sequencing, which is impractical in a timely manner. Biologically relevant proteins and microorganisms are then applied to the shotgun microarray, and the binding data are used to select biologically relevant glycans for further structural analysis. This microarray-facilitated approach addresses the fact that neither practical high-throughput synthesis nor practical high-throughput sequencing is available for glycomic analysis and allows the glycobiologist to focus his or her efforts only on functionally relevant structures. The application of this approach to a ganglioside mixture helped to define an interesting antibody to a novel glycan epitope that may be associated with Lyme disease. We anticipate that shotgun glycomics as a true functional glycomics approach (Song et al. 2011; Xia et al. 2005) will be more widely used in the future.

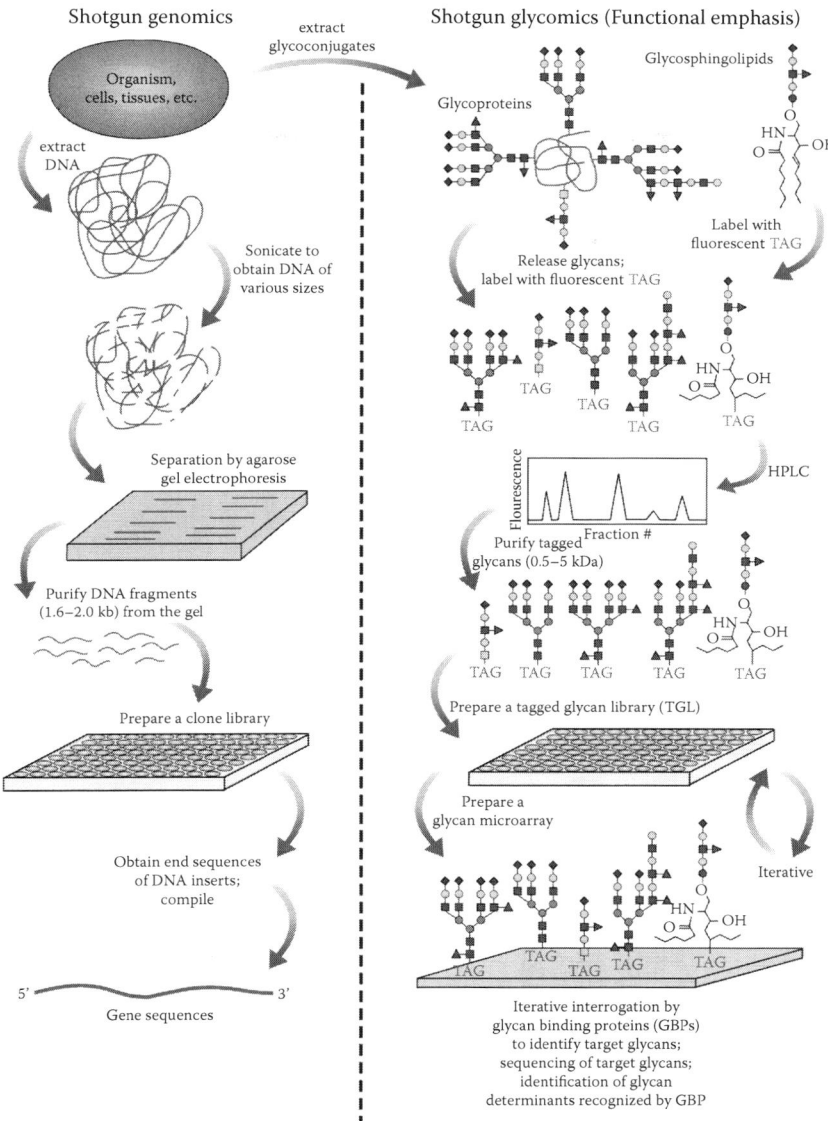

FIGURE 7.5 (See color insert.) Strategy of shotgun glycomics: In the left panel, shotgun genomics is described, in which gene sequences are obtained from cells or tissues as the starting material. Analogous to this method, we have developed shotgun glycomics (right panel). In this approach, natural glycans released from biological sources are fluorescently and functionally derivatized. These derivatives are separated by multidimensional HPLC techniques to near homogeneous fractions, and all fractions are subjected to MS to obtain compositional data and partial sequencing. A TGL is created from this material and the TGL is directly printed as shotgun glycan microarray slides for interrogation with biologically relevant proteins and microorganisms. Binding data are used to identify biologically relevant glycans for further structural analysis.

7.9 SUMMARY

Microarrays of defined glycans as a high-throughput tool for screening carbohydrate-binding specificities of proteins and microorganisms are very successful and continue to evolve in complexity. The utility of glycan microarrays is highly dependent on the size, diversity, and biological relevance of the glycans involved. The access to large numbers of complex glycans, through synthesis and/or isolation from natural sources, is essential for preparation of glycan microarrays and subsequent analyses. Shotgun glycomics, as a strategy to bypass laborious synthesis and sequencing of oligosaccharides, puts emphasis on the structures and functions of selected glycans. The preparation of glycan microarrays will continue to be an important tool in the study of functional glycomics.

REFERENCES

Alvarez, R. A. and Blixt, O. 2006. Identification of ligand specificities for glycan-binding proteins using glycan arrays. *Methods Enzymol*, 415, 292–310.

Blixt, O., Head, S., Mondala, T. et al. 2004. Printed covalent glycan array for ligand profiling of diverse glycan binding proteins. *Proc Natl Acad Sci USA*, 101, 17033–17038.

Bryan, M. C., Fazio, F., Lee, H. K. et al. 2004. Covalent display of oligosaccharide arrays in microtiter plates. *J Am Chem Soc*, 126, 8640–8641.

de Boer, A. R., Hokke, C. H., Deelder, A. M., and Wuhrer, M. 2007. General microarray technique for immobilization and screening of natural glycans. *Anal Chem*, 79, 8107–8113.

de Paz, J. L., Noti, C., and Seeberger, P. H. 2006. Microarrays of synthetic heparin oligosaccharides. *J Am Chem Soc*, 128, 2766–2767.

Fazio, F., Bryan, M. C., Blixt, O., Paulson, J. C., and Wong, C. H. 2002. Synthesis of sugar arrays in microtiter plate. *J Am Chem Soc*, 124, 14397–14402.

Feizi, T., Fazio, F., Chai, W., and Wong, C. H. 2003. Carbohydrate microarrays—a new set of technologies at the frontiers of glycomics. *Curr Opin Struc Biol*, 13, 637–645.

Feizi, T., Stoll, M.S., Yuen, C. T., Chai, W., and Lawson, A. M. 1994. Neoglycolipids: Probes of oligosaccharide structure, antigenicity, and function. *Methods Enzymol*, 230, 484–519.

Horlacher, T. and Seeberger, P. H. 2008. Carbohydrate arrays as tools for research and diagnostics. *Chem Soc Rev*, 37, 1414–1422.

Houseman, B. T., Gawalt, E. S., and Mrksich, M. 2003. Maleimide-functionalized self-assembled monolayers for the preparation of peptide and carbohydrate biochips. *Langmuir*, 19, 1522–1531.

Krishnamurthy, V. R., Wilson, J. T., Cui, W. et al. 2010. Chemoselective immobilization of peptides on abiotic and cell surfaces at controlled densities. *Langmuir*, 26, 7675–7678.

Liu, Y., Feizi, T., Campanero-Rhodes, M. A. et al. 2007. Neoglycolipid probes prepared via oxime ligation for microarray analysis of oligosaccharide-protein interactions. *Chem Biol*, 14, 847–859.

Noti, C., de Paz, J. L., Polito, L., and Seeberger, P. H. 2006. Preparation and use of microarrays containing synthetic heparin oligosaccharides for the rapid analysis of heparin-protein interactions. *Chemistry*, 12, 8664–8686.

Park, S., Lee, M.R., Pyo, S. J., and Shin, I. 2004. Carbohydrate chips for studying high-throughput carbohydrate-protein interactions. *J Am Chem Soc*, 126, 4812–4819.

Park, S., Lee, M. R., and Shin, I. 2009. Construction of carbohydrate microarrays by using one-step, direct immobilizations of diverse unmodified glycans on solid surfaces. *Bioconjug Chem*, 20, 155–162.

Park, S. and Shin, I. 2002. Fabrication of carbohydrate chips for studying protein-carbohydrate interactions. *Angew Chem Int Ed Engl*, 41, 3180–3182.

Paulson, J. C., Blixt, O., and Collins, B. E. 2006. Sweet spots in functional glycomics. *Nat Chem Biol*, 2, 238–248.

Ratner, D. M., Adams, E. W., Su, J. et al. 2004. Probing protein-carbohydrate interactions with microarrays of synthetic oligosaccharides. *Chembiochem*, 5, 379–382.

Smith, D. F., Song, X., and Cummings, R. D. 2010. Use of glycan microarrays to explore specificity of glycan-binding proteins. *Methods Enzymol*, 480, 417–444.

Song, X., Lasanajak, Y., Rivera-Marrero, C. et al. 2009a. Generation of a natural glycan microarray using 9-fluorenylmethyl chloroformate (FmocCl) as a cleavable fluorescent tag. *Anal Biochem*, 395, 151–160.

Song, X., Lasanajak, Y., Xia, B. et al. 2011. Shotgun glycomics: A microarray strategy for functional glycomics. *Nat Methods*, 8, 85–90.

Song, X., Lasanajak, Y., Xia, B., Smith, D. F., and Cummings, R. D. 2009b. Fluorescent glycosylamides produced by microscale derivatization of free glycans for natural glycan microarrays. *ACS Chem Biol*, 4, 741–750.

Song, X., Xia, B., Stowell, S. R. et al. 2009c. Novel fluorescent glycan microarray strategy reveals ligands for galectins. *Chem Biol*, 16, 36–47.

Stevens, J., Blixt, O., Paulson, J. C., and Wilson, I. A. 2006. Glycan microarray technologies: Tools to survey host specificity of influenza viruses. *Nat Rev Microbiol*, 4, 857–864.

Xia, B., Kawar, Z. S., Ju, T. et al. 2005. Versatile fluorescent derivatization of glycans for glycomic analysis. *Nat Methods*, 2, 845–850.

Yamaguchi, K., Tamaki, H., and Fukui, S. 2006. Detection of oligosaccharide ligands for hepatocyte growth factor/scatter factor (HGF/SF), keratinocyte growth factor (KGF/FGF-7), RANTES and heparin cofactor II by neoglycolipid microarrays of glycosaminoglycan-derived oligosaccharide fragments. *Glycoconj J*, 23, 513–523.

Section IV

Carbohydrates in Medicine

8 Structural Glycobiology
Applications in Cancer Research

Inka Brockhausen and Yin Gao

CONTENTS

8.1 INTRODUCTION TO TUMOR GLYCOSYLATION

Glycoproteins on cell surfaces and in secretions are heterogeneously *N*- and *O*-glycosylated, with mucins being extremely highly glycosylated, carrying a multitude of different *O*-glycan structures. Not only the expression of specific proteins but also the patterns of their glycans are usually altered in cancer and can differ between tissue origin, cell type, progression of tumors, stage of disease, and stage and site of metastasis. There are no rules that apply to all cancer cells, and often the expression of more than one gene is affected in the development of tumors. Some alterations are frequently observed in glycoproteins produced by cancer cells, such as the overexpression of MUC1 mucin and the appearance of short, truncated O-glycans (Brockhausen and Kuhns 1997; Brockhausen 1999). Other alterations may be specific for one or more types of cancer and could be useful for diagnosis and the development of individualized, targeted treatment.

Many glycan structures or their biosynthetic enzymes, glycosyltransferases, have been suggested to be markers of specific tumor cells. Methods are currently designed to analyze very small amounts of body fluids for these markers. If markers or characteristic patterns of glycans can be shown to be specific for a certain cancer type or for a stage of the disease, they would be extremely helpful in delineating the risk and guiding timely treatment. Markers could also be exploited in developing therapy that targets mainly the tumor cell but not normal cells. Detailed structural analyses and glycomics have identified great individual variations between glycan structures and their complexity in tumor cells. Abnormal glycan patterns can arise through a multitude of mechanisms involving altered glycoprotein biosynthesis. Understanding the complex control mechanisms of glycan assembly as well as the functions of specific structures will help to identify the importance of observed alterations in cancer. In a number of in vitro and animal cancer models, the biological and physiological effects of specific glycans and biosynthetic enzymes have been determined. It remains to be shown if these findings can be translated to the pathophysiology of human tumors. The goal is to design methods to assess specific cancer-associated structures in a patient's tumor and personalize the therapeutic efforts.

8.2   FUNCTIONS OF GLYCOPROTEIN-BOUND GLYCANS

There are usually multiple changes of glycosylation in cancer, due to dysregulation of glycosyltransferase expression and activities. The alteration of one enzyme can cause competing enzymes to prevail and pathways to be reshunted, resulting in a rearrangement of glycosylation patterns. The transcriptional control mechanisms are not well understood. In addition, the roles of hundreds of different glycan structures of glycoproteins, in context with their various attachment sites, are only slowly beginning to emerge. For example, sialic acid residues, present in excess on many cancer cells, have multiple functions in cell adhesion and in the immune system, which can affect cancer cell physiology. In order to understand the significance of glycosylation changes in cancer, it is important to consider the normal functions of specific carbohydrate structures.

8.2.1   Role of Glycans in Glycoprotein Functions

On mucous membranes, highly O-glycosylated mucins serve to protect and lubricate the underlying cells and suppress tumor development. MUC1 is an O-glycosylated cell-surface-bound mucin that is overexpressed in most epithelial cancer cells and may also serve to protect the cancer cell but, in addition, has been shown to play a role in the proliferation of tumor cells. Mucins, as well as glycoproteins on cell surfaces and in the extracellular matrix, are critically involved in cell–cell interactions and cell surface functions and can be linked to signaling cascades. The roles of specific glycans are slowly emerging and may include intracellular targeting, secretion, protein conformation and stability, oligomerization, and ligand binding.

Cell surface receptors for growth factors (cytokines) and apoptosis-inducing factors are involved in regulating cell growth, proliferation, and cell death of cancer cells. Many of these receptors have been identified as glycoproteins. The N-glycans can serve in controlling the cell surface expression, conformation, and ligand binding of receptors and may regulate the induction of signaling cascades (Li et al. 2007; Amith et al. 2010). Cancer cells are often hypersialylated, and sialic acid is involved in controlling growth factor receptor endocytosis and signaling. When sialic acid was removed from the cell surface of neuroblastoma cells by transsialidase or neuraminidase treatment, signaling and neurite outgrowth was induced (Amith et al. 2010). It is thus possible that sialic acid prevents receptor oligomerization as the prerequisite for signaling and thus has an important controlling function on cell behavior.

A number of mucins and glycoproteins have been suggested as cancer markers that can be detected in serum; for example, MUC16 (CA125) (Diamandis 2010) or the pan-cancer marker CA215 (Lee and Ge 2010). Antibodies against these markers may possibly be useful in targeting tumor cells and reducing tumor burden.

8.2.2 ADHESION AND INVASION OF TUMOR CELLS

Tumors are composed of a variety of cell types including vascular and inflammatory cells, fibroblasts and stromal cells, as well as a complex extracellular matrix network. This cancer cell environment provides nutrients, growth factors, chemoattracting factors and adhesive proteins, and determines cell growth and migration. The interactions between cancer cells and their environment mediate invasion, tumor progression, and metastasis.

Tumors invade when they lose the ability for intercellular adhesion. Cancer cells then migrate through the basement membrane and extracellular matrix containing glycoproteins and proteoglycans. Those cancer cells that have detached from the primary tumor and reach the circulation survive in the blood if they have the structures that block their uptake or removal by the immune system. These cells could then attach to the endothelium in distant capillary cells, migrate through the endothelial layer and form metastases. Cancer cell glycosylation, and specifically glycans that are ligands for selectin–cancer cell interactions, are critically involved in cell migration and invasion.

Glycosaminoglycan-containing proteins (proteoglycans) on cell surfaces and in the extracellular matrix can bind and store growth factors. Tumors secrete proteases and other hydrolases, and upon degradation of proteins, proteoglycans, and glycosaminoglycans surrounding tumor cells, growth factors are released and regulate cell growth and angiogenesis (Vlodavsky et al. 1990). Excess amounts of the large polysaccharide hyaluronic acid (HA) are often found in tumor environments (Simpson and Lokeshwar 2008), which can contribute to tumor progression. HA-binding proteins such as CD44 and RHAMM have been shown to regulate cell migration, invasion, and angiogenesis (Itano and Kimata 2008). The HA-binding proteoglycan versican contributes to cell motility, invasion, and metastasis of ovarian cancer cells (Ween et al. 2011). The balance between HA synthesis by HA synthases and degradation by hyaluronidases is often shifted in tumors leading to higher amounts of HA and altered tumor growth. The specific effects of HA depend on the tumor and have been shown to vary between tumor models (Itano and Kimata 2008). Degradation of the cell surface proteoglycan heparan sulfate by heparanase contributes to neovascularization, tumor progression, and metastasis. The field of glycosaminoglycans is complex but important for cancer and metastasis, and has been dealt with in recent reviews (Itano and Kimata 2008; Barash et al. 2010).

8.2.3 Selectins in Cancer

Many mammalian lectins are involved in cell–cell interactions, in the immune system, and in tumor development. Selectins, expressed on endothelium, leukocytes, or platelets, are lectins that promote metastasis through adhesion to Sialyl-Lewis x (SLe^x), sialylα(2–3)Galβ(1–4)[Fucα(1–3)]GlcNAcβ-, and related Lewis epitopes (Läubli and Borsig 2010; Brockhausen and Kuhns 1997) that can be attached to the nonreducing termini of O- or N-glycans of glycoproteins. Leukocytes utilize the selectin-binding mechanism in the process of homing and inflammation. Invasive and metastatic cancer cells often express increased amounts of SLe^x, and the inflammatory effect of cancer cells can induce the expression of selectins. This facilitates the attachment of cancer cells to the endothelium and contributes to the invasion and metastasis of tumor cells. Increased SLe^x expression is therefore associated with a poor prognosis in colon cancer and other tumors (Nakamori et al. 1993; Fujita et al. 2011). Mucins and other glycoproteins expressed by epithelial cancer cells can carry SLe^x and thus form ligands for selectins. Selectins may also bind to sulfated glycosaminoglycan chains of cancer cells, and sulfation of Lewis antigens can increase their binding affinity to selectins (Monzavi-Karbassi et al. 2007).

The expression of Lewis antigens is often abnormal in cancer. For example, normal pancreatic tissues express Lewis y, Fucα(1–2)Galβ(1–4)[Fucα(1–3)]GlcNAcβ-, while cancer tissues express Lewis x, Galβ(1–4)[Fucα(1–3)]GlcNAcβ- (Kim et al. 1988). In breast, colon, and prostate cancer cells, the expression of the selectin ligand SLe^x is associated with the degree of aggressive disease and a poor prognosis and plays an essential role in cancer cell migration (Idikio 1997; St Hill et al. 2009; Nakamori et al. 1993). SLe^x contributes to the metastasis of B16 melanoma cells (Chen and Fukuda 2006). Metastasis could be prevented by sulfated oligosaccharides that block heparanase and P-selectin (Borsig et al. 2011). E-selectin is expressed on activated endothelium and is overexpressed during the colonization stage of metastases. Thus, a downregulation of selectin expression, expression of the selectin ligand, or inhibition of E-selectin-mediated adhesion could reduce cancer cell adhesion, invasion, and metastasis (Kojima et al. 1992; Brown et al. 2006, 2009). These approaches have been partially successful in vitro, and in animal models (see Section 8.5), but need further development to effectively target cancer cells and not immune cells.

8.2.4 Galectins

Mammalian lectins have important functions in cell–cell interactions, cell growth, and apoptosis and therefore play important roles in the biology of cancer cells. Galectins are soluble lectins with carbohydrate recognition domains that bind Gal residues, mostly at the nonreducing ends of glycan chains, exposed to the external environment of the cell. Galectins can link Gal residues due to their multivalent or oligomeric nature and can affect cell adhesion as well as the conformation and signaling of cell surface receptors. Galectins that have more than one carbohydrate recognition domain can form lattices and cross-link glycoproteins and proteoglycans, thus affecting the mobility and functions of cell surface molecules. The binding of galectins depends on the number of exposed Gal residues on N- and O-glycans available for binding. Thus, a change of exposed Gal residues due to a derangement of synthesis, masking of Gal by sialic acid residues, or altered expression of highly glycosylated proteins may alter the cell's ability to respond to galectins. The higher level of sialylation in cancer therefore diminishes galectin binding (Toscano et al. 2007). The apoptosis-inducing galectin-1 may eliminate immune cells that control tumor growth. In mice, galectin-1 contributes to the escape of tumor cells from the immune system by downregulating the immune response and leads to increased tumor survival and aggressive metastases (Stannard et al. 2010). Thus, blocking galectin binding could be an additional therapeutic strategy for cancer.

Galectin-3 is a chimeric monomeric lectin and a promising anticancer target. It is widely distributed among normal and neoplastic cell types; it interacts with glycoproteins and glycolipids containing Gal exposed extracellularly, and its expression correlates with metastasis (Takenaka et al. 2004; Yu et al. 2007). Galectin-3 can bind to MUC1 carrying the cancer-associated T antigen, Galβ(1–3)GalNAc, and facilitate selectin-dependent adhesion to endothelial cells. Serum levels of galectin-3 are increased severalfold in serum from breast, gastrointestinal, and lung cancer patients (Yu et al. 2007). Its expression correlates with tumor progression in anaplastic large cell lymphoma, head and neck squamous cell carcinoma, and thyroid and gastric carcinoma. However, its expression has also been reported to negatively correlate with breast, ovarian, and prostate cancer (Takenaka et al. 2004). These apparent discrepancies may be linked to a wide range of cellular effects of galectin-3 that include cell growth and differentiation, cell adhesion, angiogenesis, and apoptosis. Galectin-3 appears to bind to the Gal residues of N-glycans (Srinivasan et al. 2009) and promote adhesion of B16 melanoma cells to fibronectin and lung metastasis in mice. Thus, metastasis of B16 melanoma cells may follow selectin-independent paths (Zhang et al. 2002).

Galectin-8 is a tandem repeat-containing lectin with two Gal-binding domains. It is also widely distributed and occurs as a number of different isoforms. In some cancer types, its expression correlates with malignancy or the degree of differentiation. Galectin-8 expression in colon cancer has been shown to be lower than normal, but the expression in other tumors varies greatly and depends on the cancer cell type or tumor type (Bidon-Wagner and Le Pennec 2004). Targeting specific galectin–glycoprotein interactions may be successful in controlling tumor growth, based on our knowledge of galectins and the biosynthesis of Gal-containing glycoconjugates.

8.3 CANCER-SPECIFIC GLYCAN STRUCTURES AND MARKERS

Many of the glycoprotein-bound extended and terminal glycan structures and epi-topes are shared between *N*- and *O*-glycans of glycoproteins and glycolipids; for example, Lewis epitopes (Figure 8.1). The core sugar residues, however, are specific for *O*-glycans or *N*-glycans. The expression of epitopes and structures of many gly-cans have been reported to be altered in cancer cells and tumors, as described below.

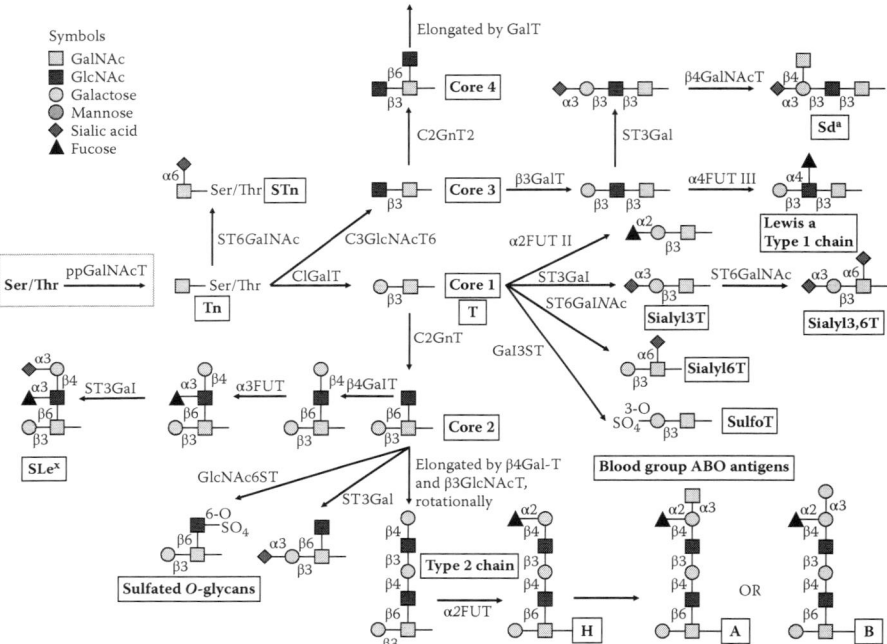

FIGURE 8.1   (See color insert.) Biosynthesis pathways of cancer-associated *O*-glycan struc-tures. The pathways of *O*-glycosylation are a complex, amazing maze, and start with GalNAc addition to Ser/Thr residues of glycoproteins. The figure shows only a few of the pathways rel-evant to cancer. Four major core structures (1–4) can be extended in various ways. Most of the glycosyltransferases that catalyze individual conversion steps (black arrows) exist as families of enzymes, with slightly different acceptor substrate specificities. This apparent redundant supply of enzymes ensures the synthesis of glycan determinants that could be functionally important. The expression levels of these enzymes, their relative activities, and localization in the Golgi and many other control factors determine the relative levels of glycans synthesized on a gly-coprotein. ppGalNAcT, polypeptide GalNAc-transferases; ST6GalNAc, α6-sialyltransferases; ClGalT, core 1 β3-Gal-transferase; C3GlcNAcT6, core 3 β3-GlcNAc-transferase, C2GnT, core 2 β6-GlcNAc-transferases, ST3Gal, α3-sialyltransferases; ST6GalNAc, α6-sialyltransferases; FUT, fucosyltransferases, Gal3ST and GlcNAc6ST, sulfotransferases; β3GlcNAcT, β3-GlcNAc-transferases;   β3GalT;   β3-Gal-transferases;   β4GalT,   β4-Gal-transferases;   β4GalNAcT, β4-GalNAc-transferases. Changes in the activities of early acting enzymes can drastically alter the patterns of *O*-glycans in cancer. Incomplete structures are found if any of the intermediate acting enzymes have low activities.

Although some of these have been labeled "markers," there is still an intensive search for markers or patterns of markers that are specific for a cancer cell type or stage of the disease.

A single GlcNAc residue linked to cytoplasmic and nuclear proteins (O-GlcNAc) has also received much attention since it may be critical for the regulation of gene transcription in cancer. Epithelial–mesenchymal transition of cancer cells is a differentiation event that leads to the characteristics of an aggressive cancer, and a poor prognosis, for example, in pancreatic cancer (Maupin et al. 2010). This transition is accompanied by many changes in the expression of glycoproteins, as well as glycosidases and glycosyltransferases, that shift the patterns of glycans of secreted and cell membrane-bound glycoproteins.

Essentially, all cancer types are associated with glycosylation changes, especially O- and N-glycans of glycoproteins and the O-GlcNAc modification of intracellular proteins. For example, the prostate produces many glycoproteins, and altered glycoprotein structures have been found in prostate adenocarcinomas (Peracaula et al. 2003, 2008; Tabarés 2006, 2007; Sayat et al. 2008; Meany et al. 2009; Dwek et al. 2010; Yoon et al. 2010; Sarrats et al. 2010; Fukushima et al. 2010; Premaratne et al. 2011). Abnormal glycans can change cancer cell antigenicity, facilitate the escape of cancer cells from the immune response, and increase the potential for metastasis (Hagisawa et al. 2005; Valenzuela et al. 2007).

N-Glycolyl-neuraminic acid (Neu5Gc) is known as a cancer-associated antigen. It cannot be synthesized in humans due to lack of CMP-Neu5Ac hydroxylase, but this sugar can be acquired from the consumption of animal products. Neu5Gc is antigenic in humans, and in cancer patients, increased levels of anti-Neu5Gc antibodies are found in the blood. In tumors, higher amounts of Neu5Gc are present on glycoproteins and glycolipids, possibly due to the fast growth of cancer cells that utilize the sialyltransferase donor substrate CMP-Neu5Gc originating from exogenous Neu5Gc (Devine et al. 1991; Malykh et al. 2001). Both the level of Neu5Gc in glycoconjugates and the corresponding antibodies may be useful markers for cancer.

8.3.1 O-GLYCAN STRUCTURES IN CANCER

O-Glycan structures in cancer cells are based on four major core structures (Figure 8.1) with cores 1 and 2 being the most common. Cores can be extended and terminated with a great variety of glycan structures. Alterations of O-glycans in cancer have been summarized in previous reviews (Brockhausen and Kuhns 1997; Brockhausen 1999). The Tn (GalNAcα-), sialyl-Tn (sialylα6GalNAc-, STn), and (sialyl)T antigens (Figure 8.1) are common in cancer cells. These small or truncated O-glycans are especially increased in advanced tumors (Itzkowitz et al. 1990; Springer et al. 1995) and are associated with a poor prognosis. T and Tn antigens are expressed in a tissue-specific fashion along the sections of the intestines and the colonic crypts. While sialyl-Tn increases in cancer, the mechanism may not necessarily involve increased synthesis but may be due to increased exposure due to decreased O-acetylation of sialic acids. A high proportion of sialic acids is normally O-acetylated, but in cancer the amount of O-acetylation is reduced. This leads to a higher degree of nonacetylated sialic acids and unmasking of epitopes that are normally recognized by sialic-acid-binding proteins or antibodies (Jass and Walsh 2001). Similarly, increased SLe^x expression in cancer may be due to decreased O-acetylation of sialic acids, or conversely, increased removal of the 9-O-acetyl ester by sialate-O-acetylesterase (Shen et al. 2004).

Simple O-glycans, Tn, T, and sialyl-T antigens are variably expressed during human sperm development (Rajpert-De Meyts et al. 2007). Thus, spermatids and spermatozoa contain Tn but not sialyl-Tn antigens. The Tn antigen appears to be converted to sialyl-Tn in adult testes. The pattern of antigen expression is different in testicular germ cell tumors, and both sialyl-Tn and sialyl-T are abundant. STn expression plays an important role in gastric cancer progression and the development of gastric carcinoma cells to malignant cell types (Pinho et al. 2007). STn may also protect cancer cells from natural killer cell lysis (Ogata et al. 1992; Blottiere et al. 1992).

Mucin-type O-glycans are the main carriers of blood group ABO antigens, and changes in these blood group structures are common in cancer (Brockhausen and Kuhns 1997). Terminal blood group epitopes in glycoconjugates may be lacking in cancer, or incompatible blood groups may be expressed. Changes in blood groups in cancer are inconsistent and include the ABO antigens, P, Sd^a/Cad, and Lewis antigens. For example, the blood group Sd^a epitope, sialylα(2–3)[GalNAcβ(1–4)] Galβ-, attached to core 3 (Figure 8.1), is missing in some colon cancer cells (Capon et al. 2001). In gastric cancer, there is a dramatic downregulation of the β4GalNAcT that synthesizes the Sd^a epitope (Dohi et al. 1996; Dohi and Kawamura 2008). Differentiated colon cancer cells Caco-2 secrete a soluble form of the enzyme.

Leukemia cells are often highly sialylated (Brockhausen and Kuhns 1997), which alters their cell surface functions, adhesive properties, and survival in the blood. Cancer glycoproteins, and especially those from metastatic cells, generally have higher amounts of sialic acid (Bresalier et al. 1996), and the ratio of 2,3- to 2,6-linked sialic acids is often changed. Many tumors, for example, from colon, liver, cervix, pancreas, and prostate are associated with increased amounts of SLe^x, but the levels of other Lewis antigens may also be altered. These selectin ligands may be attached to the O-glycan core 2 structure of mucins or mucin-like molecules and are instrumental in attaching cancer cells to the endothelium. Metastatic cells especially can express high levels of selectin ligands. These changes are expected to have a significant impact on the biology of the cancer cells.

8.3.2 Mucins in Cancer

MUC1 is a major mucin, overexpressed on the cell surfaces of epithelial cancer cells. Because of its very large size, MUC1 may interfere with cell–cell recognition events, although the O-glycans may possess ligands for cell–cell recognition and adhesion. The glycosylation patterns of MUC1 and other mucins have been shown to be abnormal in cancer (Mungul et al. 2004; Hollingsworth and Swanson 2004; Marcos et al. 2004) and often consist of simpler O-glycans (T and Tn antigens; Figure 8.1). MUC1 is involved in immunosuppression, cell adhesion and other cell surface functions (Hollingsworth and Swanson 2004; Tarp and Clausen 2008). Other mucins can also be expressed at abnormal levels in cancer and may serve functions that are similar to those of MUC1. Since the O-glycans are the predominant and exposed structures, roles of mucins are mainly attributed to the properties of O-glycans.

The altered expression, glycosylation patterns, or antigenicity of tumor mucins are often related to disease progression. MUC1, MUC3, MUC4, MUC5B, and MUC5AC are highly expressed in pancreatic cancer tissue (Balagué et al. 1995; Hollingsworth and Swanson 2004; Tarp and Clausen 2008). Mucin antigens CA19-9 and DU-PAN-2 are found in a number of tumors, including pancreatic cancer (Magnani et al. 1983). Mucin antigens are found in the serum (Metzgar et al. 1984) but may originate from cell surface mucins. For example, MUC1 not only is highly expressed but also has abnormal antigenic properties in pancreatic cancer cells, and is cleaved from the cell surface. The cell surface mucin MUC4 that is overexpressed in pancreatic cancer interacts with receptor tyrosine kinase HER2 in pancreatic adenocarcinoma cells, thus regulating cell signaling and potentially influencing the progression of cancer (Chaturvedi et al. 2008).

Prostate adenocarcinomas produce membrane-bound and secreted mucins that are altered both in expression levels and in structure (Cozzi et al. 2005; Li and Cozzi 2007; Premaratne et al. 2011). MUC1 is underglycosylated in prostate cancer but is overexpressed and accelerates cancer cell progression (Burke et al. 2006; Li and Cozzi 2007; Garbar et al. 2008). O-Glycan structures of MUC1 in prostate cancer cells include core 2 oligosaccharides (Figure 8.1) (Premaratne et al. 2011). Core 2 forms the scaffold structure for SLe^x and may play a critical role in controlling the adhesion of cancer cells to the endothelium (Hagisawa et al. 2005). A high content of shorter core 1 structures, as well as sialylated core 1 structures, were typically found in breast cancer MUC1 (Premaratne et al. 2011). The gel-forming secreted mucin MUC2 is normally highly glycosylated and expressed in the colon and other tissues. Its expression is significantly lower in nonmucinous adenocarcinoma than in mucinous adenocarcinomas of the prostate (Osunkoya et al. 2008).

Mucins in gastrointestinal cancers appear to contain less carbohydrate and increased amounts of Tn, STn, and T antigens. The degree of sulfation is also decreased in colon cancer (Brockhausen 1999), and instead, there is increased sialylation of mucins (Yamori et al. 1987). The reciprocal relationship between sialylation and sulfation may be explained by the competition of sialyl- and sulfotransferases for the same acceptor substrates. Altered sulfation may also indicate changes in sulfate transport, synthesis and transport of sulfate donor substrate PAPS, or may be due to altered glycosylation and changes in substrate structures. In human colonic adenoma cells, mucin sulfation and sulfotransferase activities toward O-glycans were lower in cells progressing to the tumorigenic phenotype (Vavasseur et al. 1994). In colon tumors, sulfotransferase activities were also lower than in normal colon tissue, which suggests that sulfotransferases are downregulated in colon cancer (Yang et al. 1994).

The expression of mucin antigens such as Lewis antigens has a gradient along the length of the normal intestines, and along the colonic crypts (Orntoft et al. 1990). These antigens are markers of poorly differentiated adenocarcinomas, as well as metastatic tumors, and are related to a poor clinical outcome. The antigenicity of MUC1 is also altered in ovarian cancer, and MUC1 is overexpressed. CA125, isolated from ovarian tumor cells, is another glycoprotein that has mucin-type O-glycans with core 2 structures and Lewis x and y determinants. Although CA125 is considered a pan-tumor marker, it is also found in benign ovarian disease (Lloyd et al. 1997). These highly glycosylated proteins are useful for diagnosis and have shown some success in cancer vaccines and to specifically target cancer cells that overexpress and abnormally glycosylate mucins.

8.3.3 O-GlcNAc in Cancer

A single GlcNAc residue (O-GlcNAc) β-linked to the hydroxyl of Ser/Thr residues of cytoplasmic and nuclear proteins is an important regulator of protein stability and function (Mi et al. 2011; Hart et al. 2010; Slawson et al. 2010). Phosphorylation of Ser/Thr residues competes with O-GlcNAc modification; thus, O-GlcNAc regulates phosphorylation, gene transcription and cell growth. Variable levels of O-GlcNAc modification of intracellular proteins have been found in cancer. O-GlcNAc can be detected with anti-O-GlcNAc antibody, and the amounts of O-GlcNAc are increased in breast cancers and other tumors and are especially high in metastatic tissues. O-GlcNAc was shown to stabilize proteins such as p53 tumor suppressor (Yang et al. 2006), which could be beneficial for the induction of apoptosis and cell growth arrest in tumors. However, tumor invasion and metastasis may be enhanced via the O-GlcNAc modification of transcription factors (Caldwell et al. 2010; Gu et al. 2010). The transcriptional regulator C-Myc is glycosylated with O-GlcNAc mainly at Ser58, which is also a phosphorylation site and a common site of mutation in lymphomas (Chou et al. 1995). In Ewing's sarcoma, O-GlcNAc may be involved in transcriptional activation of the oncogene *EWS-FLI1*, which appears to have the O-GlcNAc modification at the amino-terminal transcriptional activation domain (Bachmaier et al. 2009). Beta-catenin is involved in the regulation of intracellular Wnt signaling pathways and gene expression and is also modified by both O-GlcNAc and phosphorylation. The amount of O-GlcNAc is inversely related to the transcriptional activity of beta-catenin in prostate cancer cells (Sayat et al. 2008). The anchorage-independent growth of lung and colon cancer cells is enhanced by the O-GlcNAc modification of lung cancer cells A549, H1299, and colon cancer cells HT29. O-GlcNAc contributes to increased breast tumor growth, invasion, and metastasis (Mi et al. 2011; Caldwell et al. 2010).

The enzyme that adds O-GlcNAc to proteins is O-GlcNAc-transferase (OGT). Breast cancer tissues show increased OGT expression (Caldwell et al. 2010). The balance between synthesis and degradation appears to be disturbed in cancer and may be due to increased expression of OGT (Mi et al. 2011). While OGT adds GlcNAc to Ser or Thr residues of proteins in the cytoplasmic compartment, an N-acetylglucosaminidase (OGA) specifically removes the O-GlcNAc residue. OGA inhibitors (e.g., PUGNAC) can therefore achieve high intracellular levels of O-GlcNAc. Staining of lung and colon cancer tissues with an anti-OGT antibody showed that the expression of OGT was increased in tumors, while OGA showed no significant change of expression (Mi et al. 2011). OGT may therefore be a useful therapeutic target in some cancer types. The significance in cancer of site-specific O-GlcNAc residues is not yet known, but knowledge of the attachment and dynamics of O-GlcNAc at specific sites may add to our understanding of the regulation of tumor growth.

8.3.4 N-Glycan Structures in Cancer

Blood groups and other epitopes can be attached to the nonreducing end of *N*-glycan antennae of glycoproteins, and these epitopes can be overexpressed in cancer. The number of antennae directs the lengths, structures, and functional properties of *N*-glycans (Brockhausen, Schutzbach et al. 1998) (Figure 8.2). *N*-Glycans are important in the regulation of cell adhesion. The display of multiple possible ligands can increase the affinity and lattice formation of lectins that regulate cell adhesion, motility, cell growth, and apoptosis (Lagana et al. 2006). The most commonly observed changes in cancer cell glycoproteins are in the ratios of bi-, tri- and tetraantennary structures, as well as in core α6-fucosylation (Li, Song et al. 2010) (Figure 8.2). For example, the common serum cancer marker carcinoembryonic antigen, which is involved in cell adhesion, has highly branched and extended *N*-glycans. The antennary GlcNAc residues can all be extended and carry similar glycans that may be involved in regulating cell adhesion. The β1,6 antenna of tri- and tetraantennary

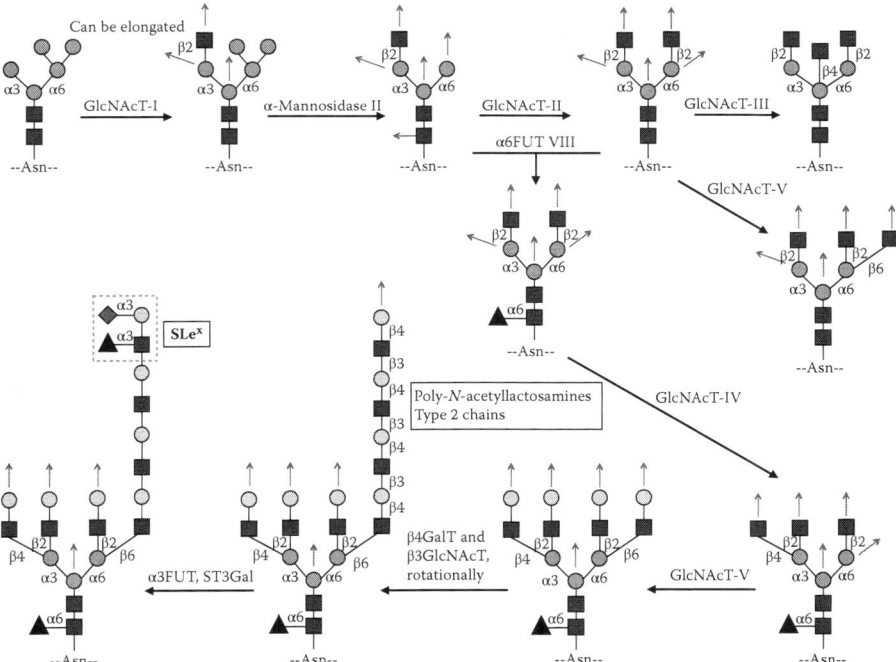

FIGURE 8.2 (See color insert.) *N*-Glycosylation pathways in cancer cells. After the assembly of a large oligosaccharide on dolichol in the cytoplasmic and endoplasmic reticulum (ER) compartments and the transfer to Asn residues of proteins in the ER, sugar residues are removed by processing glucosidases and mannosidases in the ER and Golgi. The figure shows pathways of interest in cancer. SLe^x or other terminal epitopes can be synthesized similarly to the assembly of *O*-glycans. Terminal sialylα(2–6)Gal is found in increased amounts in colon cancer and is added by ST6Gal. These termini are not found on *O*-glycans. The symbols and enzyme names are explained in Figure 8.1. Short gray arrows indicate possibilities of sugar additions and extensions.

N-glycans may differ from other antennae in their flexibility due to rotation of the 1,6 bond. Having an additional third or fourth antenna may make a significant difference in the bulkiness of glycans that may be antiadhesive and may mask or stabilize the underlying protein structure.

The extensions of the β1,6 antenna appear to block cell adhesion and are related to metastasis in human hepatoma cells (Guo et al. 2000) and in other types of cancer cells. The β1,6 antenna also contributes to loss of contact inhibition and reduced cell–extracellular matrix interactions in melanoma cells (Przybylo and Litynska 2011).

The N-glycans of the cell adhesion molecule E-cadherin regulate its adhesion function. While bisected N-glycans (Figure 8.2) caused increased cell–cell adhesion, consistent with decreased metastatic properties, the β1,6 antennae appear to cause the opposite effect, thus contributing to decreased cell adhesion and tumor cell invasiveness (Pinho et al. 2009). The mechanism of the effect of bisected N-glycans may involve control of glycoprotein expression and turnover, while the long extended glycans based on the β1,6 antenna have a direct antiadhesive effect. Other adhesion molecules are also affected by the structures of their N-glycans. For example, the adhesive functions and ligand binding of integrins are controlled by the structures of their N-glycans. In B16 melanoma cells, bisected GlcNAc residues reduced integrin $\alpha_5\beta_1$-mediated adhesion to fibronectin (Isaji et al. 2004).

In several cancer types, increased N-glycan core α6-fucosylation is seen (Li, Song et al. 2010). The Fucα(1–6)GlcNAc structure is a marker for hepatocellular carcinomas although other cancer types also show an increase in core fucosylation (Lin et al. 2011; Miyoshi, Shinzaki et al. 2010; Miyoshi, Ito et al. 2010). In mice, core Fuc appears to regulate antigenicity, growth factor receptor functions, and cell growth (Zhao et al. 2008). N-Glycans from metastatic tumors of lung and breast express high amounts of SLe^x (Arnold et al. 2011). Abnormalities can also be detected by glycomic and glycoproteomic analysis of patient's serum. Serum IgG from gastric cancer patients was shown to carry triantennary N-glycans with high amounts of SLe^x, as well as core Fucα(1–6) and agalactosyl-biantennary N-glycans, compared to normal serum IgG (Bones et al. 2011). These changes could be related to disease progression.

The highly sialylated glycans in lung cancer (Arnold et al. 2011) can be part of lectin or selectin ligands. Alternatively, sialic acid can mask ligands for galectins or other mammalian lectins involved in cell surface interactions. Terminal sialylα(2–6) residues on N-glycans are increased in colon cancer (Dall'Olio et al. 1989, 2000; Dall'Olio and Chiricolo 2001) and in other cancers and are associated with a poor prognosis. The role of α6-linked sialic acid has been studied in cultured colon cancer cells HT29 and was shown to involve cell migration (Zhu et al. 2001). Highly α6-sialylated cells were more invasive through the extracellular matrix and may therefore have an increased potential as invasive cancer cells. In contrast, in a lectin-binding study of gastric tumors with high metastatic potential, sialic acid residues in α2,3-linkage were found to be increased and to correlate with invasive depth and lymph node metastasis (Wang et al. 2009). The highly metastatic cell line SGC-7901 was particularly reactive with sialylα(2–3)-binding *Maackia amurensis* lectin.

Prostate-specific antigen (PSA), a biomarker used for prostate cancer diagnosis, contains one N-linked oligosaccharide. Normal serum PSA has mainly biantennary complex chains, while PSA from prostate cancer patients has more tri- and tetraantennary *N*-glycans (Tabarés et al. 2006; De Leoz et al. 2008; Tajiri et al. 2008). Significant changes were also seen at the nonreducing ends of the oligosaccharide chains. Sialic acid residues were found in increased amounts on prostate cancer cells and are functionally important (Suriano et al. 2005, 2009; Valenzuela et al. 2007). The ratio of α2,3- to α2,6-linked sialic acids is increased, as well as fucosylation at the nonreducing end of *N*-glycan chains (Ohyama et al. 2004; Peracaula et al. 2003; Meany et al. 2009; Fukushima et al. 2010).

The neural cell adhesion molecule (NCAM) is involved in neural recognition events, neural development, cell adhesion, and migration. NCAM has *N*-glycans carrying a high amount of polysialic acids in α2,8-linkages that control cell adhesion and are formed by polysialyltransferases ST8SiaII and ST8SiaIV. The amount of polysialic acid is significantly increased in Wilm's tumor and small lung cell cancer, especially in late stages and metastases (Tanaka et al. 2000). However, while metastases from neuroblastoma were rich in polysialic acids, NCAM was deficient in polysialic acid in patients with advanced disease accompanied by an unfavorable prognosis (Cheung et al. 2006; Korja et al. 2009). Neuroblastoma cell lines also differ in their expression in NCAM and polysialic acid (Valentiner et al. 2011). The tumors formed in SCID mice by neuroblastoma cells expressing NCAM and polysialic acid showed a disseminated pattern of small metastases. Polysialic acid therefore can reduce adhesiveness of neuroblastoma cells and promote dissemination; it regulates cell adhesion and migrations and contributes to metastasis. A number of NCAM glycoforms also express the sulfated human natural killer-1 (HNK-1) (CD57) epitope (sulfo-3-*O*-GlcAβ(1–3)Galβ(1–4)GlcNAc), which can be linked to *N*-glycans or to *O*-glycans with the core 2 structure in a cell type-specific fashion. The HNK-1 epitope is displayed on many neuroblastoma, lymphoma, leukemia, and melanoma cells and may contribute to cell adhesion and metastatic potential (Brockhausen and Kuhns 1997; Casado et al. 2008). Because of the heterogeneity and complexity of *N*-glycan structures and their cell-specific expression, highly sensitive methods to accurately define minute alterations as markers of cancer are extremely useful.

8.4 BIOSYNTHESIS AND GLYCODYNAMICS OF TUMOR-ASSOCIATED STRUCTURES

8.4.1 GENERAL MECHANISMS OF GLYCOPROTEIN BIOSYNTHESIS

Glycoprotein-bound glycans are extremely heterogeneous in structure and are assembled and modified according to specific pathways and the activities of their biosynthetic enzymes, including glycosyltransferases and sulfotransferases (Brockhausen 2006, 2007, 2010). The heterogeneity of glycan structures in glycoproteins includes incomplete or intermediate structures or highly extended chains. Intermediate structures are often substrates for several enzymes; thus, the relative activities determine the relative amounts of glycans produced in the final glycoprotein. Truncated structures of cancer glycoproteins could arise by several possible mechanisms. For example, a glycosyltransferase activity involved in chain extension or branching (Figure 8.1) may be downregulated, leading to the accumulation of intermediate structures, which then can become substrates for competing enzymes. Alternatively, specific glycosyltransferase and sulfotransferase genes could be up- or down-regulated in tumors, thus producing high amounts of core structures, extensions, or terminal epitopes that would block the synthesis of competing structures and shift the overall balance of glycans. SLe^x is assembled by enzymes from four different glycosyltransferase families, and shifts in the expression of any of these enzymes in cancer have the potential to alter the expression of SLe^x. For example, sialyltransferase ST3Gal3, which acts on N-acetyllactosamine extensions of N- and O-glycans, was shown to enhance the metastatic potential and migration of pancreatic cancer cells Capan-1 and MDA Panc-28, which was related to increased synthesis of SLe^x (Pérez-Garay et al. 2010).

A number of factors are thought to regulate the complex biosynthetic pathways. Many of the enzymes require divalent metal ions for their activities. Enzymes also appear to be stimulated by interactions or complex formation with other proteins and the membrane environment that may affect enzyme protein conformation and activity. For example, the activity of the extension enzyme β3GlcNAcT8 involved in the synthesis of N-acetyllactosamine chains in colon cancer (Ishida et al. 2005) was increased when the two members of the β3GlcNAcT family, β3GlcNAcT2 and β3GlcNAcT8, were mixed in vitro (Seko and Yamashita 2005). This indicates that the complex formation with other enzymes may be an important activating factor in the biosynthesis of glycoproteins. The proper transport, localization, and topology of enzymes as well as the concentrations of enzymes and substrates within the intracellular compartments are essential for efficient biosynthesis. The kinetic properties, specificities, and access of enzymes to their substrates and cofactors determine the final glycan structures. Under the conditions of the endoplasmic reticulum (ER) and Golgi environment, competing enzymes may differ in their kinetic properties. From studies of the peptide site specificity of glycosyltransferases, the enzymes that assemble O-glycan core structures clearly distinguish between glycosylation sites (Brockhausen et al. 2009) and may thus have different reactivities toward individual glycoproteins. Thus, an abnormal expression of mucins and glycoproteins in cancer may in part determine their glycosylation patterns. The functional consequences of alterations depend on the glycoproteins expressed in a particular cell. The control mechanisms in cancer cells may become clear upon understanding these complex regulatory factors of glycoprotein biosynthesis (Brockhausen, Schutzbach et al. 1998; Brockhausen 1999). Little is known about the transcriptional regulation of glycosyltransferase genes and differential splicing in cancer although a number of glycosyltransferase genes have been shown to contain binding sites for transcription factors and other regulatory factors. For example, the extension enzyme β4GalT5, which contributes to the growth of glioma cells (Jiang and Gu 2010), is regulated by RAS-MAPK and PI3K-AKT signaling pathways, involving E1AF and Sp1 transcription factors, which may be involved in the uncontrolled growth of cancer cells.

8.4.2 ROLE OF ABERRANT O-GLYCOSYLATION PATHWAYS IN CANCER

In cancer cells, typically, truncated, immature O-glycans, as well as the completed Lewis antigen structures, are present (Figure 8.1). Truncated O-glycans such as Tn and T antigens (GalNAc- and Galβ3GalNAc-) may be exposed when there is excessive O-glycosylation by overexpression of members of the large family of polypeptide GalNAc-transferases (ppGalNAcTs). The expression of polypeptide GalNAc-transferases is restricted and cell type specific, but ppGalNAcT 1, 2, 3, 4, 6, and 11 are expressed in many neoplastic cells. The ppGalNAcT 3 expression was found to be a negative indicator of disease aggressiveness and tumor differentiation (Table 8.1). A strong expression correlated with a good prognosis for 5 year survival in colorectal carcinoma patients (Shibao et al. 2002). The reason for this association with cancer is not clear. In pancreatic cancer tissue, the expression of ppGalNAcT 3 was shown to be increased (Taniuchi et al. 2011), and the suppression of the ppGalNAcT 3 expression in pancreatic cell lines by small interfering RNA decreased cancer cell growth and induced apoptosis. The differences between the colon and pancreatic tumors may be related to the O-glycosylation of specific proteins involved in growth regulation in these cells.

The expression of another enzyme of the ppGalNAcT family, ppGalNAcT 6, was found to be increased in pancreatic tissues and related to a better outcome in patients (Li et al. 2011). Similarly, ppGalNAcT 6 expression was shown to be upregulated in the majority of breast cancers but was not detected in other tissues (Park et al. 2010, 2011). When the expression of the enzyme was suppressed by treatment of metastatic breast cancer cells T47D with small interfering RNA, cell adhesion was enhanced while cell proliferation was suppressed (Table 8.2). This suggests a role of the enzyme in mammary carcinogenesis and growth of breast cancer cells, but the specific glycoprotein substrates remain to be identified. Breast tumors showed an unusually high but heterogeneous expression of ppGalNAcT 14 (Wu et al. 2010). The expression in tumors showed a significant association with the histological grades, invasiveness, mucinous adenocarcinomas, and ductal carcinomas in situ types.

The lack of further processing of GalNAc residues can lead to accumulation of the Tn antigen. Thus, lack of core 1 β3-Gal-transferase (C1GalT) activity in human colon cancer cells LSC (Brockhausen, Yang et al. 1998) leads to Tn expression (Figure 8.1; Table 8.1). The enzyme is unique in that a chaperone "Cosmc" has to be coexpressed. Due to a mutation in the *cosmc* gene, C1GalT is degraded in LSC cells (Ju et al. 2008).

The expression of the STn antigen can be a result of lack of O-glycan core synthesizing enzymes, allowing GalNAc to be sialylated instead of being converted to core structures 1 or 3 (Figure 8.1). STn levels may be excessive in cases where there is a high expression of ST6GalNAc-I or -II (Brockhausen et al. 2001; Marcos et al. 2004). ST6GalNAc-I is the main enzyme responsible for the synthesis of the STn antigen in the intestines and a restricted number of other tissues. Although the mRNA is not expressed in the normal mammary gland, it appears to be switched on in 30% of breast carcinoma cells, and this correlates with STn expression. The broad distribution of ST6GalNAc-I in the Golgi enables it to compete with C1GalT for the common GalNAc substrate (Sewell et al. 2006). Once STn is formed, C1GalT can no longer synthesize core 1. ST6GalNAc-I transfection resulted in increased STn in metastatic breast cancer cells T47D (Figure 8.1) (Julien et al. 2005, 2006) and led to a decrease in cell adhesion and an increase in cell migration. In addition, STn-expressing clones of human breast cancer cells MDA-MB-231 showed enhanced tumor growth in mice. In gastric carcinoma cells MKN45, the expression of STn was shown to modify cell cycle and apoptosis, adhesion, and motility (Pinho et al. 2007). These observations strongly implicate STn in tumorigenicity.

TABLE 8.1
Aberrant Glycosyltransferase and Sulfotransferase Patterns in Tumor Cells

Enzyme Abnormality	Occurrence	References
Decreased ppGalNAcT 3 expression	Aggressive colon cancer	Shibao (2002)
Increased ppGalNAcT 3 expression	Pancreatic cancer tissue	Taniuchi et al. (2011)
Increased ppGalNAcT 6 expression	Breast cancer	Park et al. (2010, 2011)
ppGalNAcT 14 expression	Breast cancer	Wu et al. (2010)
Mutation in *cosmc* and loss of C1GalT activity	Colon cancer cells LSC	Brockhausen, Yang et al. (1998) Ju et al. (2008)
Decreased core 3 β3GlcNAcT (C3GlcNAcT6) activity	Colon cancer tissues and cells	Yang et al. (1994) and King et al. (1994)
	Tumorigenic polyposis coli cells	Vavasseur et al. (1994, 1995)
Decreased core 3 β3GlcNAcT (C3GlcNAcT6) protein	Colon cancer, gastric cancer	Iwai et al. (2005)
Increased C2GnT1 expression	Progressing prostate cancer	Hagisawa et al. (2005)
	Prostate cancer cells	Gao, Cheng et al. (2010)
	Leukemia cells	Brockhausen and Kuhns (1997)
Lack of C2GnT1 expression	Variable in breast, colonic cancer cells	Brockhausen et al. (1995) and Vavasseur et al. (1995)
	Variable in prostate cancer cells	Gao, Cheng et al. (2010)
Increased FUT I, FUT IV, and ST6Gal-I expression	Colorectal adenomas	Petretti et al. (2000)
Decreased FUT III expression	Liver metastases	Petretti et al. (2000)
Increased FUT III–VII expression	Lung adenocarcinoma cells	Martín-Satué et al. (1998) and Martín-Satué et al. (1999)
Increased FUT VI expression	Breast cancer	Matsuura et al. (1998)
High ST3Gal1 activity and expression	Breast tumors	Burchell et al. (1999)
	Variable in breast cancer cells	Brockhausen et al. (1995) Gao, Cheng et al. (2010)
	Variable in prostate cancer cells	
High ST6GalNAc-II expression	Colon cancer	Schneider et al. (2001)
Increased GlcNAcT-V expression and activity	Breast tumors	Guo et al. (2010)
	Colorectal adenomas	Petretti et al. (2000)
	Liver metastases	Petretti et al. (2000)
	Bladder cancer	Ohyama (2008)
	Prostate cancer cells	Gao, Cheng et al. (2010)
	Melanoma	Przybylo and Lityńska (2011)
Increased ST6Gal-I expression	Colon cancer	Lise et al. (2000)
	Many cancers	
Increased β3GlcNAcT8 expression	Colon cancer, many cancers	Ishida et al. (2005)
Increased β4GalT5 expression	Glioma	Jiang and Gu (2010)
Loss of β4GalNAcT, Sd[a]	Gastric cancer	Dohi et al. (1996) and Dohi and Kawamura (2008)

TABLE 8.1 (*Continued*)
Aberrant Glycosyltransferase and Sulfotransferase Patterns in Tumor Cells

Enzyme Abnormality	Occurrence	References
Decreased sulfotransferase activities	Colon cancer Tumorigenic polyposis cells	Vavasseur et al. (1994); Yang et al. (1994)
Increased *O*-GlcNAc-transferase expression	Breast cancer	Caldwell et al. (2010) and Mi et al. (2011)

There are three versions of sialylated T antigens, which are found in both normal and cancer cells, as a result of α3- and α6-sialyltransferase activities (Figure 8.1). The mRNA expression of ST6GalNAc-II that acts on core 1 to synthesize sialyl6T is increased in colorectal cancer with lymph node metastases, correlating with a shorter survival (Schneider et al. 2001). The expression and activity of ST3Gal1 that synthesizes sialyl3T is increased in colon carcinomas (Table 8.1), as well as in breast tumors (Brockhausen et al. 1995; Burchell et al. 1999). ST3Gal1 competes with C2GnT1 (Dalziel et al. 2001) for the common core 1 substrate, and the resulting structures depend on the relative activities of these two enzymes. The intracellular distribution of C2GnT1 and ST3Gal1 in T47D cells is partially overlapping, which is critical for this competition. Once core 1 is sialylated, it cannot be converted to the core 2 structure, but the reverse pathway is possible; core 2 can be a substrate for ST3Gal1. Studies of mammary tumorigenesis in mice that spontaneously develop mammary tumors showed that tumor formation was significantly faster in ST3Gal1 transgenic mice. This suggests a role of sialylation in tumorigenesis, possibly by giving an advantage to cancer cells in the circulation or during tumor development (Picco et al. 2010) (Table 8.2).

The branched and complex core 2 structures are synthesized by a family of C2GnTs. We have shown that the activity and mRNA expression levels of C2GnT1 are variable among breast cancer cells (Brockhausen et al. 1995), colon cancer cells (Vavasseur et al. 1994, 1995), and prostate cancer cells (Gao, Cheng et al. 2010). While some cells have a high activity, other cells completely lack C2GnT1 activity. Normal human colon has a high activity of the C2GnT that synthesizes core 4 (C2GnT2) (Figure 8.1). However, the enzyme is variably active or absent from cultured colon cancer cells (Vavasseur et al. 1995). C2GnT2 expression in human colon cancer HCT116 cells suppressed cell growth, cell adhesion, and invasive properties. It also depressed tumor growth in mice (Huang et al. 2006) (Table 8.2). In contrast to the apparent beneficial effect of C2GnT2, the expression of C2GnT1 has been correlated with vessel invasion of colon cancer cells (Shimodaira et al. 1997). The metastatic potential of testicular germ cell tumors as well as aggressive growth is also related to C2GnT1 (Hatakeyama et al. 2010). This may be due to the fact that SLe^x structures are often attached to the GlcNAc residues of core 2.

Human and rat colonic mucins carry many *O*-glycans with core 3 structures and have the enzyme (C3GlcNAcT6) that synthesizes core 3 from GalNAc-Thr/Ser-peptide, but the activity is low in colon tumor tissues (Yang et al. 1994; King et al. 1994). In contrast to C2GnT1, which can be upregulated in cancer cells, C3GlcNAcT6

TABLE 8.2
Roles of Glycosyltransferases and Glycosylation in Cancer

Glycoconjugate, Enzyme, Lectin	Biological Role	References
ppGalNAcT 3	Increases pancreatic cancer cell growth	Taniuchi et al. (2011)
ppGalNAcT 6	Contributes to mammary carcinogenesis	Park et al. (2010, 2011)
STn	Contributes to progression of gastric cancer	Pinho et al. (2007)
	Protects cancer cells from NK cell lysis	Blottière et al. (1992)
C2GnT1	Induces aggressive growth, invasion	Hatakeyama et al. (2010)
	Metastatic potential of testicular germ cell tumors	Shimodaira et al. (1997)
C2GnT2	Decreases colon cancer cell growth and invasiveness	Huang et al. (2006)
C3GlcNAcT6	Decreases migration of HT1080 FP-10 cells	Iwai et al. (2005)
	Suppresses lung metastases	
	Decreases tumor growth and metastasis in PC-3 and LNCaP prostate cancer cells	Lee et al. (2009)
GlcNAcT-III, bisected N-glycans	In mice reduces growth factor signaling of mammary cells, inhibits breast tumor growth	Song et al. (2010) and Isaji et al. (2004) Yoshimura et al. (1995), Pinho et al. (2009), Kariya et al. (2008), and Zhao et al. 2008
	Decreases integrin-adhesion antimetastatic	
GlcNAcT-V	Decreases cell adhesion to substrate, metastatic phenotype	Demetriou et al. (1995) and Dennis et al. (1999) Pinho et al. (2009), Zhou et al. (2011), and Ohyama (2008)
	Increases lung cancer cell growth, motility, invasiveness, causes loss of contact inhibition related to malignant potential in bladder cancer	
	Disrupts acinar morphogenesis of human mammary cells MCA-10A	Guo et al. (2010)
β4GalT5	Regulates growth of glioma cells	Jiang and Gu (2010)
β4GalNAcT III	Enhances malignant potential of colon cancer cells	Huang et al. (2007)
FUT III	Leads to high expression of SLe[x] in PC-3 cells, large prostate tumors, increased adhesion to stromal cells	Inaba et al. (2003)

TABLE 8.2 (*Continued*)

Roles of Glycosyltransferases and Glycosylation in Cancer

Glycoconjugate, Enzyme, Lectin	Biological Role	References
FUT VII	Increases chemotactic migration, invasion of colon cancer cells LOVO	Li, Zhang et al. (2010)
	Contributes to metastasis of lung adenocarcinoma cells	Martín-Satué et al. (1999)
	Indicator of poor prognosis in lung carcinoma	Ogawa et al. (1997)
Sialyl-Lewis x, E-selectin	Contributes to metastasis (e.g., of B16 melanoma cells)	Chen and Fukuda (2006) and Nakamori et al. (1993), St.Hill et al. (2009)
	Causes cell adhesion, invasion, migration, metastasis	
Core Fucα1–6, FUT VIII	Regulates cell growth	Zhao et al. (2008)
Sialic acids	Prevent growth factor receptor oligomerization and signaling	Amith et al. (2010)
ST3Gal1	Increases breast cancer cell survival in mice	Picco et al. (2010)
ST3Gal3	Enhances metastatic potential and migration of pancreatic cancer cells Capan-1 and MDAPanc-28	Pérez-Garay et al. (2010)
Sialic acid α2,6	Contributes to cell migration, invasiveness	Zhu et al. (2001)
ST6Gal-I	Decreases Fas-apoptosis	Swindall and Bellis (2011)
O-GlcNAc-transferase	Regulation of protein phosphorylation, gene transcription, contributes to breast tumor growth and invasion	Caldwell et al. (2010), Mi et al. (2011), and Hart et al. (2010)
Galectin-1	In mice, evasion of tumor cells from immune system, increased tumor survival and aggressive metastases	Stannard et al. (2010)
Galectin-3	Cell adhesion (e.g., in melanoma cells)	Zhang et al. (2002)
Heparanase	Neovascularization, tumor progression, metastasis	Barash et al. (2010)
Hyaluronan	Contributes to tumor progression	Itano and Kimata (2008)
Versican	Metastasis of ovarian cancer	Ween et al. (2011)
MUC1	Adhesion, immune suppression	Hollingsworth and Swanson (2004)

and C2GnT2 activities are often lacking in cultured colon, prostate, and other cancer cells (Vavasseur et al. 1994, 1995; Gao, Cheng et al. 2010; Yang et al. 2008). These latter two enzymes (Brockhausen et al. 1985) may, therefore, contribute to the maintenance and functions of a normal mucosa. Lower levels of the C3GlcNAcT6 protein were detected in gastric cancer compared to normal gastric mucosa (Iwai et al. 2005). In the absence of core 3 synthesis, GalNAc is available as a substrate for C1GalT, allowing a higher production of O-glycan core 1, the T antigen (Figure 8.1). Studies of core 3 synthesis in colon carcinoma (Iwai et al. 2005) suggested that core 3 has a protective function (Table 8.2). Transfection of human fibrosarcoma cells HT1080 FP-10 with the C3GlcNAcT6 gene decreased migration of cells and drastically suppressed lung metastases. Lee et al. (2009) investigated the role of core 3 in prostate cancer cells. C3GlcNAcT6 transfected PC-3, and LNCaP cells showed reduced ability of migration and invasion through extracellular matrix components, and suppressed tumor formation and tumor metastasis in mice.

The E-selectin ligand SLe^x is synthesized by β3GlcNAc-transferases (β3GlcNAcT), β4Gal-transferases (β4GalT), α3Fuc-transferases (mainly FUT IV or VII), and α3sialyltransferases (ST3Gal) (Figure 8.1). The expression of the individual members of these glycosyltransferase families is cell type specific and can be altered in tumor cells, for example, in prostate cancer cells (Inaba et al. 2003; Barthel et al. 2008, 2009; Gao, Cheng et al. 2010; Matsuura et al. 1998; Kudo et al. 1998; Trinchera et al. 2011; Martín-Satué et al. 1998). Colorectal carcinomas showed increased FUT I and IV expression, while liver metastases showed decreased FUT III expression (Petretti et al. 2000). FUT III is the only FUT that synthesizes the α1,4-linkage, which is the basis for Lewis a (Fucα(1–4)[Galβ(1–3)]GlcNAcβ-) and Lewis b (Fucα(1–4)[Fucα(1–2)Galβ(1–3)]GlcNAcβ-) epitopes. Transfection of FUT III gene into prostate cancer cells PC-3 led to high levels of SLe^x and in mice to increased adhesion to stromal cells and large tumors (Inaba et al. 2003). FUT VII, which adds Fuc to sialylated substrates, is a major contributor to metastasis of HAL-24Luc cells (Martín-Satué et al. 1999) and may be a marker indicating a poor prognosis in lung carcinoma (Ogawa 1997). The effects of FUT VII in colon cancer cells LOVO include increased chemotactic migration and invasion (Li, Zhang et al. 2010). The extensions of glycan chains are important in forming the scaffold for glycosylated epitopes and can direct the biology of a cell. For example, β4-GalNAc-transferase III (Table 8.2) forms GalNAcβ(1–4)GlcNAc- chain extensions and is upregulated in colonic tumors; its overexpression enhances tumor growth, invasion, and metastasis of colon cancer cells (Huang et al. 2007). Thus, a multitude of cancer-related changes affect enzymes involved in the synthesis of O-glycan cores, their extensions, and terminal antigens. The underlying molecular mechanisms remain to be shown. We also need to clearly delineate how this affects the functions of glycoproteins.

8.4.3 Synthesis of *N*-Glycans in Cancer Cells

Asn-linked N-glycans are preassembled on dolichol-phosphate, transferred to peptides within the Asn-X-Ser/Thr sequon, and processed involving a number of cytoplasmic reactions and ER membrane-bound glycosyltransferases (Brockhausen, Schutzbach et al. 1998). Little is known about the relationship of early acting enzymes to cancer. After removal of mannose residues in the ER and Golgi, N-glycans can be converted to complex-type chains by the sequential action of GlcNAcT-I, mannosidase II, and GlcNAcT-II (Figure 8.2). The enzyme activities of GlcNAcT-II to -V, which affect the overall N-glycan branching patterns and structures, are upregulated during colon cancer cell differentiation (Brockhausen et al. 1991). The expression of GlcNAcT-III or -V (Figure 8.2) is often altered in cancer (Li, Song et al. 2010) with implications in cell migration and carcinogenesis.

GlcNAcT-III adds the bisecting GlcNAc residue to N-glycans after the action of GlcNAcT-I (Brockhausen, Schutzbach et al. 1998). GlcNAcT-III can act on bi-, tri- and tetraantennary N-glycans and decreases the possibility of further extension and branching, probably due to the crowded configuration or distorted conformation of the bisected N-glycan together with the substrate specificities of GlcNAcTs.

GlcNAcT-III is thought to have antimetastatic properties (Zhao et al. 2008), and GlcNAcT-III transfected into mouse B16 melanoma cells suppressed tumor formation in mice (Yoshimura et al. 1995). GlcNAcT-III was also shown to inhibit breast tumor growth in mice, possibly by reducing platelet-derived growth factor signaling (Song et al. 2010). While bisected N-glycans caused increased cell–cell adhesion of B16 melanoma cells (Yoshimura et al. 1995) consistent with decreased metastatic properties, GlcNAcT-V appeared to cause the opposite effect, thus contributing to decreased cell adhesion and to tumor cell invasiveness. In B16 melanoma cells, the bisecting GlcNAc residue reduced integrin $\alpha_5\beta_1$-mediated adhesion to fibronectin (Isaji et al. 2004). When human gastric carcinoma cells MKN45 were transfected with GlcNAcT-V, increased cell migration on laminin resulted, while GlcNAcT-III had the opposite effect, possibly by impairing integrin clustering (Kariya et al. 2008). Thus, in most models studied, GlcNAcT-III caused a decrease in cancer cell migration, while GlcNAcT-V promoted cancer cell motility.

The GlcNAcT-V gene has several binding sites for transcription factors (Saito et al. 1995), and it has been shown that the gene is regulated through the RAS–RAF–MAPK signal transduction pathways in cancer cells (Dennis et al. 1999). GlcNAcT-V expression was also shown to be regulated by oncogenes such as *src* (Buckhaults et al. 1997), and this may play an important role in cancer cell growth and metastasis (Lau and Dennis 2008). The display of additional carbohydrate epitopes due to GlcNAcT-V action could increase the affinity and lattice formation of lectins that regulate cell adhesion, motility, cell growth, and apoptosis (Lagana et al. 2006). The $\beta1,6$ antenna of N-glycans synthesized by GlcNAcT-V may differ from the other antennae in its flexibility due to rotation of the 1,6 bond. The long extensions often found on the $\beta1,6$ antennae reduce adherence of cancer cells, and this phenotype is related to metastasis in human hepatoma and other cancer cells (Guo et al. 2000). In prostate cancer cells LNCaP, overexpression of GlcNAcT-V was shown to stabilize the activity of the cell surface protease matriptase (Tsui et al. 2008). The invasive and metastatic properties of metastatic prostate cancer cells were also increased by GlcNAcT-V.

A high GlcNAcT-V expression in melanoma contributes to loss of contact inhibition and reduced cell–extracellular matrix interactions (Przybylo and Litynska 2011). Increased GlcNAcT-V expression was also found in gastric carcinomas, colorectal adenocarcinomas, and their liver metastases (Petretti et al. 2000), as well as in bladder cancer, where it correlates with malignant potential (Ohyama 2008). Using human mammary cells MCF-10A in a HER2 transgenic mammary tumor mouse model, GlcNAcT-V was shown to disrupt acinar morphogenesis (Guo et al. 2010) and to be associated with higher amounts of tumor-initiating cells. HER2-induced tumors were significantly delayed with GlcNAcT-V deficient cells, confirming the critical role of GlcNAcT-V in tumorigenesis. This role was also demonstrated in human lung adenocarcinoma cells that express high amounts of GlcNAcT-V and were treated with small inhibitory RNA to decrease the expression of GlcNAcT-V, which resulted in decreased cell growth in culture and tumor growth in mice (Zhou et al. 2011). Tumor growth and metastasis induced by polyomavirus middle T antigen in mice were considerably lower when GlcNAcT-V was deficient compared to GlcNAcT-V expressing control mice (Granovsky et al. 2000). In cultured mink lung epithelial cells, overexpression of GlcNAcT-V was associated with increased cell motility and loss of contact inhibition, which is consistent with a metastatic phenotype (Demetriou et al. 1995).

Not only the antennary and extended structures of *N*-glycans but their terminal epitopes play roles in cancer development. The α6-sialyltransferase responsible (ST6Gal-I) is increased in many cancers, including colon carcinoma (Dall'Olio et al. 1989, 2000), and its activity and expression correlate with a shortened survival and poor prognosis (Lise et al. 2000). The enzyme was shown to block apoptosis by modifying Fas, thus inhibiting Fas internalization as well as its binding to Fas-associated protein with death domain (FADD) (Swindall and Bellis 2011). The cancer phenotype includes a characteristic number of different transcripts of ST6Gal-I (Dall'Olio et al. 2000). The role of α6-linked sialic acid has been studied in cultured colon cancer cells HT29. Antisense DNA to ST6Gal-I resulted in a decreased ability of cells to migrate through extracellular matrix, suggesting a role of the enzyme in invasion (Zhu et al. 2001). These functional studies are useful to define potential anticancer targets.

8.5 DEVELOPMENT OF CANCER DIAGNOSIS AND THERAPY TARGETING GLYCOSYLATION

Cancer-associated structures may be useful for the diagnosis of a specific cancer type or to identify stages of the disease. While cancer markers such as CA125 are found in a variety of different cancer types, assessing a combination of markers may result in more specific diagnosis (Diamandis 2010). However, clinical studies and rigorous validation of these markers need to be carried out in human patients.

8.5.1 TUMOR VACCINES

Cells at the center of the tumor are protected from immune attacks, while cells in the blood stream are exposed to anticancer antibodies, lectins, and the cellular immune system. Circulating tumor cells may survive because their glycosylation shields them from immune recognition and degradation. Antimucin MUC1 antibodies have been detected in sera from breast, pancreatic, and colon cancer patients (Kotera et al. 1994). This suggests that an immune response to cancer mucins can be raised in cancer patients. MUC1 is a preferred therapeutic target in several cancer types including prostate cancer (Li and Cozzi 2007). Peptide-based vaccines to target MUC1 were successful in eliciting humoral and cellular immune responses

(Tarp and Clausen 2008; Huang and Wu 2010). Vaccines using glycopeptides with the cancer-associated STn, Tn, and T antigens have been shown to stimulate the formation of antibodies that recognize MUC1-expressing tumor cells and may be successful clinically (Kovjazin et al. 2011; Tarp and Clausen 2008). Sheep submaxillary mucin carrying Tn and STn glycans also induced IgM and IgG antibody responses in colon cancer patients (O'Boyle et al. 1992). Springer (1997) showed that conventional treatment for breast cancer was significantly improved using vaccinations with glycophorin carrying T and Tn antigens. Ideally, MUC1 vaccines should contain the specific aberrant glycosylation found on patient's cancer cells.

Synthetic glycoconjugates are useful in preparing structurally well-defined vaccines (Fung and Wu 1990) and those containing clusters of tumor-associated antigens. The carrier protein or peptide is also important, and both mucins and keyhole limpet hemocyanin, peptides derived from poliovirus CD4+ T cell epitope, or tetanus toxin have been effective (Guo and Wang 2009). A new synthetic approach is based on labeling cancer cells with unnatural antigens. Due to their enhanced growth, cancer cells incorporate synthetic sugars at a faster rate and display the antigen on the cell surface as part of a glycoconjugate. An example is N-phenyl-acetyl-mannosamine that is converted to an antigenic form of sialic acid. Antibodies to this antigen were shown to have strong complement-dependent cytotoxicity to cancer cells (Wang et al. 2007). Further development in this area has significant potential, using vaccines targeting either specific natural or unnatural tumor epitopes, with a combination of carriers and adjuvants, aimed at both eliciting antibody and cell-mediated responses.

8.5.2 Glycosylation Inhibitors

Inhibitors of glycosyltransferases or N-glycan-processing glycosidases can influence cellular behavior, growth, migration, and apoptosis, and some of these have been shown to reduce cancer cell invasion and metastasis (Figure 8.3). For example, swainsonine (Figure 8.4a) is an inhibitor of processing α-mannosidase II that removes Man residues before conversion of N-glycans to the complex-type chains in the Golgi (Brockhausen, Schutzbach et al. 1998) (Figure 8.3) and has the ability of attenuating glycoprotein functions attributed to complex N-glycans in cancer cells. Thus, cancer cell invasion, colonization, and growth can be inhibited by swainsonine (Jacob 1995). For example, swainsonine treatment of human gastric carcinoma cells reduced cell growth significantly (Sun et al. 2007). Nontoxic derivatives of swainsonine and other N-glycan processing inhibitors also have potential as anticancer drugs.

GalNAcα-benzyl (Figure 8.4b) is a substrate for both C1GalT and C3GlcNAcT6, which synthesize core 1 and core 3 O-glycans, respectively (Figure 8.3). GalNAcα-benzyl can penetrate into cells and block O-glycan extension in the Golgi by serving as a substrate for the second reaction in the O-glycosylation pathways. In colon cancer HT29 cells, GalNAcα-benzyl also blocked α3-sialylation of glycoproteins (Delannoy et al. 1996). The expression of Lewis antigens in pancreatic cell lines (SW1990, CAPAN-2, and PANC-1) can be reduced by treatment with GalNAcα-benzyl but not with the N-glycosylation inhibitor tunicamycin (Sawada et al. 1994). Treatment of human colonic cancer cells with GalNAcα-benzyl reduced their ability to bind to E-selectin (Kojima et al. 1992). In addition, metastases in nude

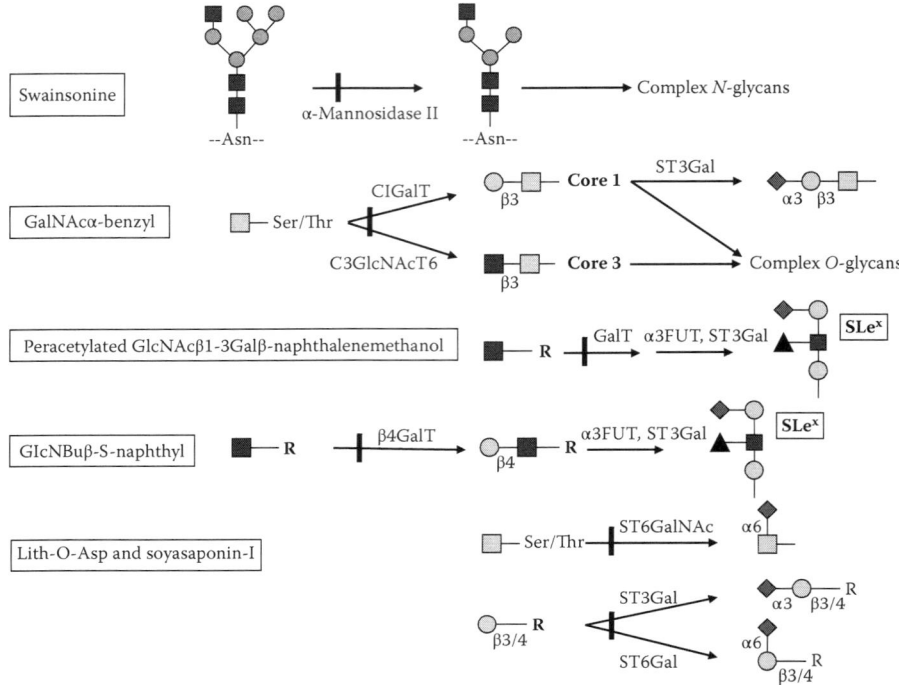

FIGURE 8.3 (See color insert.) Strategies targeting cancer cells by glycosylation inhibitors. The symbols are explained in Figure 8.1. Black bars through reaction arrows indicate the site of inhibition.

mice were significantly reduced. This indicates that GalNAcα-benzyl can suppress the synthesis of selectin ligands expressed on *O*-glycans (Bresalier et al. 1991). Other strategies to block selectin-mediated adhesion are the use of small analogs or mimetics of the glycan epitope, antibodies that target selectins, or soluble forms of glycoprotein ligands that compete with natural cell adhesion events. GalNAcα-benzyl also induced apoptosis in colon cancer cells (Patsos et al. 2009), in prostate cancer cells, and in cells derived from normal prostate (Gao, Cheng et al. 2010), as well as in HeLa cells (Li et al. 2007). However, the mechanisms of apoptosis remain to be clarified.

Brown et al. (2009) have developed per-acetylated disaccharide inhibitors of glycan extension, which can be metabolized to serve as primers for glycosylation and compete with natural substrates in the Golgi. The per-acetylated disaccharide GlcNAcβ(1–3)Galβ-naphthalenemethanol (Figure 8.4c) is taken up by cells, and after deacetylation, it prevents the synthesis of SLe^x on glycoproteins by providing a substrate for GalT (Figure 8.3) (Brown et al. 2009). In a mouse model, the peracetylated disaccharide was able to reduce SLe^x expression, as well as metastases from human colon cancer cells LS180 (Brown et al. 2006; Fuster et al. 2003).

Because of the high activities of β4GalTs in cancer cells and the involvement of β4GalT in the synthesis of the type-2 backbone structure, which is the scaffold for

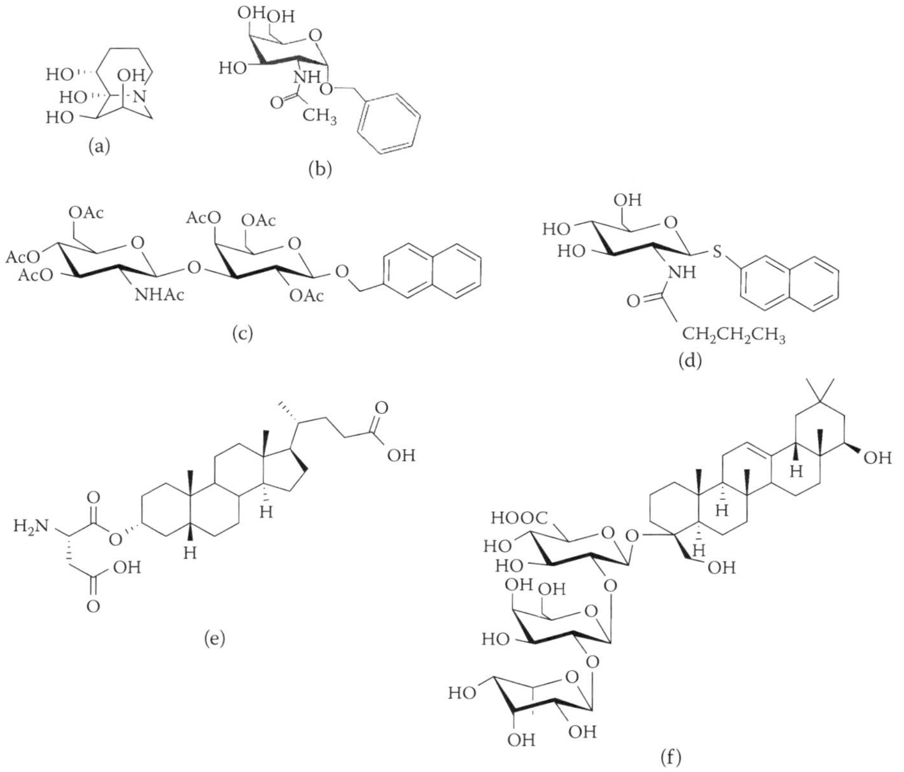

FIGURE 8.4 Structures of glycosylation inhibitors. (a) Swainsonine; (b) GalNAcα-benzyl; (c) Peracetylated GlcNAcβ(1–3)Galβ-naphthalenemethanol; (d) GlcNBuβ-S-naphthyl; (e) Lith-*O*-Asp; and (f) Soyasaponin-I.

SLe[x] (Figure 8.1), we have developed specific inhibitors for β4GalT (Brockhausen et al. 2006; Gao et al. 2010). These inhibitors have the potential to reduce the high GalT activity, the abundance of SLe[x], and galectin binding sites in cancer cells. We synthesized the substrate analog *N*-butyrylglucosamineβ-thio-naphthyl (GlcNBu-*S*-naphthyl) (Figure 8.4d) (Brockhausen et al. 2006), which potently inhibited β4GalT but not β3GalT (Gao et al. 2010) and blocked β4GalT activity more than 90% in lung, colon, and prostate cancer cell homogenates. In lung cancer cell cultures, it also reduced SLe[x] expression (Figure 8.3).

Sialyltransferase inhibitors are also valuable tools to reduce the metastatic abilities of cancer cells. Lith-*O*-Asp (Figure 8.4e) inhibits sialyltransferases (Figure 8.3) and reduced angiogenesis of human umbilical vein endothelial cells. The inhibitor also decreased sialylation of integrin-β1 and drastically reduced metastasis of highly metastatic mouse mammary cancer cells 4T1-Luc (Chen et al. 2011). Another nonspecific inhibitor of sialyltransferases, Soyasaponin-I (Figure 8.4f), was shown to inhibit ST3Gal activity. The inhibitor stimulated the adhesion of nonmetastatic human mammary cancer cells MCF-7 to extracellular matrix and decreased cell migration of highly metastatic mammary cancer cells

MDA-MB-231 (Hsu et al. 2005). Sialyltransferase inhibitor AL10 derived from lithocholic acid also penetrates into cells and reduces cell surface sialylation. It inhibited cell adhesion, migration, and invasion of ST3Gal overexpressing human lung cancer cells A549 and CL1-5 and also potently blocked lung cancer metastasis (Chiang et al. 2010). These inhibitors are useful for the study of the role of sialic acids in cancer cell behavior and are potential therapeutic tools. A major challenge is to design methods to deliver inhibitors into cells to reach the preferred site of their action (i.e., the Golgi), and to target them specifically to tumor cells, possibly by linking them to carriers that recognize structures on cancer cell surfaces.

8.6 CONCLUSIONS AND FUTURE APPLICATIONS IN CANCER RESEARCH

It seems that in cancer, infinite numbers of alterations occur. In this chapter, we have focused on only a small part of the reported structural and enzymatic aberrations, which does not mean that these are the only important ones. During the past years, it was possible to carry out many functional studies based on cloned genes of glycosyltransferases and specific inhibitors. The fields of glycoproteomics; DNA, glycan, and lectin arrays; and mass spectrometry have made invaluable contributions to our knowledge of cancer cells and development of cancer markers and sensitive diagnostic tests. A number of glycoproteins and glycan structures have been claimed to be cancer markers, but this needs to be much more rigorously validated in models and in patients. Combinations and patterns of markers may better confirm a cancer cell type and the stage of the disease and predict outcome. In order to personalize approaches to eliminating cancer cells, we need to analyze these markers with highly sensitive, reliable, and fast methods of detection. Markers may be individually different in patients and may be highly variable according to the pathophysiology of the disease.

There are many open questions in the field of the role of glycan structures in cancer that remain to be answered. Why are there so many changes? What mechanisms control the expression of so many enzymes in a cell-type-specific fashion? Why does the same change have a different effect in different cancer cells? Some of the answers lie in the cell-type-specific production of proteins. Because one enzyme can act on several glycoproteins, the protein expression levels may direct the glycosylation. Other enzymes may be specific for their glycoprotein substrates. This may explain some of the apparent discrepancies noted such as the roles of sialic acid in $\alpha 2,3$- or $\alpha 2,6$-linkages. Thus, the altered glycosylation potential in cancer affects the cells through the functions of specific glycoproteins, and enzyme expression has to be seen in context with protein expression. Clearly, the glycans affect glycoprotein function, and depending on the glycoprotein, it may affect conformation, stability, transport and secretion, cell surface expression, accessibility, ligand binding, and oligomerization, all with a potential impact on the growth of tumors. GlcNAcT-III and -V are examples of how the role of an oligosaccharide can be altered, and how different glycoproteins are affected differently by these enzymes. Therefore, we need to explore the functions of specific glycan structures on individual molecules such as growth factor receptors. For example, the role of sialic acid in receptor signaling is a very exciting area worth exploring in much more detail. The details of glycan function in cell death also remain to be determined.

Glycolipids also carry many cancer antigens such as SLe^x, extending into the extracellular space with important roles in cancer cell behavior. However, their detailed discussion is beyond the scope of this chapter. The actual display of glycolipids and glycoproteins on cell surface in context with the membrane, their concentrations in rafts, and clusters may be critical for the functions of the cell surface. The glycans of both glycolipids and glycoproteins may interact with each other, as well as with components of the cell membrane with functional implications that affect cancer cell growth and cell death.

ACKNOWLEDGMENTS

The financial support of the Prostate Cancer Fight Foundation—Motorcycle Ride for Dad and a Discovery Grant by the Natural Science and Engineering Council of Canada is acknowledged.

REFERENCES

Amith S. R., Jayanth P., Franchuk S. et al. 2010. Neu1 desialylation of sialyl alpha-2,3-linked beta-galactosyl residues of TOLL-like receptor 4 is essential for receptor activation and cellular signaling. *Cell Signal*, 22, 314–324.

Arnold J. N., Saldova R., Galligan M. C. et al. 2011. Novel glycan biomarkers for the detection of lung cancer. *J Proteome Res*, 10, 1755–1764.

Bachmaier R., Aryee D. N., Jug G. et al. 2009. O-GlcNAcylation is involved in the transcriptional activity of EWS-FLI1 in Ewing's sarcoma. *Oncogene*, 28, 1280–1284.

Balagué C., Audié J. P., Porchet N., and Real F. X. 1995. In situ hybridization shows distinct patterns of mucin gene expression in normal, benign, and malignant pancreas tissues. *Gastroenterology*, 109, 953–964.

Barash U., Cohen-Kaplan V., Dowek I. et al. 2010. Proteoglycans in health and disease: New concepts for heparanase function in tumor progression and metastasis. *FEBS J*, 277, 3890–3903.

Barthel S. R., Gavino J. D., Wiese G. K. et al. 2008. Analysis of glycosyltransferase expression in metastatic prostate cancer cells capable of rolling activity on microvascular endothelial (E)-selectin. *Glycobiology*, 18, 806–817.

Barthel S. R., Wiese G. K., Cho J. et al. 2009. Alpha 1,3 fucosyltransferases are master regulators of prostate cancer cell trafficking. *Proc Natl Acad Sci USA*, 106, 19491–19496.

Bidon-Wagner N. and Le Pennec J. P. 2004. Human galectin-8 isoforms and cancer. *Glycoconj J*, 19, 557–563.

Blottière H. M., Burg C., Zennadi R. et al. 1992. Involvement of histo-blood-group antigens in the susceptibility of colon carcinoma cells to natural killer-mediated cytotoxicity. *Int J Cancer*, 52, 609–618.

Bones J., Byrne J. C., O'Donoghue N. et al. 2011. Glycomic and glycoproteomic analysis of serum from patients with stomach cancer reveals potential markers arising from host defense response mechanisms. *J Proteome Res*, 10, 1246–1265.

Borsig L., Vlodavsky I., Ishai-Michaeli R., Torri G., and Vismara E. 2011. Sulfated hexasaccharides attenuate metastasis by inhibition of P-selectin and heparanase. *Neoplasia*, 13, 445–452.

Bresalier R. S., Niv Y., Byrd J. C. et al. 1991. Mucin production by human colonic carcinoma cells correlates with their metastatic potential in animal models of colon cancer metastasis. *J Clin Invest*, 87, 1037–1045.

Bresalier R., Ho S., Schoeppner H. et al. 1996. Enhanced sialylation of mucin-associated carbohydrate structures in human colon cancer metastasis. *Gastroenterology*, 110, 1354–1367.

Brockhausen I., Matta K. L., Orr J., and Schachter H. 1985. Mucin synthesis. UDP-GlcNAc:GalNAc-R beta 3-N-acetylglucosaminyltransferase and UDP-GlcNAc:GlcNAc beta 1-3GalNAc-R (GlcNAc to GalNAc) beta 6-N-acetylglucosaminyltransferase from pig and rat colon mucosa. *Biochemistry*, 24, 1866–1874.

Brockhausen I., Romero P., and Herscovics A. 1991. Glycosyltransferase changes upon differentiation of CaCo-2 human colonic adenocarcinoma cells. *Cancer Res*, 51, 3136–3142.

Brockhausen I., Yang J., Burchell J., Whitehouse C., and Taylor-Papadimitriou J. 1995. Mechanism underlying aberrant glycosylation of the MUC1 mucin in breast cancer cells. *Eur J Biochem*, 233, 607–617.

Brockhausen I. and Kuhns W. 1997. *Glycoproteins and human disease, Medical Intelligence Unit.* CRC Press and Mosby Year Book. New York: Chapman & Hall.

Brockhausen I., Schutzbach J., and Kuhns W. 1998. Glycoproteins and their relationship to human disease. *Acta Anat (Basel)*, 161, 36–78.

Brockhausen I., Yang J., Dickinson N., Ogata S., and Itzkowitz S. H. 1998. Enzymatic basis for sialyl-Tn expression in human colon cancer cells. *Glycoconj J*, 15, 595–603.

Brockhausen, I. 1999. Pathways of O-glycan biosynthesis in cancer cells. *Biochim Biophys Acta*, 1473, 67–95.

Brockhausen I., Yang J., Lehotay M., Ogata S., and Itzkowitz S. H. 2001. Pathways of mucin O-glycosylation in normal and malignant rat colonic epithelial cells reveal a mechanism for cancer-associated Sialyl-Tn antigen expression. *Biol Chem*, 382, 219–232.

Brockhausen I., Benn M., Bhat S. et al. 2006. UDP-Gal: GlcNAc-R β1,4-Galactosyltransferase a target enzyme for drug design. Acceptor specificity and inhibition of the enzyme. *Glycoconj J*, 23, 525–541.

Brockhausen I. 2006. Mucin-type O-glycans in human colon and breast cancer: Glycodynamics and functions. *EMBO Rep*, 7, 599–604.

Brockhausen I. 2007. Biochemical aspects: Biosynthesis of mucin-type O-glycans. In *Comprehensive Glycosciences (From Chemistry to Systems Biology)*, Kamerling J. P. (ed). pp. 33–59. Burlington, MA: Acad. Press/Elsevier.

Brockhausen I., Dowler T., and Paulsen H. 2009. Site directed processing: Role of amino acid sequences and glycosylation of acceptor glycopeptides in the assembly of extended mucin type O-glycan core 2. *Biochim Biophys Acta*, 1790, 1244–1257.

Brockhausen I. 2010. Biosynthesis of complex mucin-type O-glycans. In *Comprehensive Natural Products II Chemistry and Biology*, Mander L., Lui H. -W., and Wang P. G. (eds). pp. 315–350. Oxford: Elsevier.

Brown J. R., Fuster M. M., Li R. et al. 2006. A disaccharide based inhibitor of glycosylation attenuates metastatic tumor cell dissemination. *Clin Cancer Res*, 12, 2894–2901.

Brown J. R., Yang F., Sinha A. et al. 2009. Deoxygenated disaccharide analogs as specific inhibitors of beta1-4-galactosyltransferase 1 and selectin-mediated tumor metastasis. *J Biol Chem*, 284, 4952–4959.

Buckhaults P., Chen L., Fregien N., and Pierce M. 1997. Transcriptional regulation of N-acetylglucosaminyltransferase V by the src oncogene. *J Biol Chem*, 272, 19575–19581.

Burchell J., Poulsom R., Hanby A. et al. 1999. An alpha 2,3 sialyltransferase (ST3Gal I) is elevated in primary breast carcinomas. *Glycobiology*, 9, 1307–1311.

Burke P. A., Gregg J. P., Bakhtiar B. et al. 2006. Characterization of MUC1 glycoprotein on prostate cancer for selection of targeting molecules. *Int J Oncol*, 29, 49–55.

Caldwell S. A., Jackson S. R., Shahriari K. S. et al. 2010. Nutrient sensor O-GlcNAc transferase regulates breast cancer tumorigenesis through targeting of the oncogenic transcription factor FoxM1. *Oncogene*, 29, 2831–2842.

Capon C., Maes E., Michalski J. C., Leffler H., and Kim Y.S. 2001. Sd(a)-antigen-like structures carried on core 3 are prominent features of glycans from the mucin of normal human descending colon. *Biochem J*, 358, 657–664.

Casado J. G., Delgado, E., Patsavoudi E. et al. 2008. Functional implications of HNK-1 expression on invasive behaviour of melanoma cells. *Tumour Biol*, 29, 304–310.

Chaturvedi P., Singh A. P., Chakraborty S. et al. 2008. MUC4 mucin interacts with and stabilizes the HER2 oncoprotein in human pancreatic cancer cells. *Cancer Res*, 68, 2065–2070.

Chen S. and Fukuda M. 2006. Cell type-specific roles of carbohydrates in tumor metastasis. *Methods Enzymol*, 416, 371–380.

Chen J. Y., Tang Y. A., Huang S.M. et al. 2011. A novel sialyltransferase inhibitor suppresses FAK/paxillin signaling and cancer angiogenesis and metastasis pathways. *Cancer Res*, 71, 473–483.

Cheung I. Y., Vickers A., and Cheung N. K. 2006. Sialyltransferase STX (ST8SiaII): A novel molecular marker of metastatic neuroblastoma. *Int J Cancer*, 119, 152–156.

Chiang C. H., Wang C. H., Chang H. C. et al. 2010. A novel sialyltransferase inhibitor AL10 suppresses invasion and metastasis of lung cancer cells by inhibiting integrin-mediated signaling. *J Cell Physiol*, 223, 492–499.

Chou T. Y., Hart G. W., and Dang C. V. 1995. c-Myc is glycosylated at threonine 58, a known phosphorylation site and a mutational hot spot in lymphomas. *J Biol Chem*, 270, 18961–18965.

Cozzi P. J., Wang J., Delprado W. et al. 2005. MUC1, MUC2, MUC4, MUC5AC and MUC6 expression in the progression of prostate cancer. *Clin Exp Metastasis*, 22, 565–573.

Dall'Olio F., Malagolini N., di Stefano G. et al. 1989. Increased CMP-NeuAc:Gal beta 1,4GlcNAc-R alpha 2,6 sialyltransferase activity in human colorectal cancer tissues. *Int J Cancer*, 44, 434–439.

Dall'Olio F., Chiricolo M., Ceccarelli C. et al. 2000. Beta-galactoside alpha 2,6 sialyltransferase in human colon cancer: Contribution of multiple transcripts to regulation of enzyme activity and reactivity with Sambucus nigra agglutinin. *Int J Cancer*, 88, 58–65.

Dall'Olio F. and Chiricolo M. 2001. Sialyltransferases in cancer. *Glycoconj J*, 18, 841–850.

Dalziel M., Whitehouse C., McFarlane I. et al. 2001. The relative activities of the C2GnT1 and ST3Gal-I glycosyltransferases determine O-glycan structure and expression of a tumor-associated epitope on MUC1. *J Biol Chem*, 276, 11007–11015.

De Leoz M. L., An H. J., Kronewitter S. et al. 2008. Glycomic approach for potential biomarkers on prostate cancer: Profiling of N-linked glycans in human sera and pRNS cell lines. *Dis Markers*, 25, 243–258.

Delannoy P., Kim I., Emery N. et al. 1996. Benzyl-N-acetyl-alpha-D-galactosaminide inhibits the sialylation and the secretion of mucins by a mucin secreting HT-29 cell subpopulation. *Glycoconj J*, 13, 717–726.

Demetriou M., Nabi I. R., Coppolino M., Dedhar S., and Dennis J. W. 1995. Reduced contact-inhibition and substratum adhesion in epithelial cells expressing GlcNAc-transferase V. *J Cell Biol*, 130, 383–392.

Dennis J. W., Granovsky M., and Warren C. E. 1999. Glycoprotein glycosylation and cancer progression. *Biochim Biophys Acta*, 1473, 21–34.

Devine P. L., Clark B. A., Birrell G. W. et al. 1991. The breast tumor-associated epitope defined by monoclonal antibody 3E1.2 is an O-linked mucin carbohydrate containing N-glycolylneuraminic acid. *Cancer Res*, 51, 5826–5836.

Diamandis E. P. 2010. Cancer biomarkers: Can we turn recent failures into success? *J Natl Cancer Inst*, 102, 1462–1467.

Dohi T., Yuyama Y., Natori Y. et al. 1996. Detection of N-acetylgalactosaminyltransferase mRNA which determines expression of Sda blood group carbohydrate structure in human gastrointestinal mucosa and cancer. *Int J Cancer*, 67, 626–631.

Dohi T. and Kawamura Y. I. 2008. Incomplete synthesis of the Sda/Cad blood group carbohydrate in gastrointestinal cancer. *Biochim Biophys Acta*, 1780, 467–471.

Dwek M. V., Jenks A., and Leathem A. J. 2010. A sensitive assay to measure biomarker glycosylation demonstrates increased fucosylation of prostate specific antigen (PSA) in patients with prostate cancer compared with benign prostatic hyperplasia. *Clin Chim Acta*, 411, 1935–1939.

Fujita T., Murayama K., Hanamura T. et al. 2011. CSLEX (Sialyl Lewis X) is a useful tumor marker for monitoring of breast cancer patients. *Jpn J Clin Oncol*, 41, 394–399.

Fukushima K., Satoh T., Baba S., and Yamashita K. 2010. {alpha}1,2-fucosylated and {beta}-N-acetylgalactosaminylated prostate specific antigen as an efficient marker of prostatic cancer. *Glycobiology*, 21, 452–460.

Fung P. Y. S., Madej M., Koganty R. R., and Longenecker B. M. 1990. Active specific immunotherapy of a murine mammary adenocarcinoma using a synthetic tumor-associated glycoconjugate. *Cancer Res*, 50, 4308–4314.

Fuster M. M., Brown J. R., Wang L., and Esko J. D. 2003. A disaccharide precursor of sialyl Lewis X inhibits metastatic potential of tumor cells. *Cancer Res*, 63, 2775–2781.

Gao Y., Lazar C., Szarek W. A., and Brockhausen, I. 2010. Specificity of β4galactosyltransferase inhibitor 2-naphthyl 2-butanamido-2-deoxy-1-thio-beta-D-glucopyranoside. *Glycoconj J*, 27, 673–684.

Gao Y., Chen P. W., Chachadi V., Wang Y., Anastassiades T., and Brockhausen, I. 2010. The Characteristic Glycosylation Potential of Human Prostate Cancer Cells. *Glycobiology*, 20, 1491. Abstract 131.

Garbar C., Mascaux C., and Wespes E. 2008. Expression of MUC1 and sialyl-Tn in benign prostatic glands, high-grade prostate intraepithelial neoplasia and malignant prostatic glands: A preliminary study. *Anal Quant Cytol Histol*, 30, 71–77.

Granovsky M., Fata J., Pawling J. et al. 2000. Suppression of tumor growth and metastasis in Mgat5-deficient mice. *Nat Med*, 6, 306–312.

Gu Y., Mi W., Ge Y. et al. 2010. GlcNAcylation plays an essential role in breast cancer metastasis. *Cancer Res*, 70, 6344–6351.

Guo H. B., Zhang Q. S., and Chen H. L. 2000. Effects of H-ras and v-sis overexpression on N-acetylglucosaminyltransferase V and metastasis-related phenotypes in human hepatocarcinoma cells. *J Cancer Res Clin Oncol*, 126, 263–270.

Guo H. B., Johnson H., Randolph M. et al. 2010. Specific posttranslational modification regulates early events in mammary carcinoma formation. *Proc Natl Acad Sci USA*, 107, 21116–21121.

Guo Z. and Wang Q. 2009. Recent development in carbohydrate-based cancer vaccines. *Curr Opin Chem Biol*, 13, 608–617.

Hagisawa S., Ohyama C., Takahashi T. et al. 2005. Expression of core 2 beta1, 6-N-acetyl glucosaminyltransferase facilitates prostate cancer progression. *Glycobiology*, 15, 1016–1024.

Hart, G. W., Slawson C., Ramirez-Correa G., and Lagerlof O. 2010. Cross talk between O-GlcNAcylation and phosphorylation: Roles in signaling, transcription, and chronic disease. *Annu Rev Biochem*, 80, 825–858.

Hatakeyama S., Kyan A., Yamamoto H. et al. 2010. Core 2 N-acetylglucosaminyltransferase-1 expression induces aggressive potential of testicular germ cell tumor. *Int J Cancer*, 127, 1052–1059.

Hollingsworth M. A. and Swanson B. J. 2004. Mucins in cancer: Protection and control of the cell surface. *Nat Rev Cancer*, 4, 45–60.

Hsu C. C., Lin T. W., Chang W. W. et al. 2005. Soyasaponin-I-modified invasive behavior of cancer by changing cell surface sialic acids. *Gynecol Oncol*, 96, 415–422.

Huang M. C., Chen H. Y., Huang H. C. et al. 2006. C2GnT-M is downregulated in colorectal cancer and its re-expression causes growth inhibition of colon cancer cells. *Oncogene*, 25, 3267–3276.

Huang J., Liang J. T., Huang H. C. et al. 2007. Beta1, 4-N-acetylgalactosaminyltransferase III enhances malignant phenotypes of colon cancer cells. *Mol Cancer Res*, 5, 543–552.

Huang Y. L. and Wu C. Y. 2010. Carbohydrate-based vaccines: Challenges and opportunities. *Expert Rev Vaccines*, 9, 1257–1274.

Idikio H. A. 1997. Sialyl-lewis-X, Gleason grade and stage in non-metastatic human prostate cancer. *Glycoconj J*, 14, 875–877.

Inaba Y., Ohyama C., Kato T. et al. 2003. Gene transfer of alpha1,3-fucosyltransferase increases tumor growth of the PC-3 human prostate cancer cell line through enhanced adhesion to prostatic stromal cells. *Int J Cancer*, 107, 949–957.

Isaji T., Gu J., Nishiuchi R. et al. 2004. Introduction of bisecting GlcNAc into integrin alpha-5beta1 reduces ligand binding and down-regulates cell adhesion and cell migration. *J Biol Chem*, 279, 19747–19754.

Ishida H., Togayachi A., Sakai T. et al. 2005. A novel beta1,3-N-acetylglucosaminyltransferase (beta3Gn-T8), which synthesizes poly-N-acetyllactosamine, is dramatically upregulated in colon cancer. *FEBS Lett*, 579, 71–8.

Itano N. and Kimata, K. 2008. Altered hyaluronan biosynthesis in cancer progression. *Semin Cancer Biol*, 18, 268–274.

Itzkowitz S. H., Bloom E. J., Kokal W. A. et al. 1990. Sialosyl-Tn: A novel mucin antigen associated with prognosis in colorectal cancer patients. *Cancer*, 66, 1960–1966.

Iwai T., Kudo T., Kawamoto R. et al. 2005. Core 3 synthase is down-regulated in colon carcinoma and profoundly suppresses the metastatic potential of carcinoma cells. *Proc Natl Acad Sci USA*, 102, 4572–4577.

Jacob G. S. 1995. Glycosylation inhibitors in biology and medicine. *Curr Opin Struct Biol*, 5, 605–611.

Jass J. R. and Walsh M. D. 2001. Altered mucin expression in the gastrointestinal tract: A review. *J Cell Mol Med*, 5, 327–351.

Jiang J. and Gu J. 2010. Beta1,4-galactosyltransferase V A growth regulator in glioma. *Methods Enzymol*, 479, 3–23.

Ju T., Lanneau G. S., Gautam T. et al. 2008. Human tumor antigens Tn and sialyl Tn arise from mutations in Cosmc. *Cancer Res*, 68, 1636–1646.

Julien S., Lagadec C., Krzewinski-Recchi M. et al. 2005. Stable expression of sialyl-Tn antigen in T47-D cells induces a decrease of cell adhesion and an increase of cell migration. *Breast Cancer Res Treat*, 90, 77–84.

Julien S., Adriaenssens E., Ottenberg K. et al. 2006. ST6GalNAc I expression in MDA-MB-231 breast cancer cells greatly modifies their O-glycosylation pattern and enhances their tumourigenicity. *Glycobiology*, 16, 54–64.

Kariya Y., Kato R., Itoh S. et al. 2008. N-Glycosylation of laminin-332 regulates its biological functions. A novel function of the bisecting GlcNAc. *J Biol Chem*, 283, 33036–33045.

Kim Y. S., Itzkowitz S. H., Yuan M. et al. 1988. Lex and Ley antigen expression in human pancreatic cancer. *Cancer Res*, 48, 475–482.

King M. J., Chan A., Roe R. et al. 1994. Two different glycosyltransferase defects that result in GalNAc alpha-O-peptide (Tn) expression. *Glycobiology*, 4, 267–279.

Kojima N., Handa K., Newman W., and Hakomori S. 1992. Inhibition of selectin-dependent tumor cell adhesion to endothelial cells and platelets by blocking O-glycosylation of these cells. *Biochem Biophys Res Commun*, 182, 1288–1295.

Korja M., Jokilammi A., Salmi T. T. et al. 2009. Absence of polysialylated NCAM is an unfavorable prognostic phenotype for advanced stage neuroblastoma. *BMC Cancer*, 9, 57.

Kotera Y., Fontenot J. D., Pecher G., Metzgar R. S., and Finn O. J. 1994. Humoral immunity against a tandem repeat epitope of human mucin MUC-1 in sera from breast, pancreatic, and colon cancer patients. *Cancer Res*, 54, 2856–2860.

Kovjazin R., Volovitz I., Kundel Y. et al. 2011. ImMucin: A novel therapeutic vaccine with promiscuous MHC binding for the treatment of MUC1-expressing tumors. *Vaccine*, 29, 4676–4686.

Kudo T., Ikehara Y., Togayachi A. et al. 1998. Up-regulation of a set of glycosyltransferase genes in human colorectal cancer. *Lab Investig*, 78, 797–811.

Läubli H. and Borsig L. 2010. Selectins promote tumor metastasis. *Semin Cancer Biol*, 20, 169–177.

Lagana A., Goetz J. G., Cheung P. et al. 2006. Galectin binding to Mgat5-modified N-glycans regulates fibronectin matrix remodeling in tumor cells. *Mol Cell Biol*, 26, 3181–3193.

Lau K. S. and Dennis J. W. 2008. N-Glycans in cancer progression. *Glycobiology*, 18, 750–760.

Lee G. and Ge B. 2010. Inhibition of in vitro tumor cell growth by RP215 monoclonal antibody and antibodies raised against its anti-idiotype antibodies. *Cancer Immunol Immunother*, 59, 1347–1356.

Lee S. H., Hatakeyama S., Yu S. Y. et al. 2009. Core3 O-glycan synthase suppresses tumor formation and metastasis of prostate carcinoma PC3 and LNCaP cells through down-regulation of alpha2beta1 integrin complex. *J Biol Chem*, 284, 17157–17169.

Li M., Song L., and Qin X. 2010. Glycan changes: Cancer metastasis and anti-cancer vaccines. *J Biosci*, 35, 665–673.

Li W., Zhang W., Luo J. et al. 2010. Alpha1,3 fucosyltransferase VII plays a role in colorectal carcinoma metastases by promoting the carbohydration of glycoprotein CD24. *Oncol Rep*, 23, 1609–1617.

Li Y. and Cozzi P. J. 2007. MUC1 is a promising therapeutic target for prostate cancer therapy. *Curr Cancer Drug Targets*, 7, 259–271.

Li Y., Yang X., Nguyen A. H., and Brockhausen I. 2007. Requirement of N-glycosylation for the secretion of recombinant extracellular domain of human Fas in HeLa cells. *Int J Biochem Cell Biol*, 39, 1625–1636.

Li Z., Yamada S., Inenaga S. et al. 2011. Polypeptide N-acetylgalactosaminyltransferase 6 expression in pancreatic cancer is an independent prognostic factor indicating better overall survival. *Br J Cancer*, 104, 1882–1889.

Lin Z., Simeone D. M., Anderson M. A. et al. 2011. Mass spectrometric assay for analysis of haptoglobin fucosylation in pancreatic cancer. *J Proteome Res*, 10, 2602–2611.

Lise M., Belluco C., Perera S. P. et al. 2000. Clinical correlations of alpha 2,6-sialyltransferase expression in colorectal cancer patients. *Hybridoma*, 19, 281–286.

Lloyd K. O., Yin B. W., and Kudryashov V. 1997. Isolation and characterization of ovarian cancer antigen CA 125 using a new monoclonal antibody (VK-8): Identification as a mucin-type molecule. *Int J Cancer*, 71, 842–850.

Magnani J. L., Steplewski Z., Koprowski H., and Ginsburg V. 1983. Identification of the gastrointestinal and pancreatic cancer-associated antigen detected by monoclonal antibody 19-9 in the sera of patients as a mucin. *Cancer Res*, 43, 5489–5492.

Malykh Y. N., Schauer R., and Shaw L. 2001. N-Glycolylneuraminic acid in human tumours. *Biochimie*, 83, 623–634.

Marcos N. T., Pinho S., Grandela C. et al. 2004. Role of the human ST6GalNAc-I and ST6GalNAc-II in the synthesis of the cancer-associated sialyl-Tn antigen. *Cancer Res*, 64, 7050–7057.

Martín-Satué M., Marrugat R., Cancelas J. A., and Blanco J. 1998. Enhanced expression of alpha(1,3)-fucosyltransferase genes correlates with E-selectin-mediated adhesion and metastatic potential of human lung adenocarcinoma cells. *Cancer Res*, 58, 1544–1550.

Martín-Satué M., de Castellarnau C., and Blanco J. 1999. Overexpression of alpha(1,3)-fucosyltransferase VII is sufficient for the acquisition of lung colonization phenotype in human lung adenocarcinoma HAL-24Luc cells. *Br J Cancer*, 80, 1169–1174.

Matsuura N., Narita T., Hiraiwa N. et al. 1998. Gene expression of fucosyl- and sialyltransferases which synthesize sialyl Lewis x, the carbohydrate ligands for E-selectin, in human breast cancer. *Int J Oncol*, 12, 1157–1164.

Maupin K. A., Sinha A., Eugster E. et al. 2010. Glycogene expression alterations associated with pancreatic cancer epithelial-mesenchymal transition in complementary model systems. *PLoS One*, 5, e13002.

Meany D. L., Zhang Z., Sokoll L. J., Zhang H., and Chan D. W. 2009. Glycoproteomics for prostate cancer detection: Changes in serum PSA glycosylation patterns. *J Proteome Res*, 8, 613–619.

Metzgar R. S., Rodriguez N., Finn O. J. et al. 1984. Detection of a pancreatic cancer-associated antigen (DU-PAN-2 antigen) in serum and ascites of patients with adenocarcinoma. *Proc Natl Acad Sci USA*, 81, 5242–5246.

Mi W., Gu Y., Han C. et al. 2011. O-GlcNAcylation is a novel regulator of lung and colon cancer malignancy. *Biochim Biophys Acta*, 1812, 514–519.

Miyoshi E., Shinzaki S., Moriwaki K., and Matsumoto H. 2010. Identification of fucosylated haptoglobin as a novel tumor marker for pancreatic cancer and its possible application for a clinical diagnostic test. *Methods Enzymol*, 478, 153–164.

Miyoshi E., Ito Y., and Miyoshi Y. 2010. Involvement of aberrant glycosylation in thyroid cancer. *J Oncol*, 2010: 816595.

Monzavi-Karbassi B., Stanley J. S., Hennings L. et al. 2007. Chondroitin sulfate glycosaminoglycans as major P-selectin ligands on metastatic breast cancer cell lines. *Int J Cancer*, 120, 1179–1191.

Mungul A., Cooper L., Brockhausen I. et al. 2004. Sialylated core 1 based O-linked glycans enhance the growth rate of mammary carcinoma cells in MUC1 transgenic mice. *Internat J Oncol*, 25, 937–943.

Nakamori S., Kamyama M., Imaoka S. et al. 1993. Increased expression of sialyl Lewisx antigen correlates with poor survival in patients with colorectal carcinoma: Clinicopathological and immunohistochemical study. *Cancer Res*, 53, 3632–3637.

O'Boyle K. P., Zamore R., Adluri S. et al. 1992. Immunization of colorectal cancer patients with modified ovine submaxillary gland mucin and adjuvants induces IgM and IgG antibodies to sialylated Tn. *Cancer Res*, 52, 5663–5667.

Ogata S., Maimonis P. J., and Itzkowitz S. H. 1992. Mucins bearing the cancer-associated sialosyl-Tn antigen mediate inhibition of natural killer cell cytotoxicity. *Cancer Res*, 52, 4741–4746.

Ogawa J. I., Inoue H., and Koide S. 1997. Alpha-2, 3-Sialyltransferase type 3N and alpha-1,3-fucosyltransferase type VII are related to sialyl Lewis(x) synthesis and patient survival from lung carcinoma. *Cancer*, 79, 1678–1685.

Ohyama C., Hosono M., Nitta. K. et al. 2004. Carbohydrate structure and differential binding of prostate specific antigen to Maackia amurensis lectin between prostate cancer and benign prostate hypertrophy. *Glycobiology*, 14, 671–679.

Ohyama C. 2008. Glycosylation in bladder cancer. *Int J Clin Oncol*, 13, 308–313.

Orntoft T. F., Harving N., and Langkilde N. C. 1990. O-Linked mucin-type glycoproteins in normal and malignant colon mucosa: Lack of T-antigen expression and accumulation of Tn and sialosyl-Tn antigens in carcinomas. *Int J Cancer*, 45, 666–672.

Osunkoya A. O., Adsay N. V., Cohen C., Epstein J. I., and Smith S. L. 2008. MUC2 expression in primary mucinous and nonmucinous adenocarcinoma of the prostate: An analysis of 50 cases on radical prostatectomy. *Mod Pathol*, 21, 789–794.

Park J. H., Nishidate T., Kijima K. et al. 2010. Critical roles of mucin 1 glycosylation by transactivated polypeptide N-acetylgalactosaminyltransferase 6 in mammary carcinogenesis. *Cancer Res*, 70, 2759–2769.

Park J. H., Katagiri T., Chung S., Kijima K., and Nakamura Y. 2011. Polypeptide N-acetylgalactosaminyltransferase 6 disrupts mammary acinar morphogenesis through O-glycosylation of fibronectin. *Neoplasia*, 13, 320–326.

Patsos G., Hebbe-Viton V., Robbe-Masselot C. et al. 2009. O-Glycan inhibitors generate aryl-glycans, induce apoptosis and lead to growth inhibition in colorectal cancer cell lines. *Glycobiology*, 19, 382–398.

Peracaula R., Tabarés G., Royle L. et al. 2003. Altered glycosylation pattern allows the distinction between prostate-specific antigen (PSA) from normal and tumor origins. *Glycobiology*, 13, 457–470.

Peracaula R., Barrabés S., Sarrats A., Rudd P. M., and de Llorens R. 2008. Altered glycosylation in tumours focused to cancer diagnosis. *Dis Markers*, 25, 207–218.

Pérez-Garay M., Arteta B., Pagès L. et al. 2010. Alpha2,3-sialyltransferase ST3Gal III modulates pancreatic cancer cell motility and adhesion in vitro and enhances its metastatic potential in vivo. *PLoS One*, 5, e12524.

Petretti T., Kemmner W., Schulze B., and Schlag P. M. 2000. Altered mRNA expression of glycosyltransferases in human colorectal carcinomas and liver metastases. *Gut*, 46, 359–366.

Picco G., Julien S., Brockhausen I. et al. 2010. Over-expression of ST3Gal-I promotes mammary tumorigenesis. *Glycobiology*, 20, 1241–1250.

Pinho S., Marcos N. T., Ferreira B. et al. 2007. Biological significance of cancer-associated sialyl-Tn antigen: Modulation of malignant phenotype in gastric carcinoma cells. *Cancer Lett*, 249, 157–170.

Pinho S. S., Osório H., Nita-Lazar M. et al. 2009. Role of E-cadherin N-glycosylation profile in a mammary tumor model. *Biochem Biophys Res Commun*, 379, 1091–1096.

Premaratne P., Welén K., Damber J. E., Hansson G. C., and Backstrom M. 2011. O-Glycosylation of MUC1 mucin in prostate cancer and the effects of its expression on tumor growth in a prostate cancer xenograft model. *Tumour Biol*, 32, 203–213.

Przybylo M. and Lityńska A. 2011. Glycans in melanoma screening. Part 1. The role of β1,6-branched N-linked oligosaccharides in melanoma. *BiochemSoc Trans*, 39, 370–373.

Rajpert-De Meyts E., Poll S. N., Goukasian I. et al. 2007. Changes in the profile of simple mucin-type O-glycans and polypeptide GalNAc-transferases in human testis and testicular neoplasms are associated with germ cell maturation and tumour differentiation. *Virchows Arch*, 451, 805–814.

Saito H., Gu J., Nishikawa A. et al. 1995. Organization of the human N-acetylglucosaminyltransferase V gene. *Eur J Biochem*, 233, 18–26.

Sarrats A., Saldova R., Comet J. et al. 2010. Glycan characterization of PSA 2-DE subforms from serum and seminal plasma. *OMICS*, 14, 465–474.

Sawada T., Ho J. J., Chung Y. S., Sowa M., and Kim Y. S. 1994. E-selectin binding by pancreatic tumor cells is inhibited by cancer sera. *Int J Cancer*, 57, 901–907.

Sayat R., Leber B., Grubac V., Wiltshire L., and Persad S. 2008. O-GlcNAc-glycosylation of beta-catenin regulates its nuclear localization and transcriptional activity. *Exp Cell Res*, 314, 2774–2787.

Schneider F., Kemmer W., Haensch W. et al. 2001. Overexpression of sialyltransferase CMP-sialic acid: Galbeta1,3GalNAc-R alpha6-sialyltransferase is related to poor patient survival in human colorectal carcinomas. *Cancer Res*, 61, 4605–4611.

Seko A. and Yamashita K. 2005. Characterization of a novel galactose beta1, 3-N-acetylglucosaminyltransferase (beta3Gn-T8): The complex formation of beta3Gn-T2 and beta3Gn-T8 enhances enzymatic activity. *Glycobiology*, 15, 943–951.

Sewell R., Baeckstroem M., Dalziel M. et al. 2006. The ST6GalNAc-I sialyltransferase localises throughout the Golgi and is responsible for the synthesis of the tumour-associated sialyl Tn O-glycan in human breast cancer. *J Biol Chem*, 281, 3586–3594.

Shen Y., Kohla G., Lrhorfi A. L. et al. 2004. O-Acetylation and de-O-acetylation of sialic acids in human colorectal carcinoma. *Eur J Biochem*, 271, 281–290.

Shibao K., Izumi H., Nakayama Y. et al. 2002. Expression of UDP-N-acetyl-alpha-D-galactosamine-polypeptide GalNAc N-acetylgalactosaminyl transferase-3 in relation to differentiation and prognosis in patients with colorectoral carcinoma. *Cancer*, 94, 1939–1946.

Shimodaira K., Nakayama J., Nakamura N. et al. 1997. Carcinoma-associated expression of core 2 beta-1,6-N-acetylglucosaminyltransferase gene in human colorectal cancer: Role of O-glycans in tumor progression. *Cancer Res*, 57, 5201–5206.

Simpson M. A. and Lokeshwar V. B. 2008. Hyaluronan and hyaluronidase in genitourinary tumors. *Front Biosci*, 13, 5664–5680.

Slawson C., Copeland R. J., and Hart G. W. 2010. O-GlcNAc signaling: A metabolic link between diabetes and cancer? *Trends Biochem Sci*, 35, 547–555.

Song Y., Aglipay J. A., Bernstein J. D., Goswami S., and Stanley P. 2010. The bisecting GlcNAc on N-glycans inhibits growth factor signaling and retards mammary tumor progression. *Cancer Res*, 70, 3361–3371.

Springer G. F., Desai P. R., Spencer B. D. et al. 1995. T/Tn antigen vaccine is effective and safe in preventing recurrence of advanced breast carcinoma. *Cancer Detect Prev*, 19, 374–380.

Springer G. F. 1997. Immunoreactive T and Tn epitopes in cancer diagnosis, prognosis, and immunotherapy. *J Mol Med*, 75, 594–602.

Srinivasan N., Bane S. M., Ahire S. D., Ingle A. D., and Kalraiya R.D. 2009. Poly N-acetyllactosamine substitutions on N- and not O-oligosaccharides or Thomsen-Friedenreich antigen facilitate lung specific metastasis of melanoma cells via galectin-3. *Glycoconj J*, 26, 445–456.

Stannard K. A., Collins P. M., Ito K. et al. 2010. Galectin inhibitory disaccharides promote tumour immunity in a breast cancer model. *Cancer Lett*, 299, 95–110.

St. Hill C. A., Farooqui M., Mitcheltree G. et al. 2009. The high affinity selectin glycan ligand C2-O-sLex and mRNA transcripts of the core 2 beta-1,6-N-acetylglucosaminyltransferase (C2GnT1) gene are highly expressed in human colorectal adenocarcinomas. *BMC Cancer*, 9, 79.

Sun J. Y., Zhu M. Z., Wang S. W. et al. 2007. Inhibition of the growth of human gastric carcinoma in vivo and in vitro by swainsonine. *Phytomedicine*, 14, 353–359.

Suriano R., Ghosh S. K., Ashok B. T. et al. 2005. Differences in glycosylation patterns of heat shock protein, gp96: Implications for prostate cancer prevention. *Cancer Res*, 65, 6466–6475.

Suriano R., Ghosh S. K., Chaudhuri D. et al. 2009. Sialic acid content of tissue-specific gp96 and its potential role in modulating gp96-macrophage interactions. *Glycobiology*, 19, 1427–1435.

Swindall A. F. and Bellis S. L. 2011. Sialylation of the Fas death receptor by ST6Gal-I provides protection against Fas-mediated apoptosis in colon carcinoma cells. *J Biol Chem*, 286, 22982–22990.

Tabarés G., Radcliffe C.M., Barrabés S. et al. 2006. Different glycan structures in prostate-specific antigen from prostate cancer sera in relation to seminal plasma PSA. *Glycobiology*, 16, 132–145.

Tabarés G., Jung K., Reiche J. et al. 2007. Free PSA forms in prostatic tissue and sera of prostate cancer patients: Analysis by 2-DE and western blotting of immunopurified samples. *Clin Biochem*, 40, 343–350.

Tajiri M., Ohyama C., and Wada Y. 2008. Oligosaccharide profiles of the prostate specific antigen in free and complexed forms from the prostate cancer patient serum and in seminal plasma: A glycopeptide approach. *Glycobiology*, 18, 2–8.

Takenaka Y., Fukumori T., and Raz A. 2004. Galectin-3 and metastasis. *Glycoconj J*, 19, 543–549.

Tanaka F., Otake Y., Nakagawa T. et al. 2000. Expression of polysialic acid and STX, a human polysialyltransferase, is correlated with tumor progression in non-small cell lung cancer. *Cancer Res*, 60, 3072–3080.

Taniuchi K., Cerny R. L., Tanouchi A. et al. 2011. Overexpression of GalNAc-transferase GalNAc-T3 promotes pancreatic cancer cell growth. *Oncogene*, 30, 4843–4854.

Tarp M. A. and Clausen H. 2008. Mucin-type O-glycosylation and its potential use in drug and vaccine development. *Biochim Biophys Acta*, 1780, 546–563.

Toscano M. A., Bianco G. A., Ilarregui J. M. et al. 2007. Differential glycosylation of TH1, TH2 and TH-17 effector cells selectively regulates susceptibility to cell death. *Nat Immunol*, 8, 825–834.

Trinchera M., Malagolini N., Chiricolo M. et al. 2011. The biosynthesis of the selectin-ligand sialyl Lewis x in colorectal cancer tissues is regulated by fucosyltransferase VI and can be inhibited by an RNA interference-based approach. *Int J Biochem Cell Biol*, 43, 130–139.

Tsui K. H., Chang P. L., Feng T. H. et al. 2008. Evaluating the function of matriptase and N-acetylglucosaminyltransferase V in prostate cancer metastasis. *Anticancer Res*, 28, 1993–1999.

Valentiner U., Mühlenhoff M., Lehmann U., Hildebrandt H., and Schumacher U. 2011. Expression of the neural cell adhesion molecule and polysialic acid in human neuroblastoma cell lines. *Int J Oncol*, 39, 417–424.

Valenzuela H. F., Pace K. E., Cabrera P. V. et al. 2007. O-Glycosylation regulates LNCaP prostate cancer cell susceptibility to apoptosis induced by galectin-1. *Cancer Res*, 67, 6155–6162.

Vavasseur F., Dole K., Yang J. et al. 1994. O-Glycan biosynthesis in human colorectal adenoma cells during progression to cancer. *Eur J Biochem*, 222, 415–424.

Vavasseur F., Yang J., Dole K., Paulsen H., and Brockhausen I. 1995. Synthesis of O-glycan core 3: Characterization of UDP-GlcNAc: GalNAc-R beta3-N-acetyl-glucosaminyl-transferase activity from colonic mucosal tissues and lack of the activity in human cancer cell lines. *Glycobiology*, 5, 351–357.

Vlodavsky I., Korner G., Ishai-Michaeli R. et al. 1990. Extracellular matrix-resident growth factors and enzymes: Possible involvement in tumor metastasis and angiogenesis. *Cancer Metastasis Rev*, 9, 203–226.

Wang F. L., Cui S. X., Sun L. P. et al. 2009. High expression of alpha 2, 3-linked sialic acid residues is associated with the metastatic potential of human gastric cancer. *Cancer Detect Prev*, 32, 437–443.

Wang Q., Zhang J., and Guo Z. 2007. Efficient glycoengineering of GM3 on melanoma cell and monoclonal antibody-mediated selective killing of the glycoengineered cancer cells. *Bioorg Med Chem*, 15, 7561–7567.

Ween M. P., Hummitzsch K., Rodgers R. J., Oehler M. K., and Ricciardelli C. 2011. Versican induces a pro-metastatic ovarian cancer cell behavior which can be inhibited by small hyaluronan oligosaccharides. *Clin Exp Metastasis*, 28, 113–125.

Wu C., Guo X., Wang W. et al. 2010. N-Acetylgalactosaminyltransferase-14 as a potential biomarker for breast cancer by immunohistochemistry. *BMC Cancer*, 10, 123.

Yamori T., Kimura H., Stewart K. et al. 1987. Differential production of high molecular weight sulfated glycoproteins in normal colonic mucosa, primary colon carcinoma, and metastases. *Cancer Res*, 47, 2741–2747.

Yang J. M., Byrd J. C., Siddiki B. B. et al. 1994. Alterations of O-glycan biosynthesis in human colon cancer tissues. *Glycobiology*, 4, 873–884.

Yang W. H., Kim J. E., Nam H. W. et al. 2006. Modification of p53 with O-linked N-acetylglucosamine regulates p53 activity and stability. *Nat Cell Biol*, 8, 1074–1083.

Yang X., Yip J., Harrison M., and Brockhausen I. 2008. Primary human osteoblasts and bone cancer cells as models to study glycodynamics in bone. *Internat J Biochem Cell Biol*, 40, 471–483.

Yoon S. J., Park S. Y., Pang P. C. et al. 2010. N-Glycosylation status of beta-haptoglobin in sera of patients with prostate cancer vs. benign prostate diseases. *Int J Oncol*, 36, 193–203.

Yoshimura M., Nishikawa A., Ihara Y., Taniguchi S., and Taniguchi N. 1995. Suppression of lung metastasis of B16 mouse melanoma by N-acetylglucosaminyltransferase III gene transfection. *Proc Natl Acad Sci USA*, 92, 8754–8758.

Yu L. G., Andrews N., Zhao Q. et al. 2007. Galectin-3 interaction with Thomsen-Friedenreich disaccharide on cancer-associated MUC1 causes increased cancer cell endothelial adhesion. *J Biol Chem*, 282, 773–781.

Zhang J., Nakayama J., Ohyama C. et al. 2002. Sialyl Lewis X-dependent lung colonization of B16 melanoma cells through a selectin-like endothelial receptor distinct from E- or P-selectin. *Cancer Res*, 62, 4194–4198.

Zhao Y. Y., Takahashi M., Gu J. G. et al. 2008. Functional roles of N-glycans in cell signaling and cell adhesion in cancer. *Cancer Sci*, 99, 1304–1310.

Zhou X., Chen H., Wang Q., Zhang L., and Zhao J. 2011. Knockdown of Mgat5 inhibits CD133+ human pulmonary adenocarcinoma cell growth in vitro and in vivo. *Clin Invest Med*, 34, E155.

Zhu Y., Srivatana U., Ullah A. et al. 2001. Suppression of a sialyltransferase by antisense DNA reduces invasiveness of human colon cancer cells in vitro. *Biochim Biophys Acta*, 1536, 148–160.

9 Structural Glycobiology
Applications in Organ Transplantation

Dale Christiansen, Mark Agostino, Elizabeth Yuriev, Paul A. Ramsland, and Mauro S. Sandrin

CONTENTS

9.1 INTRODUCTION

Naturally occurring antibodies (i.e., those existing without known exogenous antigen stimulation) are fundamental in the response to foreign organisms and function in both innate and adaptive immune responses. Indeed, natural preformed antibodies in a recipient are the first immunological barrier for clinical transplantation. To overcome the chronic shortage of organs available for transplantation, the practice of live donation across the ABO blood group barrier is now performed in many countries (Nydegger et al. 2005; Wu et al. 2003). Furthermore, pig-to-human xenotransplantation is also being investigated as a viable solution. However, without intervention, the recipient's natural antibodies would cause hyperacute rejection of the graft. These natural antibodies recognize carbohydrate epitopes on the donor tissue (A or B blood group antigens for allotransplantation; Galα(1-3)Gal [αGal] for xenotransplantation), which are synthesized by closely related members of family 6 glycosyltransferases. In this chapter, we highlight the biochemical, genetic, and immunological features of these clinically important carbohydrate antigens from the enzymes responsible for their synthesis to their molecular interactions with the immune system.

9.2 NATURAL ANTIBODIES AND THEIR ANTIGENS

Natural antibodies are found in the serum of healthy individuals and are produced in the absence of obvious immunization. It is currently believed that such antibodies are produced as a consequence of exposure to environmental microbes (Sandrin and McKenzie 1994; Sandrin, Vaughan et al. 1994). However, low amounts of natural antibodies (mainly of immunoglobulin M [IgM] isotype) have been detected without any known cause of stimulation in germ-free, antigen-free, and maternal antibody–deprived animals (Bos et al. 1989). It is believed that these antibodies, along with the innate immune system, play a major role as a first line of defense against infection.

9.2.1 THE ABO BLOOD GROUPS

By mixing sera and erythrocytes from unrelated nonimmunized individuals, Landsteiner (1962) discovered and defined the ABO blood groups. Two blood group antigens, A and B, were identified with individuals having antibodies to the non-self-antigens; that is, individuals with blood group A have anti-B antibodies and individuals with blood group B have anti-A antibodies. In addition, two other phenotypes were recognized. The AB phenotype exhibits neither anti-A nor anti-B antibodies as they express both the antigens. The O phenotype lacks both A and B antigens and has both anti-A and anti-B antibodies. Most of these naturally occurring anti-AB antibodies are IgM (Mollison et al. 1997), which is the most important class of immunoglobulin involved in hyperacute graft rejection (Ierino and Sandrin 2007; Sandrin and McKenzie 1994), although in many instances immunoglobulin G (IgG) is also present (Mollison et al. 1997). For allotransplantation, clinical hyperacute rejection has been prevented by the judicious choice of donor and recipient. However, for transplantation across ABO incompatibilities, it is essential to reduce the presence of anti-A/-B antibodies prior to transplantation. Various strategies have been developed to achieve this in recipients (Gloor and Stegall 2007; Stussi et al. 2006).

9.2.2 THE αGAL XENOANTIGEN

Current technology is such that desired traits can be readily achieved in many domestic species by genetic modification. With this in mind, reevaluation of xenotransplantation to overcome the shortfall in available organs has recommenced. As it has been known for many years, all humans have naturally occurring IgM antibodies to pig antigens; thus, pig-to-human transplantation reintroduces the problem of hyperacute rejection. A significant advance for the field was the identification of αGal as the major antigen recognized by the human natural antibodies (Cooper et al. 1993, 1994; Good et al. 1992; Sandrin and McKenzie 1994; Sandrin et al. 1993; Sandrin, Vaughan et al. 1994). Depletion of these antibodies for clinical xenotransplantation is not viable as approximately 1% of total human IgG and 2% of human IgM is directed to αGal (Galili et al. 1984). In contrast, anti-ABO antibody levels are estimated as one quarter of this value (Galili et al. 1993). Therefore, this highlights that successful xenotransplantation requires genetic modifications of the pig to eliminate the antigen.

9.2.3 TARGET ANTIGENS FOR NATURAL ANTIBODIES

Natural antibodies recognize carbohydrate antigens that are nonreducing terminal α1,3-linked galactose (Gal) or N-acetylgalactosamine (GalNAc) found on both glycoproteins (as either N- or O-linked sugars) and glycolipids formed on precursor backbones by glycosyltransferases. The A and B transferases transfer either a Gal residue (B transferase) or a GalNAc residue (A transferase) on fucosylated precursors (for a review on this topic, see the study by Yamamoto [2004]). In contrast, the αGal antigen can be synthesized by two distinct enzymes: (1) The α1,3-galactosyltransferase (αGT), which transfers Gal residues to the nonfucosylated precursor N-acetyllactosamine (LacNAc) found on both glycoproteins and glycolipids (Larsen et al. 1989; Sandrin, Dabkowski et al. 1994), and (2) the isogloboside (iGb) 3 synthase (iGb3S), which transfers Gal residues exclusively to the glycolipid lactosylceramide (LacCer) to produce iGb3 (Keusch et al. 2000; Milland et al. 2006) (Figure 9.1).

Structural studies have been performed on free and lectin-bound αGal-containing saccharides and reveal some important features of these carbohydrate xenoantigens (reviewed by Yuriev et al. [2009]). A detailed comparison of glycosidic linkages revealed similar conformational properties for the free and lectin-bound saccharides. In particular, the Galα(1-3)Gal linkages showed a broad range of glycosidic torsions, indicating that the minimal epitope recognized by natural antibodies is relatively flexible. In contrast, the subsequent Galβ(1-4)GlcNAc/Glc linkages of αGal oligosaccharides adopt a limited number of conformations, suggesting that this portion

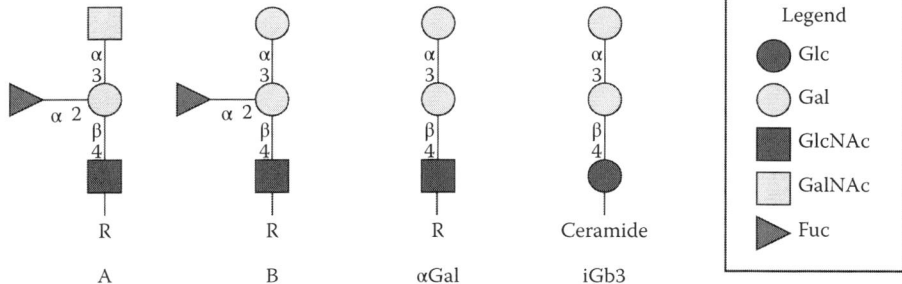

FIGURE 9.1 **(See color insert.)** Diagrammatic representation of the terminal carbohydrate structures of A and B blood group antigens, αGal and iGb3. Only the iGb3 structure is on LacCer, the others are on LacNAc. Glc = glucose; GlcNAc = N-acetylglucosamine; Fuc = fucose. Structures are represented in the Consortium for Functional Glycomics notation and were generated using GlycanBuilder (Ceroni et al. 2007). (Adapted from Milland, J. and Sandrin, M. S., *Tissue Antigens*, 68, 459–466, 2006.)

of the xenoantigens act as a rigid "stalk" for displaying the antigenic αGal epitopes. One interpretation of these conformational properties of carbohydrate xenoantigens is that the flexible terminal Galα(1-3)Gal epitopes can be recognized by a wider variety of binding sites than the rigid and less antigenic Galβ(1-4)GlcNAc/Glc stalk region. However, the presence of either a GlcNAc or a Glc at the third position may well influence the binding specificity of lectins and antibodies toward αGal carbohydrate xenoantigens.

9.2.4 ANTI-αGAL MONOCLONAL ANTIBODIES

The origin of natural antibodies and the role that gut bacteria play in the generation of anti-αGal and anti-AB antibodies are still unclear. Anti-αGal monoclonal antibodies (mAbs), generated by several laboratories including our own, have been useful in understanding the immunoglobulin gene structure, understanding the amino acid changes in the binding site of a natural antibody induced by an immune response, and determining expression of the epitope.

The first of the anti-αGal mAbs, Gal-13, was produced by immunizing BALB/c (αGal⁺) mice with rabbit erythrocytes (Galili, Basbaum et al. 1987). This antibody only recognized αGal on glycolipids (Galili, Basbaum et al. 1987; Teneberg et al. 1996).

The production of αGT knockout mice (GGTA1⁻/⁻; see Section 9.3.2) allowed us and others to develop a series of anti-αGal mAbs (Galili et al. 1998; Milland et al. 2007; Nozawa et al. 2001; Xu et al. 2001). As GGTA1⁻/⁻ mice, similar to human beings, are unable to synthesize the αGal epitope, these mAbs are likely to be structurally and functionally similar to human anti-αGal antibodies. In contrast to Gal-13, these mAbs recognize αGal on both glycolipids and glycoproteins (Galili et al. 1998; Milland et al. 2007). Furthermore, studies from our laboratory have demonstrated that the terminal carbohydrate residues determine most of the binding affinity, whereas the subsequent residues of the oligosaccharide define the fine specificity (Milland et al. 2007) (see Section 9.2.6). Indeed, one of these anti-αGal mAbs (15.101) was invaluable in determining expression of the epitope on iGb3 (Christiansen et al. 2008), and residual αGal has been reported on GGTA1⁻/⁻ mice (Milland et al. 2006) and some GGTA1⁻/⁻ pigs (Sharma et al. 2003).

9.2.5 ANTI-αGAL ANTIBODY GENES

Anti-αGal IgG and IgM antibody genes have been isolated and sequenced from patients before and after perfusion through pig hepatocytes (Kearns-Jonker et al. 1999). The immunoglobulin heavy-chain variable-region (IgV_H) genes of these antibodies demonstrated a high degree of amino acid identity. Indeed, many were derived from a single IgV_H germ line gene family, that is, IgV_H3. Moreover, the restricted number of germ line IgV_H gene sequences suggested a limited population of progenitors. Hence, polyclonal anti-αGal antibodies demonstrate little genetic or amino acid variability.

We performed similar studies in mice, using the mAbs produced in GGTA1⁻/⁻ mice (Nozawa et al. 2001). Comparisons of these IgV_H genes also demonstrated a limited number of germ line genes with the majority being V_H441 (Nozawa et al. 2001). In addition, both human beings and mice share conserved amino acid sequences at particular canonical positions within the H1 and H2 hypervariable loops of the IgV_H region, and both are therefore predicted to display similar three-dimensional structures (Nozawa et al. 2001). Similar results were found using mAbs produced from naive and xenotransplanted $GGTA1^{-/-}$ mice (Xu et al. 2001).

9.2.6 ANTI-αGAL ANTIBODY STRUCTURE

Although a crystal structure is yet to be determined for an αGal-binding antibody, we and other researchers have examined the structural basis for αGal recognition using a variety of experimental and computational approaches. In examining the crystal structures of lectins in complex with αGal carbohydrates, two major mechanisms for recognition were identified, that is, end-on insertion (Ramsland et al. 2003) and groove-type binding interactions (Yuriev et al. 2009). Kearns-Jonker and colleagues (2007) used homology modeling, docking, and mutagenesis to examine the structural basis for human αGal antibodies that were induced after exposure to pig hepatocytes following placement on a bioartificial liver. Potential amino acids were identified in the binding site and in the complementarity-determining region 3 (CDR3) region of the heavy chain. Docking of αGal carbohydrates into the binding sites of two closely related mouse mAbs, 8.17 and 15.101, demonstrated that end-on insertion was a likely mechanism for binding of antibodies to terminal Galα(1-3)Gal determinants. Furthermore, we identified that interactions with the third carbohydrate can determine the fine specificity of mAb for Galα(1-3)Galβ(1-4) GlcNAc- or Galα(1-3)Galβ(1-4)Glc-type carbohydrate xenoantigens (Milland et al. 2007). Based on these studies, our group utilized a panel of mAbs generated from $GGTA1^{-/-}$ mice (Nozawa et al. 2001) to further examine the structural basis for αGal carbohydrate recognition using *in silico* approaches, with the key findings summarized here.

By using a docking protocol that was validated and optimized against a set of high-resolution crystal structures of carbohydrate–antibody complexes (Agostino, Jene et al. 2009), a series of possible carbohydrate poses was generated for a variety of αGal carbohydrates ranging in size from monosaccharides to pentasaccharides. These ensembles of binding poses (up to 100 poses for each ligand) were used to generate "site maps" (see Chapter 5) in which hydrogen bonding and van der Waals interactions were tallied across the poses and mapped to the solvent-accessible surfaces of a panel of six αGal-binding mAbs. The site maps facilitated the identification of antibody-binding site residues that are frequently involved in the recognition of αGal carbohydrates. With this approach, we determined that a consensus set of residues (at nine structurally conserved positions) could account for the majority of predicted αGal interactions. Combined with a few additional residues, the consensus residues accounted for well over 80% of all interactions with different antibodies and the ensembles of the different αGal ligands (Agostino, Sandrin et al. 2009). The use of structurally conserved residues for binding αGal ligands across the panel of six mAbs supports the idea that a relatively conserved repertoire of antibodies (natural or elicited) recognizes αGal xenoantigens.

We next turned to examining the recognition of αGal from the point of view of the carbohydrate, where the interaction-based tallies were applied for *in silico* epitope mapping (Agostino et al. 2010). The terminal Galα(1-3)Gal disaccharide accounted for the majority of interactions with the different antibodies. In particular, the O3, O6, and C6 positions of the terminal Gal and the O6 and C6 positions of the second Gal were predominant sites on the ligands for interaction with antibodies. The third residue could also participate in contacts with the antibodies and the interactions with this residue (GlcNAc or Glc) helped to explain the fine specificity of one anti-αGal mAb, 15.101, that selectively binds to iGb3-type carbohydrates.

In addition to interaction-based selection from site and epitope mapping, we used the restricted conformational properties of Galβ(1-4)GlcNAc/Glc linkages (see Section 9.2.3) to help identify preferred binding modes for the trisaccharide or longer carbohydrate ligands (Agostino et al. 2010). Whereas different antibodies in the panel were predicted to bind αGal carbohydrates in different ways, the predominant binding mode was found to be end-on insertion with the Galα(1,3)Gal portion binding most deeply within the binding sites. The comparison of site maps, epitope maps, and preferred binding modes further revealed the structural basis for selective binding to Galα(1-3)Galβ(1-4)Glc by mAb 15.101, especially when compared with the closely related mAb 8.17 (Figure 9.2). Differences in the site maps highlight the pivotal contribution of a single residue (Gly_{109H} in 8.17 and Ser_{109H} in 15.101) in the binding site near the Galβ(1-4)GlcNAc/Glc disaccharide. In 8.17, Gly_{109H} contributes markedly to hydrogen bonding (Figure 9.2a) and only weakly to van der Waals (Figure 9.2b) site maps. In contrast, Ser_{109H} is only marginally involved in hydrogen bonding in 15.101 (Figure 9.2c), but it contributes more to van der Waals interactions (Figure 9.2d). The outcome of these differences is that Galα(1-3)Galβ(1-4)Glc can enter the binding site more deeply in 15.101 than can Galα(1-3)Galβ(1-4)GlcNAc in 8.17 (Agostino et al. 2010). This *in silico* observation is in agreement with the experimental finding of the reduced affinity of 15.101 for Galα(1-3)Galβ(1-4)GlcNAc compared with Galα(1-3)Galβ(1-4)Glc (Milland et al. 2007).

Although end-on insertion appears to dominate the binding of αGal carbohydrates by antibodies, it is possible that antibodies can discriminate between different xenoantigens using different binding strategies. Although it is not discussed in detail here, it is possible that groove-type binding modes occur to facilitate selective interactions with longer carbohydrate chains, including poly-αGal xenoantigens. In addition, preliminary structural studies of non-αGal xenoantibodies suggest that similar antibody recognition may occur with other, as yet unidentified, carbohydrate xenoantigens (Harnden et al. 2010).

9.2.7 ANTI-αGAL ANTIBODY RESPONSES

Human beings produce natural anti-αGal IgM and IgG in an age-dependent manner (Cramer 2000). Increases in both titer and affinity have been reported in sera of diabetic patients transplanted with fetal porcine islet cell clusters (Galili et al. 1995). Similarly, increases in anti-αGal IgM and IgG are seen after the implantation of porcine aortic (Konakci et al. 2005; Mangold et al. 2009) or pulmonary

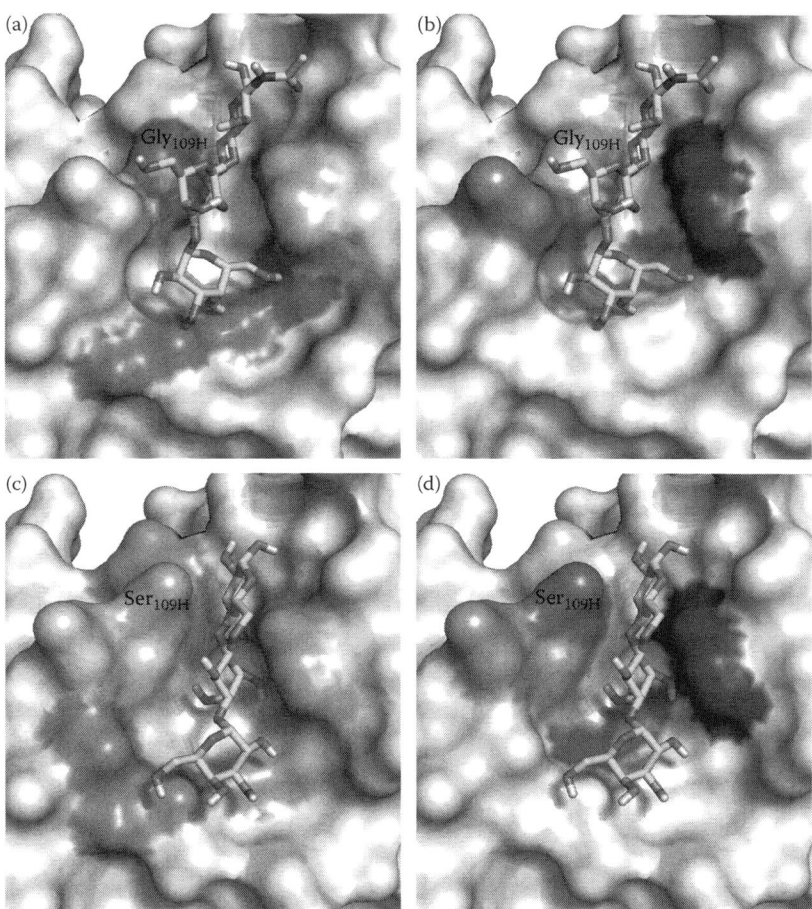

FIGURE 9.2 **(See color insert.)** Interaction-based site maps and predicted αGal carbohydrate-binding modes for two murine mAbs. Figure shows the likely binding mode obtained for Galα(1-3)Galβ(1-4)GlcNAc in complex with mAb 8.17, with hydrogen bonding (a) and van der Waals (b) site maps displayed on the solvent-accessible surfaces. It also shows the likely binding mode obtained for Galα(1-3)Galβ(1-4)Glc in complex with mAb 15.101, with hydrogen bonding (c) and van der Waals (d) site maps displayed on the solvent-accessible surfaces. Figures generated in PyMOL based on site mapping and binding-mode determination results reported previously (Agostino, Sandrin et al. 2009; Agostino et al. 2010).

(Park et al. 2010) valves. In contrast, no anti-αGal antibodies were observed when pig tendons treated with α-galactosidase were used to replace anterior cruciate ligaments (Stone et al. 2007).

The generation of GGTA1$^{-/-}$ mice has provided a very useful small-animal model to study anti-carbohydrate antibody production. Similar to human beings, naturally occurring anti-αGal antibodies are present in these mice (Thall et al. 1995) and are produced by B1-like B cells (Ohdan et al. 2000). The B-1 B cells, a subset of B cells, secrete natural serum antibodies from a restricted set of variable (V)-region genes. Furthermore, the induced anti-αGal IgM and IgG antibody responses are dependent on both CD4+ T cells (Chong et al. 2000; Cretin et al. 2002; Milland et al. 2006) and CD40 costimulation (Cretin et al. 2002; Tanemura et al. 2000). It has been suggested that xenogeneic peptides, possibly glycopeptides, activate a larger repertoire of helper T cells (Tanemura et al. 2000), resulting in stronger anti-αGal antibody responses observed with xenogeneic rather than allogeneic cells.

In contrast to the aforementioned anti-αGal antibody responses, where αGal glycopeptides can be presented by major histocompatibility complexes (MHCs) class II, CD1 molecules are generally considered as presenting glycolipids to T or natural killer T (NKT) cells to augment antibody responses. However, our recent study demonstrates that although purified neutral glycolipids (iGb3 and blood group B) require CD4+ T cells to generate anti-carbohydrate antibody responses, there is no requirement for CD1 NKT cells or peptides (Christiansen et al. 2011). Currently, it is unclear how these glycolipids are processed or what presentation pathway is used for this purpose.

9.2.8 CLINICAL RELEVANCE OF ANTI-αGAL ANTIBODIES

In general, anti-αGal antibodies do not have clinical relevance except in xenotransplantation. However, in certain circumstances they can interfere with treatments or become life threatening. The mAbs and soluble proteins produced for therapy may contain αGal residues (Ashford et al. 1993; Borrebaeck et al. 1993), the presence of which may result in reduced *in vivo* half-life due to immune complex formation (Borrebaeck et al. 1993; Sandrin and McKenzie 1994). Chung et al. (2008) reported anaphylactic reactions in cancer patients treated with a chimeric mouse–human mAb to epidermal growth factor receptor (cetuximab) due to the presence of anti-αGal IgE antibodies. Anti-αGal IgE antibodies have also been shown to be important in sensitivity to cat IgA (Gronlund et al. 2009) and meat-induced anaphylaxis (Commins et al. 2009; Jacquenet et al. 2009). In contrast to classical anaphylaxis that occurs within minutes after injection, the anti-αGal IgE meat-induced anaphylaxis typically manifests 3–6 hours after ingestion (Commins et al. 2009). Based on a previously reported association between red-meat allergies and tick bites (Van Nunen et al. 2009), Commins and coworkers (2011) demonstrated that patients with tick bites had elevated anti-αGal IgE. These authors suggest that immune response to the ectoparasite gives rise to the allergy.

9.3 GLYCOSYLTRANSFERASES

Carbohydrate antigens are typically oligosaccharides synthesized by glycosyltransferases that are Golgi resident type II integral membrane proteins. Structurally, these proteins consist of a short amino-terminal cytoplasmic tail, a transmembrane region, a stalk region, and a carboxyl-terminal catalytic domain that resides within the lumen of the Golgi (Joziasse 1992). Based on amino acid sequences, glycosyltransferases are currently grouped into 92 different families (CAZY: www.cazy.org); this grouping highlights the vast diversity of these enzymes and the reactions they catalyze. However, for optimum function, most glycosyltransferases have a pH range similar to that typically found within the secretory pathway (pH of 5.0–7.0) and require the presence of divalent cations (Mg^{2+} or Mn^{2+}). The glycosyltransferases relevant to this chapter, that is, the A and B transferases, αGT and iGb3S, all belong to family 6 based on their sequence conservation (~45% amino acid identity) (Milland and Sandrin 2006). The members of this family are further classified as retaining enzymes, as they retain the anomeric configuration that exists between the nucleoside and the sugar (α configuration) during catalysis. The exon organization is such that the cytoplasmic and the transmembrane domains are each encoded by separate exons. The stalk region is encoded by a variable number of exons and the catalytic domain is typically encoded by two exons, a small one toward the N-terminus and a larger one encoding the majority of the domain (Milland and Sandrin 2006).

9.3.1 HUMAN A AND B TRANSFERASES

The A and B blood group antigens, which are expressed on the surface of red blood cells and a variety of other cell types (Hakomori 1999), are synthesized by two glycosyltransferases that are encoded by two alleles at the ABO locus on chromosome 9q34 (Yamamoto et al. 1995). The A blood group antigen is synthesized by the A-specific α1,3-N-acetylgalactosaminyltransferase (GTA) (Yamamoto 2004). This enzyme catalyzes the transfer of a GalNAc residue in an α1,3-linkage to H substance (Fucα(1-2)Galβ(1-4)GlcNAc-R) (Seto et al. 1999). Similarly, the B blood group antigen is synthesized by the B-specific αGT (GTB) (Yamamoto 2004). The GTB uses the same substrate as GTA; however, it catalyzes the transfer of a Gal residue in an α1,3-linkage to H substance (Seto et al. 1999).

It is noted that GTA and GTB were among the first mammalian glycosyltransferases to be cloned, and the molecular basis for ABO polymorphism was established by demonstrating that GTA and GTB are indeed alleles (Yamamoto et al. 1990). Remarkably, the specificity of these enzymes is determined by amino acid differences at only four positions: (1) Arg/Gly$_{176}$, (2) Gly/Ser$_{235}$, (3) Leu/Met$_{266}$, and (4) Gly/Ala$_{268}$ (Yamamoto et al. 1990). Moreover, complementary deoxyribonucleic acid (cDNA) from individuals of blood group O has a frameshift mutation, due to deletion of a single nucleotide, resulting in a nonfunctional GTA, thus providing the genetic basis for O blood group (Yamamoto and Hakomori 1990).

Further structure–function studies were undertaken to elucidate residues involved in donor specificity. Several groups sequentially altered each of the four amino acids to that of the alternate enzyme (i.e., replacing residues in GTA with those found at the same positions in GTB) (Seto et al. 1997, 1999; Yamamoto and Hakomori 1990). Furthermore, the Gly/Ala$_{268}$ was mutated to every other amino acid (Yamamoto and McNeill 1996). The enzymes were then tested for their ability to transfer either Gal or GalNAc in an α1,3-linkage to H substance. The results of these studies showed that of the four amino acid differences, only the last two, that is, differences at positions 266 and 268, were crucial for determining substrate specificity of an enzyme (Seto et al. 2000; Yamamoto and McNeill 1996) and that Gly/Ala$_{268}$ directly interacts with the carbohydrate of the nucleotide sugar (Yamamoto and McNeill 1996). The crystal structure of the catalytic domains of both GTA and GTB confirmed that

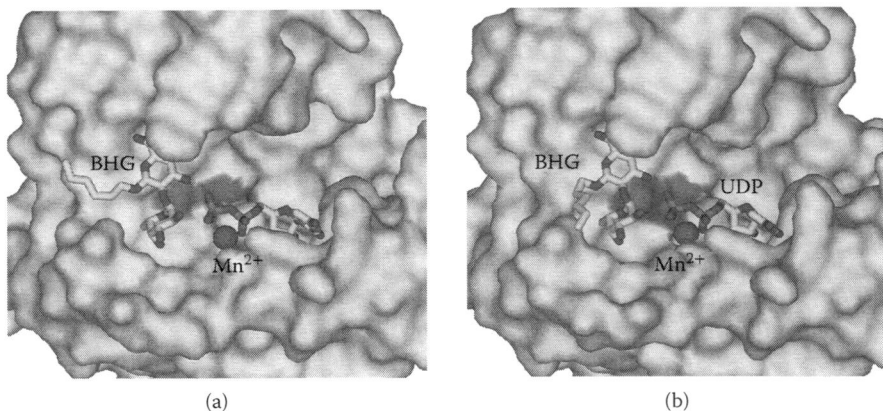

(a)   (b)

FIGURE 9.3 (**See color insert.**) Comparison of the crystal structures of A and B transferases. (a) The A transferase crystal structure (PDB ID: 1LZI) and (b) the B transferase crystal structure (PDB ID: 1LZJ) (Patenaude et al. 2002). Solvent-accessible surface is shown with key residues at positions 266 (blue) and 268 (magenta) mapped to the surface. The bound synthetic H (BHG) substance and a nucleotide (UDP) are shown as stick representations and the Mn^{2+} ion is shown as a space-filling sphere. Images were generated with DS Visualizer, version 2.5 (Accelrys, Inc., San Diego, California).

Leu/Met$_{266}$ and Gly/Ala$_{268}$ are positioned within the active sites and are in contact with the nucleotide-sugar residues (Figure 9.3) (Patenaude et al. 2002). However, these studies showed that only Leu/Met$_{266}$ can distinguish between uridine diphosphate (UDP)-GalNAc and UDP-Gal.

In humans, several natural mutations in glycosyltransferases family 6 have been characterized. There are mutations that affect the enzymatic kinetics of a transferase, such as the A2 phenotype, which has much lower amounts of A antigen expression due to an altered stop codon, causing read-through of an additional 21 amino acids (Yamamoto et al. 1992), or weak A or B blood groups (with reduced antigen expression) from amino acid substitutions (Hult et al. 2010). Other mutations generate a chimeric enzyme (GTA/B) that can synthesize both A and B antigens, which is known as the "*cis*-AB phenotype" (Yamamoto et al. 1993a; Yazer et al. 2006). The third group of mutations produces nonfunctional enzymes resulting in the O phenotype (Chester and Olsson 2001). In contrast to the most common O blood mutation that yields a truncated protein discussed above, another nondeletional O blood group allele, with a single-point mutation, synthesizes nonfunctional full-length GTA (Chester and Olsson 2001; Grunnet et al. 1994; Yamamoto et al. 1993b).

9.3.2 αGT

The αGal epitope is generated by an αGT, which transfers a terminal Gal residue to a subterminal Gal (Figure 9.1), and is expressed in most tissues (Sandrin and McKenzie 1994). The cDNA clones for this transferase have been isolated from numerous species, including mouse (Joziasse et al. 1992; Larsen et al. 1989), ox (Joziasse et al. 1989), rat (Taylor et al. 2003), and pig (Sandrin, Dabkowski et al. 1994), and are encoded by the *GGTA1* gene that comprises nine exons (Joziasse et al. 1992; Koike et al. 2000). Indeed, most mammalian species have a functional αGT, with the exception of Old World primates and human beings (Galili, Clark et al. 1987; Galili et al. 1988) in which mutations have inactivated the gene (Galili and Swanson 1991). In human beings, two αGT genes have been identified: (1) One on chromosome 9 (HGT-10, encoding only the N-terminal region of the protein) and (2) one on chromosome 12 (HGT-2, a processed pseudogene lacking introns, which contains multiple deletions and insertions) (Joziasse et al. 1991; Shaper et al. 1992). The inactivation of αGT is a relatively recent evolutionary event, which occurred after higher-order primates diverged into New and Old World monkeys (Galili 1999).

The catalytic domain of bovine αGT, the first family 6 member for which the tertiary structure was solved (Boix et al. 2001; Gastinel et al. 2001), is a globular protein comprising two central β-sheets, each comprising four β-strands, surrounded by four long α-helices (Figure 9.4). The four amino-terminal parallel β-strands, surrounded by two of the α-helices, form the nucleotide sugar–binding site. The remaining two parallel β-strands, flanked by two antiparallel β-strands and surrounded by the other two α-helices, are involved in acceptor substrate binding. The catalytic pocket is formed by portions of these two subdomains and a small β-sheet domain (comprising two β-strands), with the conserved Asp-Val-Asp (DVD) motif, characteristic of family 6, at the base.

Mutational/functional studies can be used to investigate the importance of individual amino acid residues for enzymatic function. Site-directed mutagenesis of pig αGT demonstrated that His_{271} was critical for enzyme function (Lazarus et al. 2002). In addition, molecular modeling suggests that His_{271} interacts with the Gal moiety of UDP-Gal (Lazarus et al. 2002). Enzymatic and structural analysis of bovine αGT

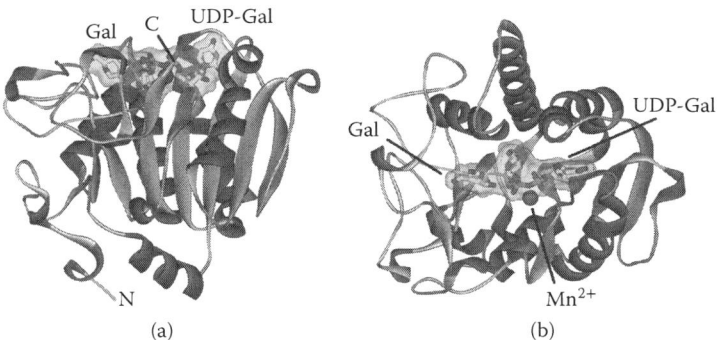

(a)   (b)

FIGURE 9.4   (See color insert.) Crystal structure of bovine αGT, a family 6 glycosyltransferase. (a) Side view and (b) end-on view of the αGT catalytic domain (PDB ID: 1G93) (Gastinel et al. 2001) are ribbon-style representations with secondary structure elements shown (β-strands in cyan; α-helices in red). The bound Gal and UDP-Gal are shown as stick representations and the Mn^{2+} ion is shown as a space-filling sphere. A solvent-accessible surface (transparent yellow) is displayed over the sugar substrate and nucleotide-sugar donor molecules. The N- and C-termini are indicated for the side view of the αGT catalytic domain. Images were generated with DS Visualizer, version 2.5 (Accelrys, Inc., San Diego, California).

mutants confirmed the importance of this residue for enzymatic function and identified the molecular interaction of His with Gal (Zhang et al. 2003). Furthermore, this study also showed contact between Gln_{238} and the acceptor substrate. Other studies have highlighted the important roles played by several other amino acid residues in enzymatic function (Tumbale et al. 2008; Zhang et al. 2004).

As discussed in Section 9.2.2, the major barrier to clinical pig-to-human xenotransplantation is the hyperacute rejection of vascularized organs due to presence of natural anti-Gal antibodies and complement activation. The most obvious way of preventing this is to develop αGal^- pigs by inactivating the *GGTA1* gene. This has been achieved in several species including mice (Tearle et al. 1996; Thall et al. 1995), pigs (Dai et al. 2002; Lai et al. 2002; Nottle et al. 2007; Phelps et al. 2003), and cattle (Sendai et al. 2006).

The *GGTA1* gene was inactivated in mice by targeting exon 9 (Tearle et al. 1996; Thall et al. 1995). As expected, these mice were viable, had marked decrease in αGal expression, and developed natural anti-αGal antibodies (Tearle et al. 1996; Thall et al. 1995). These mice have been useful in establishing small-animal models to evaluate immune responses to αGal (Christiansen et al. 2011; Gock et al. 2000; McKenzie, Koulmanda et al. 1998; McKenzie, Li et al. 1998; Pearse et al. 1998; Salvaris et al. 2000). Translation of this strategy to pigs was more problematic and required the advent of nuclear transfer technology. Several centers have produced $GGTA1^{-/-}$ pigs that show the same characteristics as $GGTA1^{-/-}$ mice (i.e., viable and show reduced αGal expression) (Dai et al. 2002; Lai et al. 2002; Nottle et al. 2007; Phelps et al. 2003). Transplantation of organs from $GGTA1^{-/-}$ pigs into immunosuppressed primates does not lead to hyperacute rejection; however, vascular injury still occurs due to the development of coagulopathy, leading to graft failure (Chen et al. 2005; Ekser et al. 2009; Hisashi et al. 2008; Kuwaki et al. 2005; Shimizu et al. 2008; Yamada et al. 2005).

9.3.3 iGb3S

Residual αGal expression has been observed in $GGTA1^{-/-}$ animals (Milland et al. 2006; Sharma et al. 2003), which demonstrates that in addition to αGT other enzymes can synthesize αGal. The most likely candidate is iGb3S (Keusch et al. 2000; Milland et al. 2006), another member of family 6. Indeed, in the rat, iGb3S was originally considered to be the αGT ortholog (Keusch et al. 2000); however, both functional αGTs (αGT and iGb3S) are present (Keusch et al. 2000; Taylor et al. 2003). Furthermore, iGb3S cDNA was cloned from the thymus of $GGTA1^{-/-}$ mice (Milland et al. 2006). The unique feature of iGb3S is that, in contrast to αGT, LacCer is the only substrate that can be utilized to synthesize the iGb series of glycolipids (Keusch et al. 2000; Taylor et al. 2003).

Several studies suggest that iGb3 is an endogenous glycolipid involved in the activation and development of NKT cells in both mice and human beings (Zhou et al. 2004). Studies have confirmed that iGb3 is an agonist for subsets of mouse and human NKT cells (Brigl et al. 2006; Cheng et al. 2007; Schumann et al. 2006; Scott-Browne et al. 2007; Wei et al. 2006; Xia et al. 2006; Zhou et al. 2004). However, normal development of NKT cells was seen in iGb3S$^{-/-}$ mice (Porubsky et al. 2007) and, furthermore, iGb3 was undetectable in mouse or human thymus (Speak et al. 2007), suggesting that iGb3 is not essential for NKT development.

To determine whether human iGb3S is enzymatically active, the catalytic domain of rat iGb3S was replaced with the analogous human domain (Christiansen et al. 2008). Unlike the rat iGb3S, the chimeric glycosyltransferase was unable to synthesize iGb3 (Christiansen et al. 2008). Moreover, substitution of single amino acids in the rat iGb3S sequence, with their human analogs, resulted in a significant reduction in activity (L$_{187}$P) or complete inactivation (Y$_{252}$N) of the rat enzyme (Christiansen et al. 2008). Thus, similar to αGT, iGb3S is inactivated in human beings.

The residual αGal on GGTA1$^{-/-}$ pigs was considered to be iGb3 (Milland et al. 2006; Milland et al. 2005; Sharma et al. 2003); however, more recent studies using mAbs and lectins (Diswall et al. 2011; Puga Yung et al. 2009) and mass spectrometry/nuclear magnetic resonance (NMR) (Diswall et al. 2007; Puga Yung et al. 2009) could not detect glycolipids terminating in αGal. Nevertheless, a fucosylated form of iGb3, originally reported in pig gut mucosa (Slomiany et al. 1974), was identified in GGTA1$^{-/-}$ and the wild-type pigs (Diswall et al. 2007). These findings suggest that iGb3 is unlikely to be involved in antibody-mediated xenograft destruction.

9.4   CONCLUSIONS

The demand for organs for transplantation has greatly outstripped the supply, leading to the consideration of pig-to-human xenotransplantation and live donation across the ABO blood group barrier. A more complete understanding of the immunobiology and structural glycobiology of carbohydrate antigens may result in the development of novel therapeutic approaches to prevent antibody-mediated damage in xenotransplantation.

REFERENCES

Agostino, M., Jene, C., Boyle, T., Ramsland, P. A., and Yuriev, E. 2009. Molecular docking of carbohydrate ligands to antibodies: Structural validation against crystal structures. *J Chem Inf Model*, 49, 2749–2760.

Agostino, M., Sandrin, M. S., Thompson, P. E., Yuriev, E., and Ramsland, P. A. 2009. In silico analysis of antibody-carbohydrate interactions and its application to xenoreactive antibodies. *Mol Immunol*, 47, 233–246.

Agostino, M., Sandrin, M. S., Thompson, P. E., Yuriev, E., and Ramsland, P. A. 2010. Identification of preferred carbohydrate binding modes in xenoreactive antibodies by combining conformational filters and binding site maps. *Glycobiology*, 20, 724–735.

Ashford, D. A., Alafi, C. D., Gamble, V. M. et al. 1993. Site-specific glycosylation of recombinant rat and human soluble CD4 variants expressed in Chinese hamster ovary cells. *J Biol Chem*, 268, 3260–3267.

Boix, E., Swaminathan, G. J., Zhang, Y. et al. 2001. Structure of UDP complex of UDP-galactose:beta-galactoside-alpha -1,3-galactosyltransferase at 1.53-A resolution reveals a conformational change in the catalytically important C terminus. *J Biol Chem*, 276, 48608–48614.

Borrebaeck, C. K., Malmborg, A. C., and Ohlin, M. 1993. Does endogenous glycosylation prevent the use of mouse monoclonal antibodies as cancer therapeutics? *Immunol Today*, 14, 477–479.

Bos, N. A., Kimura, H., Meeuwsen, C. G. et al. 1989. Serum immunoglobulin levels and naturally occurring antibodies against carbohydrate antigens in germ-free BALB/c mice fed chemically defined ultrafiltered diet. *Eur J Immunol*, 19, 2335–2339.

Brigl, M., van den Elzen, P., Chen, X. et al. 2006. Conserved and heterogeneous lipid antigen specificities of CD1d-restricted NKT cell receptors. *J Immunol*, 176, 3625–3634.

Ceroni, A., Dell, A., and Haslam, S.M. 2007. The GlycanBuilder: A fast, intuitive and flexible software tool for building and displaying glycan structures. *Source Code Biol Med*, 2, 3.

Chen, G., Qian, H., Starzl, T. et al. 2005. Acute rejection is associated with antibodies to non-Gal antigens in baboons using Gal-knockout pig kidneys. *Nat Med*, 11, 1295–1298.

Cheng, L., Ueno, A., Cho, S. et al. 2007. Efficient activation of Valpha14 invariant NKT cells by foreign lipid antigen is associated with concurrent dendritic cell-specific self recognition. *J Immunol*, 178, 2755–2762.

Chester, M. A. and Olsson, M. L. 2001. The ABO blood group gene: A locus of considerable genetic diversity. *Transfus Med Rev*, 15, 177–200.

Chong, A., Blinder, L., Ma, L. et al. 2000. Anti-galactose-alpha(1,3) galactose antibody production in alpha1,3-galactosyltransferase gene knockout mice after xeno and allo transplantation. *Transpl Immunol*, 8, 129–137.

Christiansen, D., Milland, J., Mouhtouris, E. et al. 2008. Humans lack iGb3 due to the absence of functional iGb3-synthase: Implications for NKT cell development and transplantation. *PLoS Biol*, 6, e172.

Christiansen, D., Vaughan, H. A., Miland, J. et al. 2011. Antibody responses to glycolipid borne carbohydrates require CD4+ T cells but not CD1 or NKT cells. *Immunol Cell Biol*, 89, 502–510.

Chung, C. H., Mirakhur, B., Chan, E. et al. 2008. Cetuximab-induced anaphylaxis and IgE specific for galactose-alpha-1,3-galactose. *N Engl J Med*, 358, 1109–1117.

Commins, S. P., James, H. R., Kelly, L. A. et al. 2011. The relevance of tick bites to the production of IgE antibodies to the mammalian oligosaccharide galactose-alpha-1,3-galactose. *J Allergy Clin Immunol*, 127, 1286–1293.

Commins, S. P., Satinover, S. M., Hosen, J. et al. 2009. Delayed anaphylaxis, angioedema, or urticaria after consumption of red meat in patients with IgE antibodies specific for galactose-alpha-1,3-galactose. *J Allergy Clin Immunol*, 123, 426–433.

Cooper, D. K., Good, A. H., Koren, E. et al. 1993. Identification of alpha-galactosyl and other carbohydrate epitopes that are bound by human anti-pig antibodies: Relevance to discordant xenografting in man. *Transpl Immunol*, 1, 198–205.

Cooper, D. K., Koren, E., and Oriol, R. 1994. Oligosaccharides and discordant xenotransplantation. *Immunol Rev*, 141, 31–58.

Cramer, D. V. 2000. Natural antibodies and the host immune responses to xenografts. *Xenotransplantation*, 7, 83–92.

Cretin, N., Bracy, J., Hanson, K., and Iacomini, J. 2002. The role of T cell help in the production of antibodies specific for Gal alpha 1–3Gal. *J Immunol*, 168, 1479–1483.

Dai, Y., Vaught, T. D., Boone, J. et al. 2002. Targeted disruption of the alpha1,3-galactosyltransferase gene in cloned pigs. *Nat Biotechnol*, 20, 251–255.

Diswall, M., Angstrom, J., Schuurman, H. J. et al. 2007. Studies on glycolipid antigens in small intestine and pancreas from alpha1,3-galactosyltransferase knockout miniature swine. *Transplantation*, 84, 1348–1356.

Diswall, M., Gustafsson, A., Holgersson, J., Sandrin, M. S., and Breimer, M. E. 2011. Antigen-binding specificity of anti-alphaGal reagents determined by solid-phase glycolipid-binding assays. A complete lack of alphaGal glycolipid reactivity in alpha1,3GalT-KO pig small intestine. *Xenotransplantation*, 18, 28–39.

Ekser, B., Rigotti, P., Gridelli, B., and Cooper, D.K. 2009. Xenotransplantation of solid organs in the pig-to-primate model. *Transpl Immunol*, 21, 87–92.

Galili, U. 1999. Evolution of alpha 1,3galactosyltransferase and of the alpha-Gal epitope. *Subcell Biochem*, 32, 1–23.

Galili, U., Anaraki, F., Thall, A., Hill-Black, C., and Radic, M. 1993. One percent of human circulating B lymphocytes are capable of producing the natural anti-Gal antibody. *Blood*, 82, 2485–2493.

Galili, U., Basbaum, C. B., Shohet, S. B., Buehler, J., and Macher, B. A. 1987. Identification of erythrocyte Gal alpha 1-3Gal glycosphingolipids with a mouse monoclonal antibody, Gal-13. *J Biol Chem*, 262, 4683–4688.

Galili, U., Clark, M. R., Shohet, S. B., Buehler, J., and Macher, B. A. 1987. Evolutionary relationship between the natural anti-Gal antibody and the Gal alpha 1–3Gal epitope in primates. *Proc Natl Acad Sci USA*, 84, 1369–1373.

Galili, U., LaTemple, D. C., and Radic, M. Z. 1998. A sensitive assay for measuring alpha-Gal epitope expression on cells by a monoclonal anti-Gal antibody. *Transplantation*, 65, 1129–1132.

Galili, U., Rachmilewitz, E. A., Peleg, A., and Flechner, I. 1984. A unique natural human IgG antibody with anti-α-galactosyl specificity. *J Exp Med*, 160, 1519–1531.

Galili, U., Shohet, S. B., Kobrin, E., Stults, C. L., and Macher, B. A. 1988. Man, apes, and Old World monkeys differ from other mammals in the expression of alpha-galactosyl epitopes on nucleated cells. *J Biol Chem*, 263, 17755–17762.

Galili, U. and Swanson, K. 1991. Gene sequences suggest inactivation of alpha-1,3-galactosyltransferase in catarrhines after the divergence of apes from monkeys. *Proc Natl Acad Sci USA*, 88, 7401–7404.

Galili, U., Tibell, A., Samuelsson, B., Rydberg, L., and Groth, C. G. 1995. Increased anti-Gal activity in diabetic patients transplanted with fetal porcine islet cell clusters. *Transplantation*, 59, 1549–1556.

Gastinel, L. N., Bignon, C., Misra, A. K. et al. 2001. Bovine alpha1,3-galactosyltransferase catalytic domain structure and its relationship with ABO histo-blood group and glycosphingolipid glycosyltransferases. *EMBO J*, 20, 638–649.

Gloor, J. M. and Stegall, M. D. 2007. ABO incompatible kidney transplantation. *Curr Opin Nephrol Hypertens*, 16, 529–534.

Gock, H., Salvaris, E., Murray-Segal, L. et al. 2000. Hyperacute rejection of vascularized heart transplants in BALB/c Gal knockout mice. *Xenotransplantation*, 7, 237–246.

Good, A. H., Cooper, D. K. C., Malcolm, A. J. et al. 1992. Identification of carbohydrate structures that bind human antiporcine antibodies: Implications for discordant xenografting in humans. *Transplant Proc*, 24, 559–562.

Gronlund, H., Adedoyin, J., Commins, S. P., Platts-Mills, T. A., and van Hage, M. 2009. The carbohydrate galactose-alpha-1,3-galactose is a major IgE-binding epitope on cat IgA. *J Allergy Clin Immunol*, 123, 1189–1191.

Grunnet, N., Steffensen, R., Bennett, E. P., and Clausen, H. 1994. Evaluation of histo-blood group ABO genotyping in a Danish population: Frequency of a novel O allele defined as O2. *Vox Sang*, 67, 210–215.

Hakomori, S. 1999. Antigen structure and genetic basis of histo-blood groups A, B and O: Their changes associated with human cancer. *Biochim Biophys Acta*, 1473, 247–266.

Harnden, I., Kiernan, K., and Kearns-Jonker, M. 2010. The anti-nonGal xenoantibody response to alpha1,3-galactosyltransferase gene knockout pig xenografts. *Curr Opin Organ Transplant*, 15, 207–211.

Hisashi, Y., Yamada, K., Kuwaki, K. et al. 2008. Rejection of cardiac xenografts transplanted from alpha1,3-galactosyltransferase gene-knockout (GalT-KO) pigs to baboons. *Am J Transplant*, 8, 2516–2526.

Hult, A. K., Yazer, M. H., Jorgensen, R. et al. 2010. Weak A phenotypes associated with novel ABO alleles carrying the A2-related 1061C deletion and various missense substitutions. *Transfusion*, 50, 1471–1486.

Ierino, F. L. and Sandrin, M. S. 2007. Spectrum of the early xenograft response: From hyperacute rejection to delayed xenograft injury. *Crit Rev Immunol*, 27, 153–166.

Jacquenet, S., Moneret-Vautrin, D. A., and Bihain, B. E. 2009. Mammalian meat-induced anaphylaxis: Clinical relevance of anti-galactose-alpha-1,3-galactose IgE confirmed by means of skin tests to cetuximab. *J Allergy Clin Immunol*, 124, 603–605.

Joziasse, D. H. 1992. Mammalian glycosyltransferases: Genomic organization and protein structure. *Glycobiology*, 2, 271–277.

Joziasse, D. H., Shaper, J. H., Jabs, E. W., and Shaper, N. L. 1991. Characterization of an alpha 1-3-galactosyltransferase homologue on human chromosome 12 that is organized as a processed pseudogene. *J Biol Chem*, 266, 6991–6998.

Joziasse, D. H., Shaper, J. H., Van den Eijnden, D. H., Van den Tunen, A. J., and Sharper, N. L. 1989. Bovine α1,3 galactosyltransferase: Isolation and characterisation of a cDNA clone. Identification of homologous sequences in human genomic DNA. *J Biol Chem*, 264, 14290–14297.

Joziasse, D. H., Shaper, N. L., Kim, D., Van den Eijnden, D. H., and Shaper, J. H. 1992. Murine α1,3 galactosyltransferase: A single gene locus specifies four isoforms of the enzyme by alternative spicing. *J Biol Chem*, 267, 5534–5541.

Kearns-Jonker, M., Barteneva, N., Mencel, R. et al. 2007. Use of molecular modeling and site-directed mutagenesis to define the structural basis for the immune response to carbohydrate xenoantigens. *BMC Immunol*, 8, 3.

Kearns-Jonker, M., Swensson, J., Ghiuzeli, C. et al. 1999. The human antibody response to porcine xenoantigens is encoded by IGHV3-11 and IGHV3-74 IgVH germline progenitors. *J Immunol*, 163, 4399–4412.

Keusch, J. J., Manzella, S. M., Nyame, K. A., Cummings, R. D., and Baenziger, J. U. 2000. Expression cloning of a new member of the ABO blood group glycosyltransferases, iGb3 synthase, that directs the synthesis of isoglobo-glycosphingolipids. *J Biol Chem*, 275, 25308–25314.

Koike, C., Friday, R. P., Nakashima, I. et al. 2000. Isolation of the regulatory regions and genomic organization of the porcine alpha1,3-galactosyltransferase gene. *Transplantation*, 70, 1275–1283.

Konakci, K. Z., Bohle, B., Blumer, R. et al. 2005. Alpha-Gal on bioprostheses: Xenograft immune response in cardiac surgery. *Eur J Clin Invest*, 35, 17–23.

Kuwaki, K., Tseng, Y. L., Dor, F. J. et al. 2005. Heart transplantation in baboons using alpha1,3-galactosyltransferase gene-knockout pigs as donors: Initial experience. *Nat Med*, 11, 29–31.

Lai, L., Kolber-Simonds, D., Park, K. W. et al. 2002. Production of alpha-1,3-galactosyltransferase knockout pigs by nuclear transfer cloning. *Science*, 295, 1089–1092.

Landsteiner, K. 1962. *The Specificity of Serological Reactions*. New York: Dover Publications Inc.

Larsen, R. D., Rajan, V. P., Ruff, M. M. et al. 1989. Isolation of a cDNA encoding a murine UDPgalactose:beta-D-galactosyl- 1,4-N-acetyl-D-glucosaminide alpha-1,3-galactosyltransferase: Expression cloning by gene transfer. *Proc Natl Acad Sci USA*, 86, 8227–8231.

Lazarus, B. D., Milland, J., Ramsland, P. A., Mouhtouris, E., and Sandrin, M. S. 2002. Histidine 271 has a functional role in pig alpha-1,3galactosyltransferase enzyme activity. *Glycobiology*, 12, 793–802.

Mangold, A., Szerafin, T., Hoetzenecker, K. et al. 2009. Alpha-Gal specific IgG immune response after implantation of bioprostheses. *Thorac Cardiovasc Surg*, 57, 191–195.

McKenzie, I. F., Koulmanda, M., Mandel, T. E., and Sandrin, M. S. 1998. Pig islet xenografts are susceptible to "anti-pig" but not Gal alpha(1,3)Gal antibody plus complement in Gal o/o mice. *J Immunol*, 161, 5116–5119.

McKenzie, I. F., Li, Y.Q., Patton, K., Thall, A. D., and Sandrin, M. S. 1998. A murine model of antibody-mediated hyperacute rejection by galactose-alpha(1,3)galactose antibodies in Gal o/o mice. *Transplantation*, 66, 754–763.

Milland, J., Christiansen, D., Lazarus, B. D. et al. 2006. The molecular basis for galalpha(1,3) gal expression in animals with a deletion of the alpha1,3galactosyltransferase gene. *J Immunol*, 176, 2448–2454.

Milland, J., Christiansen, D., and Sandrin, M. S. 2005. Alpha1,3-galactosyltransferase knockout pigs are available for xenotransplantation: Are glycosyltransferases still relevant? *Immunol Cell Biol*, 83, 687–693.

Milland, J. and Sandrin, M. S. 2006. ABO blood group and related antigens, natural antibodies and transplantation. *Tissue Antigens*, 68, 459–466.

Milland, J., Yuriev, E., Xing, P. X. et al. 2007. Carbohydrate residues downstream of the terminal Galalpha(1,3)Gal epitope modulate the specificity of xenoreactive antibodies. *Immunol Cell Biol*, 85, 623–632.

Mollison, P. E., Engelfreit, C. P., and Contreras, M. 1997. *Blood Transfusion in Clinical Medicine*. Oxford: Blackwell Science.

Nottle, M. B., Beebe, L. F., Harrison, S. J. et al. 2007. Production of homozygous alpha-1,3-galactosyltransferase knockout pigs by breeding and somatic cell nuclear transfer. *Xenotransplantation*, 14, 339–344.

Nozawa, S., Xing, P. X., Wu, G. D. et al. 2001. Characteristics of immunoglobulin gene usage of the xenoantibody binding to gal-alpha(1,3)gal target antigens in the gal knockout mouse. *Transplantation*, 72, 147–155.

Nydegger, U., Mohacsi, P., Koestner, S. et al. 2005. ABO histo-blood group system-incompatible allografting. *Int Immunopharmacol*, 5, 147–153.

Ohdan, H., Swenson, K. G., Kruger Gray, H. S. et al. 2000. Mac-1-negative B-1b phenotype of natural antibody-producing cells, including those responding to Gal alpha 1,3Gal epitopes in alpha 1,3-galactosyltransferase-deficient mice. *J Immunol*, 165, 5518–5529.

Park, C. S., Park, S. S., Choi, S. Y. et al. 2010. Anti alpha-gal immune response following porcine bioprosthesis implantation in children. *J Heart Valve Dis*, 19, 124–130.

Patenaude, S. I., Seto, N. O., Borisova, S. N. et al. 2002. The structural basis for specificity in human ABO(H) blood group biosynthesis. *Nat Struct Biol*, 9, 685–690.

Pearse, M. J., Witort, E., Mottram, P. et al. 1998. Anti-Gal antibody-mediated allograft rejection in alpha1,3-galactosyltransferase gene knockout mice: A model of delayed xenograft rejection. *Transplantation*, 66, 748–754.

Phelps, C. J., Koike, C., Vaught, T. D. et al. 2003. Production of alpha 1,3-galactosyltransferase-deficient pigs. *Science*, 299, 411–414.

Porubsky, S., Speak, A. O., Luckow, B. et al. 2007. Normal development and function of invariant natural killer T cells in mice with isoglobotrihexosylceramide (iGb3) deficiency. *Proc Natl Acad Sci USA*, 104, 5977–5982.

Puga Yung, G., Schneider, M. K., and Seebach, J.D. 2009. Immune responses to alpha1,3 galactosyltransferase knockout pigs. *Curr Opin Organ Transplant*, 14, 154–160.

Ramsland, P. A., Farrugia, W., Yuriev, E., Edmundson, A. B., and Sandrin, M. S. 2003. Evidence for structurally conserved recognition of the major carbohydrate xenoantigen by natural antibodies. *Cell Mol Biol (Noisy-le-grand)*, 49, 307–317.

Salvaris, E., Gock, H., Han, W. et al. 2000. Naturally acquired anti-alpha Gal antibodies in a murine allograft model similar to delayed xenograft rejection. *Xenotransplantation*, 7, 42–47.

Sandrin, M. S., Dabkowski, P. L., Henning, M. M., Mouhtouris, E., and McKenzie, I. F. C. 1994. Characterisation of cDNA clones for Porcine α(1,3)galactosyl transferase: The enzyme generating the Galα(1-3)Gal epitope. *Xenotransplantation*, 1, 81–88.

Sandrin, M. S. and McKenzie, I. F. C. 1994. Galα(1,3)Gal, the major xenoantigen(s) recognised in pigs by human natural antibodies. *Immunol Rev*, 141, 169–190.

Sandrin, M. S., Vaughan, H. A., Dabkowski, P. L., and McKenzie, I. F. 1993. Anti-pig IgM antibodies in human serum react predominantly with Gal(alpha 1-3)Gal epitopes. *Proc Natl Acad Sci USA*, 90, 11391–11395.

Sandrin, M. S., Vaughan, H. A., and McKenzie, I. F. C. 1994. Identification of Gal(α1-3) Gal as the major epitope for pig to human vascularised xenografts. *Transplant Rev* 8, 134–139.

Schumann, J., Mycko, M. P., Dellabona, P., Casorati, G., and MacDonald, H. R. 2006. Cutting edge: Influence of the TCR Vbeta domain on the selection of semi-invariant NKT cells by endogenous ligands. *J Immunol*, 176, 2064–2068.

Scott-Browne, J. P., Matsuda, J. L., Mallevaey, T. et al. 2007. Germline-encoded recognition of diverse glycolipids by natural killer T cells. *Nat Immunol*, 8, 1105–1113.

Sendai, Y., Sawada, T., Urakawa, M. et al. 2006. alpha1,3-Galactosyltransferase-gene knockout in cattle using a single targeting vector with loxP sequences and cre-expressing adenovirus. *Transplantation*, 81, 760–766.

Seto, N. O., Compston, C. A., Evans, S. V. et al. 1999. Donor substrate specificity of recombinant human blood group A, B and hybrid A/B glycosyltransferases expressed in Escherichia coli. *Eur J Biochem*, 259, 770–775.

Seto, N. O., Compston, C. A., Szpacenko, A., and Palcic, M.M. 2000. Enzymatic synthesis of blood group A and B trisaccharide analogues. *Carbohydr Res*, 324, 161–169.

Seto, N. O., Palcic, M. M., Compston, C. A. et al. 1997. Sequential interchange of four amino acids from blood group B to blood group A glycosyltransferase boosts catalytic activity and progressively modifies substrate recognition in human recombinant enzymes. *J Biol Chem*, 272, 14133–14138.

Shaper, N. L., Lin, S. P., Joziasse, D. H., Kim, D. Y., and Yang-Feng, T. L. 1992. Assignment of two human alpha-1,3-galactosyltransferase gene sequences (GGTA1 and GGTA1P) to chromosomes 9q33-q34 and 12q14-q15. *Genomics*, 12, 613–615.

Sharma, A., Naziruddin, B., Cui, C. et al. 2003. Pig cells that lack the gene for alpha1-3 galactosyltransferase express low levels of the gal antigen. *Transplantation*, 75, 430–436.

Shimizu, A., Hisashi, Y., Kuwaki, K. et al. 2008. Thrombotic microangiopathy associated with humoral rejection of cardiac xenografts from alpha1,3-galactosyltransferase gene-knockout pigs in baboons. *Am J Pathol*, 172, 1471–1481.

Slomiany, B. L., Slomiany, A., and Horowitz, M.I. 1974. Characterization of blood-group-H-active ceramide tetrasaccharide from hog-stomach mucosa. *Eur J Biochem*, 43, 161–165.

Speak, A. O., Salio, M., Neville, D. C. et al. 2007. Implications for invariant natural killer T cell ligands due to the restricted presence of isoglobotrihexosylceramide in mammals. *Proc Natl Acad Sci USA*, 104, 5971–5976.

Stone, K. R., Abdel-Motal, U. M., Walgenbach, A. W., Turek, T. J., and Galili, U. 2007. Replacement of human anterior cruciate ligaments with pig ligaments: A model for anti-non-gal antibody response in long-term xenotransplantation. *Transplantation*, 83, 211–219.

Stussi, G., West, L., Cooper, D. K. and Seebach, J. D. 2006. ABO-incompatible allotransplantation as a basis for clinical xenotransplantation. *Xenotransplantation*, 13, 390–399.

Tanemura, M., Yin, D., Chong, A. S., and Galili, U. 2000. Differential immune responses to alpha-gal epitopes on xenografts and allografts: Implications for accommodation in xenotransplantation. *J Clin Invest*, 105, 301–310.

Taylor, S. G., McKenzie, I. F., and Sandrin, M. S. 2003. Characterization of the rat alpha(1,3) galactosyltransferase: Evidence for two independent genes encoding glycosyltransferases that synthesize Galalpha(1,3)Gal by two separate glycosylation pathways. *Glycobiology*, 13, 327–337.

Tearle, R. G., Tange, M. J., Zannettino, Z. L. et al. 1996. The a-1,3-galactosyltransferase knockout mouse: Implications for xenotransplantation. *Transplantation*, 61, 13–19.

Teneberg, S., Lonnroth, I., Torres Lopez, J. F. et al. 1996. Molecular mimicry in the recognition of glycosphingolipids by Gal alpha 3 Gal beta 4 GlcNAc beta-binding Clostridium difficile toxin A, human natural anti alpha-galactosyl IgG and the monoclonal antibody Gal-13: Characterization of a binding-active human glycosphingolipid, non-identical with the animal receptor. *Glycobiology*, 6, 599–609.

Thall, A. D., Maly, P., and Lowe, J. B. 1995. Oocyte Gal alpha 1,3Gal epitopes implicated in sperm adhesion to the zona pellucida glycoprotein ZP3 are not required for fertilization in the mouse. *J Biol Chem*, 270, 21437–21440.

Tumbale, P., Jamaluddin, H., Thiyagarajan, N., Brew, K., and Acharya, K. R. 2008. Structural basis of UDP-galactose binding by alpha-1,3-galactosyltransferase (alpha3GT): Role of negative charge on aspartic acid 316 in structure and activity. *Biochemistry*, 47, 8711–8718.

Van Nunen, S. A., O'Connor, K. S., Clarke, L. R., Boyle, R. X., and Fernando, S. L. 2009. An association between tick bite reactions and red meat allergy in humans. *Med J Aust*, 190, 510–511.

Wei, D. G., Curran, S. A., Savage, P. B., Teyton, L., and Bendelac, A. 2006. Mechanisms imposing the Vbeta bias of Valpha14 natural killer T cells and consequences for microbial glycolipid recognition. *J Exp Med*, 203, 1197–1207.

Wu, A., Buhler, L. H. and Cooper, D. K. 2003. ABO-incompatible organ and bone marrow transplantation: Current status. *Transpl Int*, 16, 291–299.

Xia, C., Yao, Q., Schumann, J. et al. 2006. Synthesis and biological evaluation of alpha-galactosylceramide (KRN7000) and isoglobotrihexosylceramide (iGb3). *Bioorg Med Chem Lett*, 16, 2195–2199.

Xu, H., Sharma, A., Chen, L. et al. 2001. The structure of anti-Gal immunoglobulin genes in naive and stimulated Gal knockout mice. *Transplantation*, 72, 1817–1825.

Yamada, K., Yazawa, K., Shimizu, A. et al. 2005. Marked prolongation of porcine renal xenograft survival in baboons through the use of alpha1,3-galactosyltransferase gene-knockout donors and the cotransplantation of vascularized thymic tissue. *Nat Med*, 11, 32–34.

Yamamoto, F. 2004. Review: ABO blood group system—ABH oligosaccharide antigens, anti-A and anti-B, A and B glycosyltransferases, and ABO genes. *Immunohematology*, 20, 3–22.

Yamamoto, F., Clausen, H., White, T., Marken, J., and Hakomori, S. 1990. Molecular genetic basis of the histo-blood group ABO system. *Nature*, 345, 229–233.

Yamamoto, F. and Hakomori, S. 1990. Sugar-nucleotide donor specificity of histo-blood group A and B transferases is based on amino acid substitutions. *J Biol Chem*, 265, 19257–19262.

Yamamoto, F. and McNeill, P. D. 1996. Amino acid residue at codon 268 determines both activity and nucleotide-sugar donor substrate specificity of human histo-blood group A and B transferases. In vitro mutagenesis study. *J Biol Chem*, 271, 10515–10520.

Yamamoto, F., McNeill, P. D., and Hakomori, S. 1992. Human histo-blood group A2 transferase coded by A2 allele, one of the A subtypes, is characterized by a single base deletion in the coding sequence, which results in an additional domain at the carboxyl terminal. *Biochem Biophys Res Commun*, 187, 366–374.

Yamamoto, F., McNeill, P. D., and Hakomori, S. 1995. Genomic organization of human histo-blood group ABO genes. *Glycobiology*, 5, 51–58.

Yamamoto, F., McNeill, P. D., Kominato, Y. et al. 1993a. Molecular genetic analysis of the ABO blood group system: 2. cis-AB alleles. *Vox Sang*, 64, 120–123.

Yamamoto, F., McNeill, P. D., Yamamoto, M. et al. 1993b. Molecular genetic analysis of the ABO blood group system: 4. Another type of O allele. *Vox Sang*, 64, 175–178.

Yazer, M. H., Olsson, M. L., and Palcic, M. M. 2006. The cis-AB blood group phenotype: Fundamental lessons in glycobiology. *Transfus Med Rev*, 20, 207–217.

Yuriev, E., Agostino, M., Farrugia, W. et al. 2009. Structural biology of carbohydrate xenoantigens. *Expert Opin Biol Ther*, 9, 1017–1029.

Zhang, Y., Deshpande, A., Xie, Z. et al. 2004. Roles of active site tryptophans in substrate binding and catalysis by alpha-1,3 galactosyltransferase. *Glycobiology*, 14, 1295–1302.

Zhang, Y., Swaminathan, G. J., Deshpande, A. et al. 2003. Roles of individual enzyme-substrate interactions by alpha-1,3-galactosyltransferase in catalysis and specificity. *Biochemistry*, 42, 13512–13521.

Zhou, D., Mattner, J., Cantu, C., 3rd et al. 2004. Lysosomal glycosphingolipid recognition by NKT cells. *Science*, 306, 1786–1789.

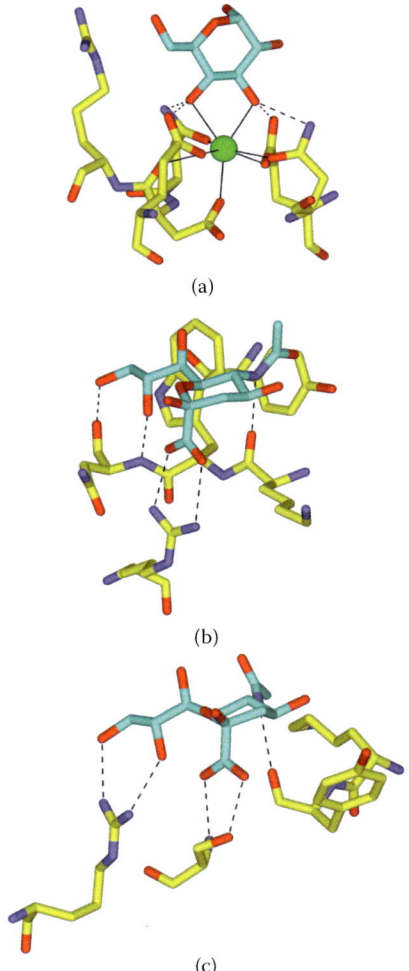

(a)

(b)

(c)

FIGURE 1.2 Anchored binding of terminal carbohydrate residues.

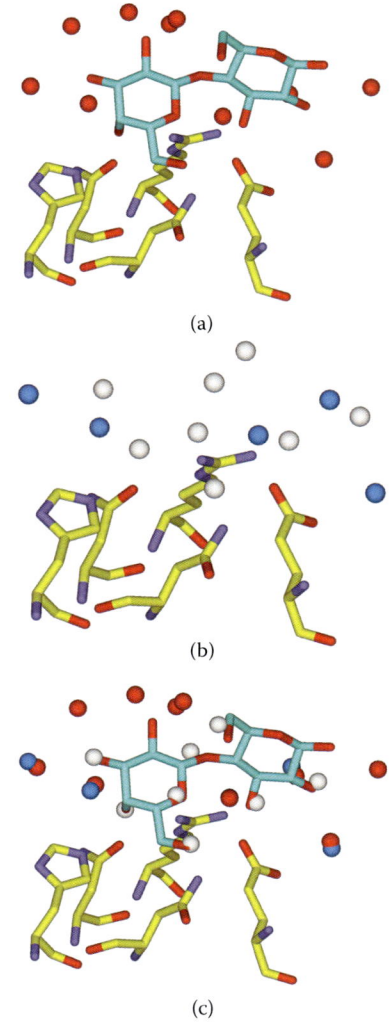

(a)

(b)

(c)

FIGURE 1.3 Central role of water in protein recognition of carbohydrates.

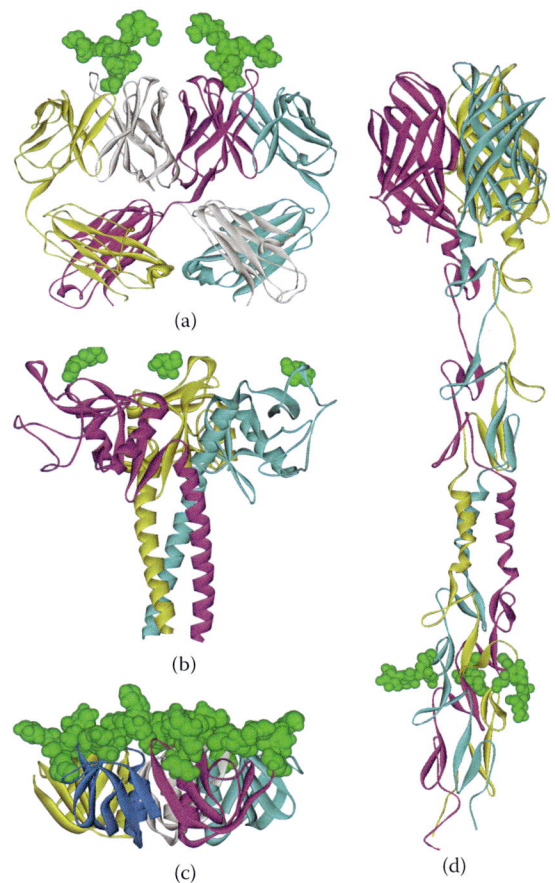

FIGURE 1.5 Examples of crystal structures of protein oligomers involved in binding to carbohydrates.

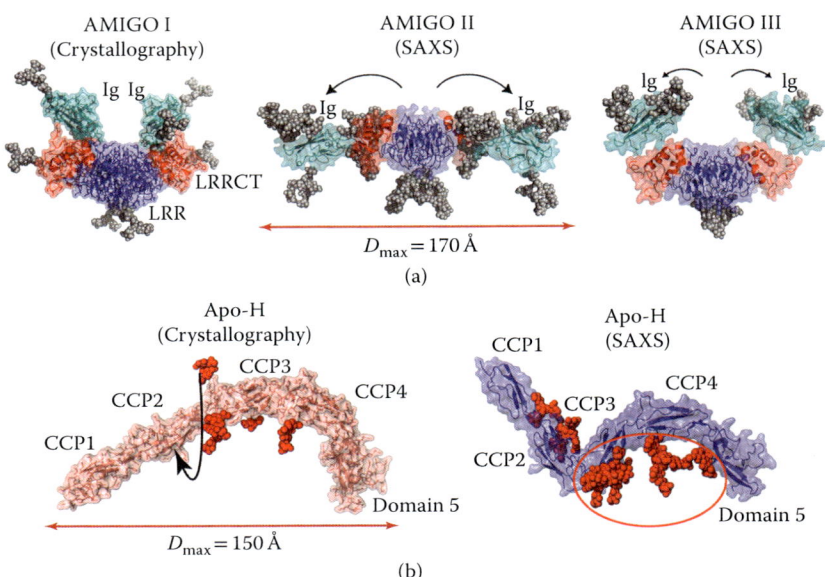

FIGURE 1.6 Crystallography in combination with solution SAXS of glycoproteins: AMIGO proteins and Apo-H.

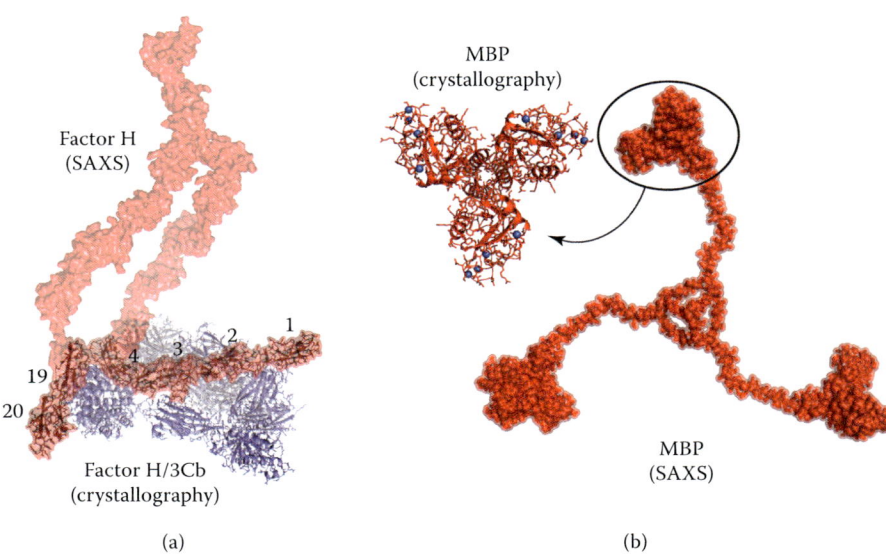

FIGURE 1.7 Building complex models from crystallographic and small-angle scattering data: Factor H and mannan-binding protein (MBP).

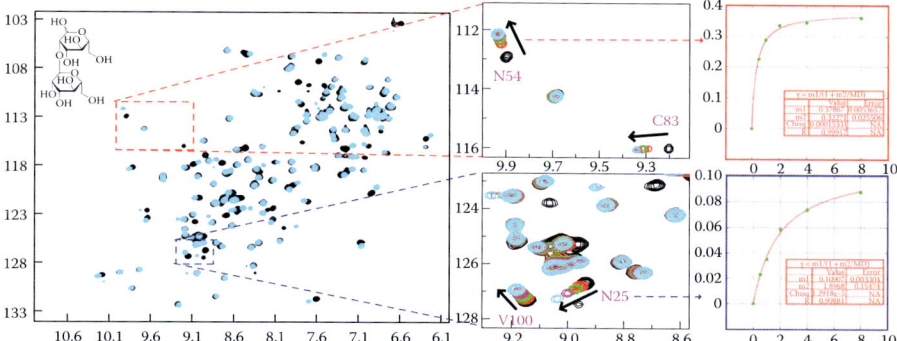

FIGURE 2.2 Superposition of the ^{1}H-^{15}N 2D HSQC spectra of free (black) and Manα(1-2) Man-bound (cyan) LKAMG.

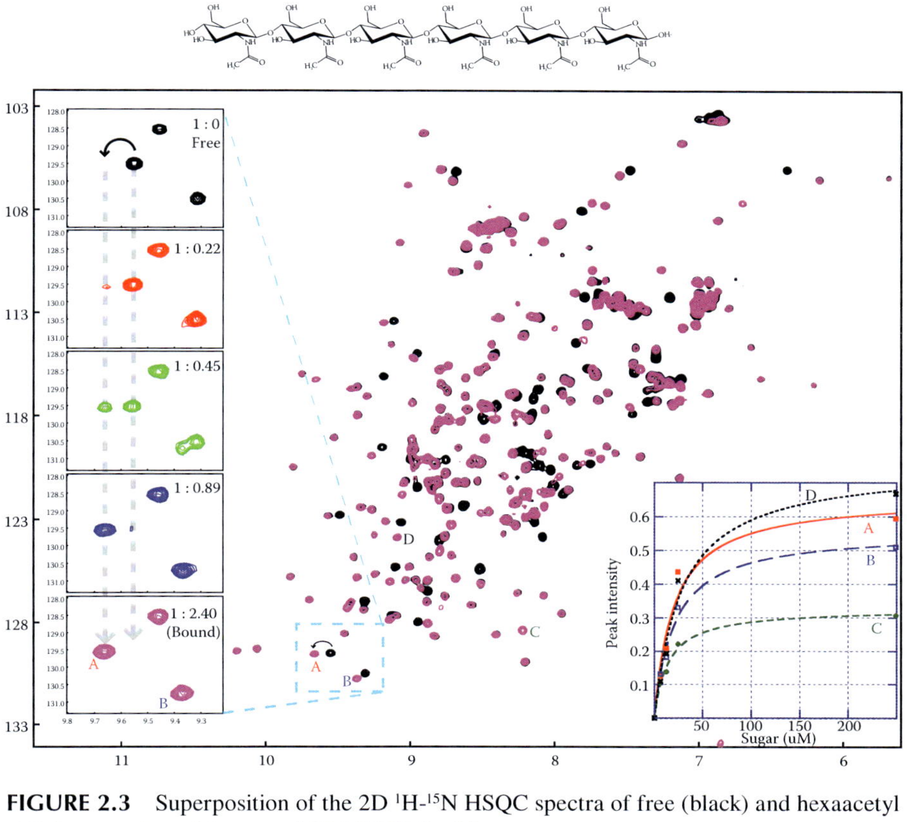

FIGURE 2.3 Superposition of the 2D ^{1}H-^{15}N HSQC spectra of free (black) and hexaacetyl chitohexaose-bound (magenta) MoCVNH-LysM.

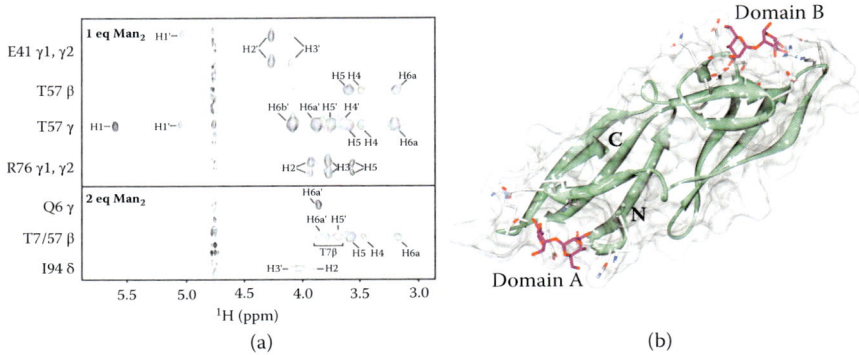

FIGURE 2.8 (a) Selected intermolecular NOEs between CV-N side chains of E41 Hγ1/γ2, T57 Hβ, T57 Hγ, R76 Hγ1/γ2, Q6 Hγ, T57 Hβ, and I94 Hδ and Manα(1-2)Man. (b) The calculated bound conformation of the dimannose in the protein binding sites.

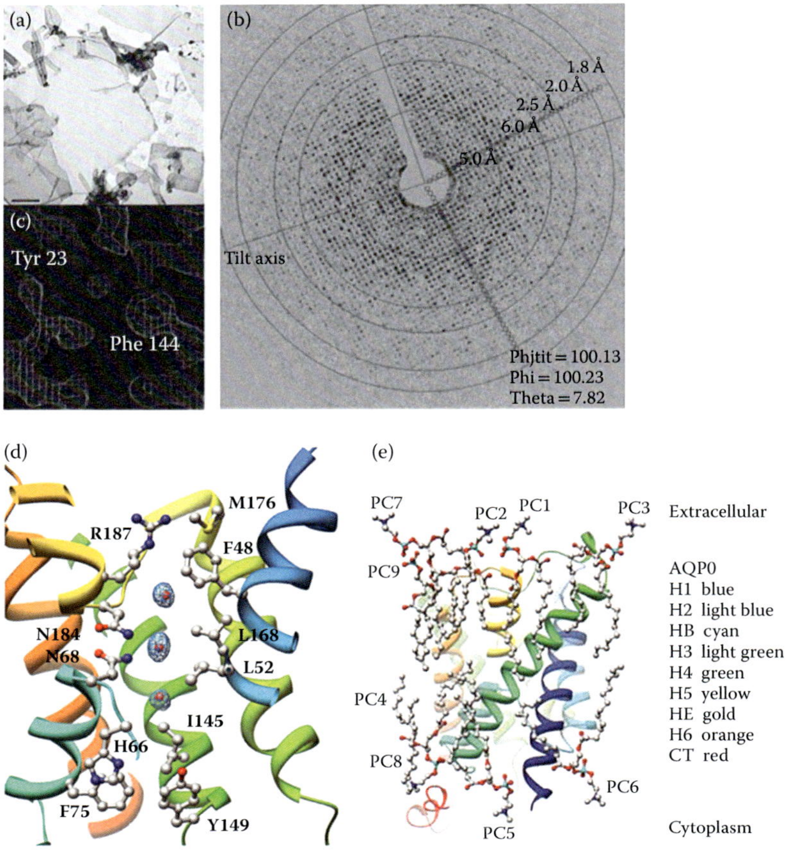

FIGURE 3.4 Structure determination of aquaporin-0 by electron crystallography.

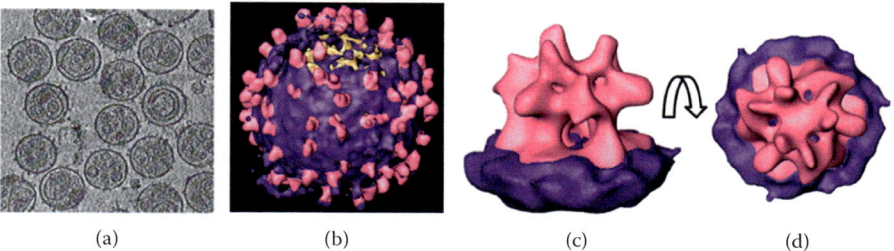

FIGURE 3.5   Analysis of intact moloney murine leukemia virus and its Env glycoprotein by cryoET.

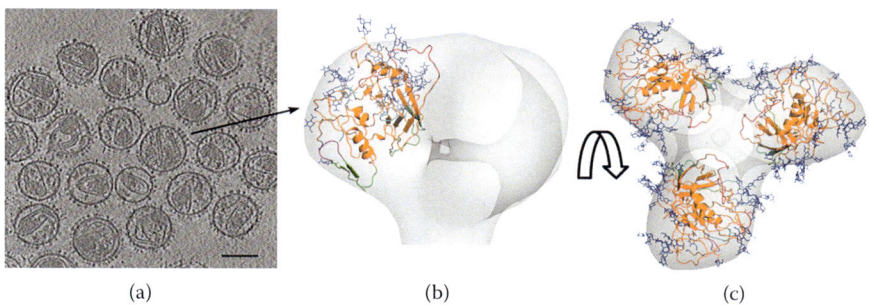

FIGURE 3.6   The structure of the simian immunodeficiency virus (SIV) Env complex by cryoET.

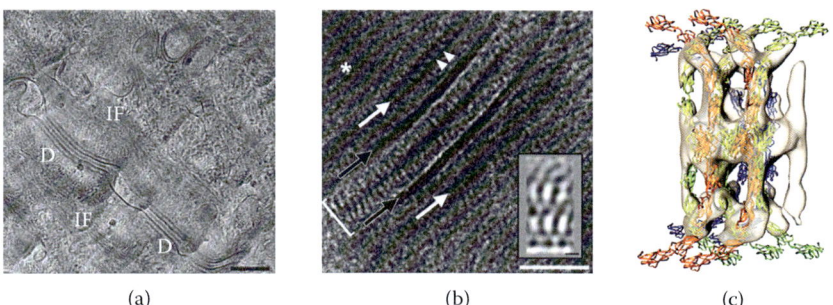

FIGURE 3.7   Molecular architecture of cadherins in native epidermal desmosomes.

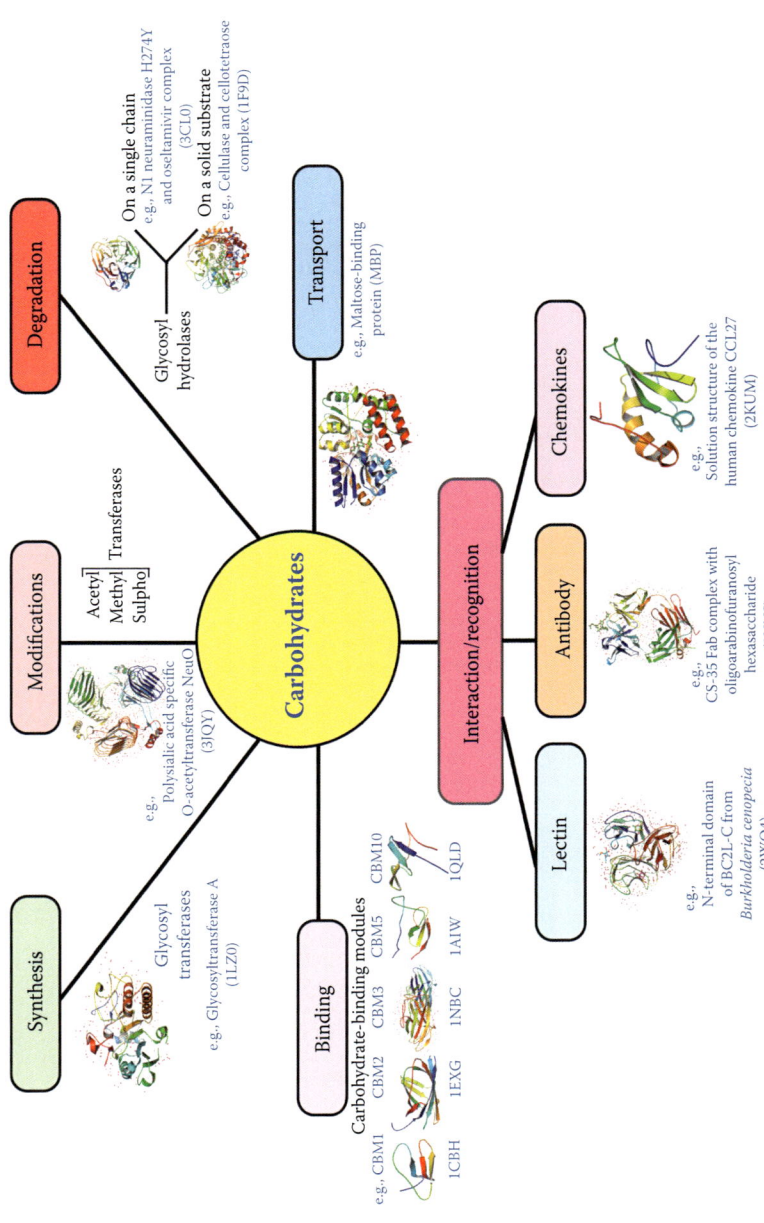

FIGURE 4.1 Synopsis of the families of proteins interacting with carbohydrates, illustrated with examples from PDB for synthesis, modifications (acetyltransferase), degradation by glycosyl hydrolases on a single chain or on a solid substrate, binding by carbohydrate-binding modules (CBMs), transport, interaction/recognition by lectins, antibodies, and chemokines.

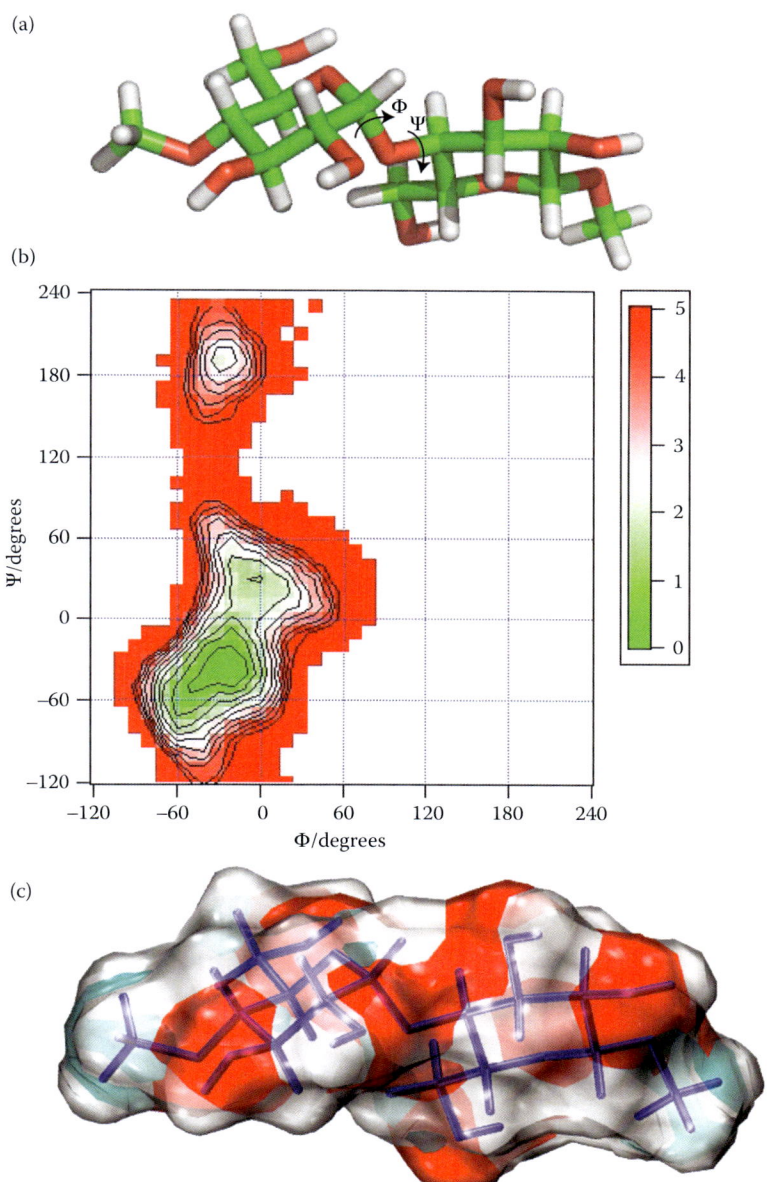

FIGURE 4.2 (a) Molecular representation of the disaccharide (Glcα(1-4)Glcβ) with the Φ and Ψ torsion angles shown on the glycosidic linkage. The potential energy surface shows conformational energy with respect to the Φ and Ψ torsion angles. (b) The favored low-energy Φ/Ψ combinations are shown in light color, while the high-energy regions are shown in red and the inaccessible regions are shown in white. (c) The surface of the disaccharide is composed of hydrophobic (green) and hydrophilic (red) patches, formed by nonpolar aliphatic protons and polar hydroxyl groups.

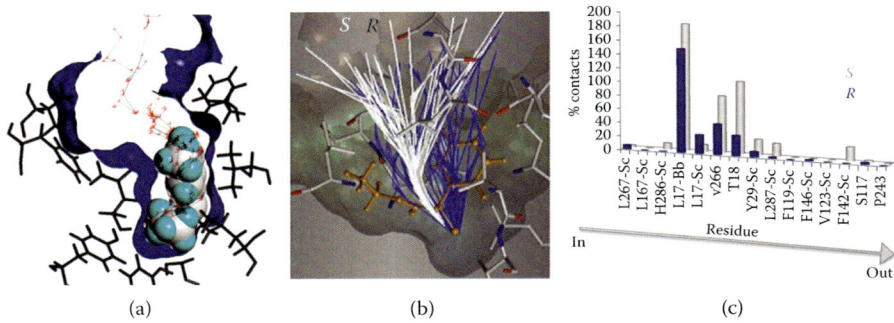

(a)   (b)   (c)

FIGURE 4.5 Illustration of the molecular robotics approach to investigate the role of substrate accessibility to the active site on *Burkholderia cepacia* lipase enantioselectivity.

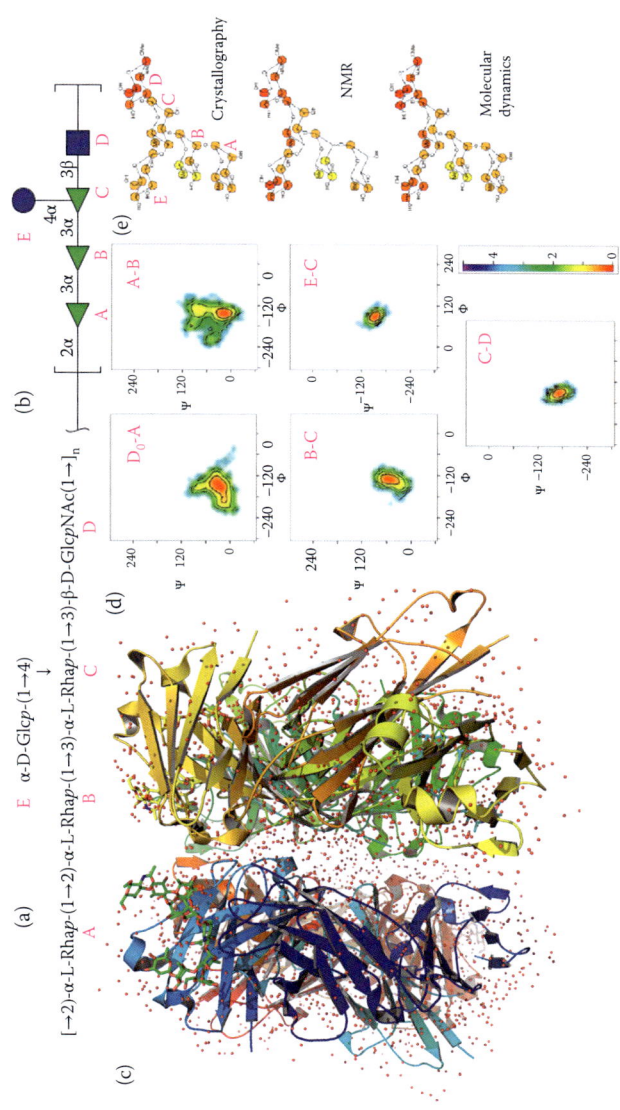

FIGURE 4.7 Features of *Shigella flexeneri* O antigen interacting with monoclonal antibody.

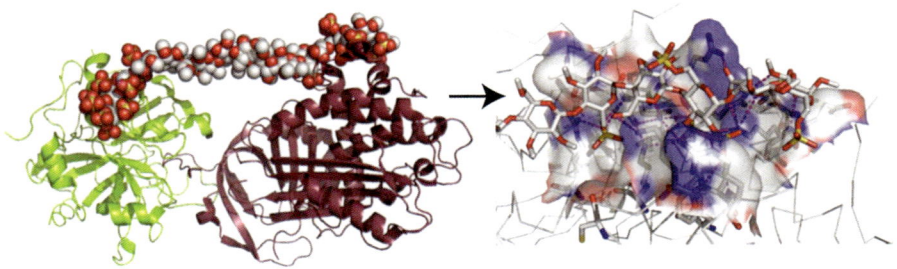

FIGURE 4.8 General view (left) of the crystal structure of ternary complexes between antithrombin (reddish-brown ribbon), thrombin (green ribbon), and heparin analog; (right) blowup of the binding site of antithrombin interacting with the specific heparin fragment.

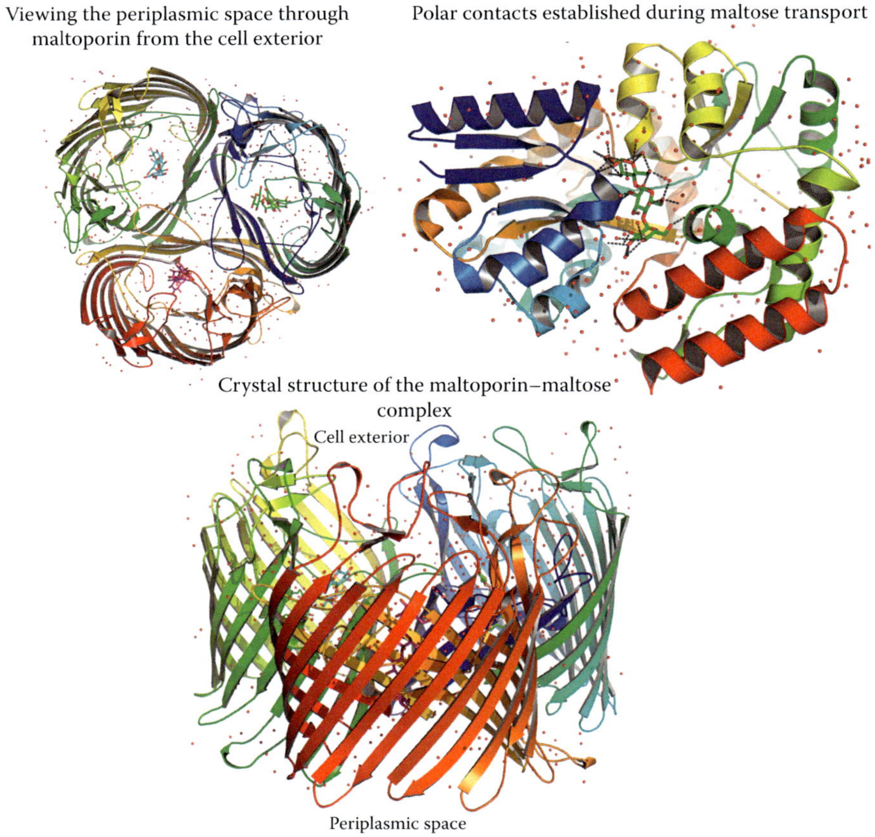

Viewing the periplasmic space through maltoporin from the cell exterior

Polar contacts established during maltose transport

Crystal structure of the maltoporin–maltose complex

Cell exterior

Periplasmic space

FIGURE 4.9 Three-dimensional structure of maltoporin along with snapshots of the interaction of maltooligosaccharides within the channel.

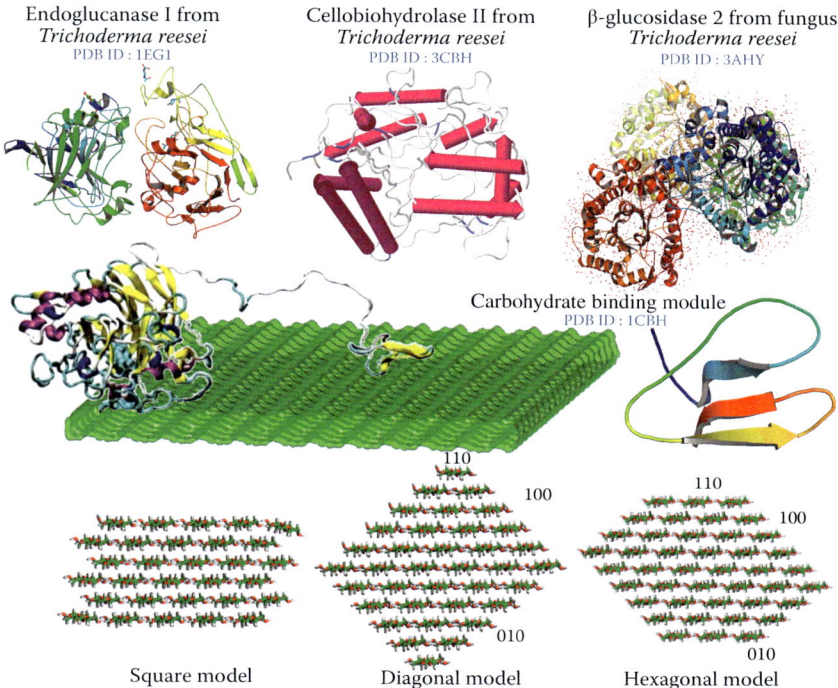

FIGURE 4.10 The enzymatic digestion of cellulose.

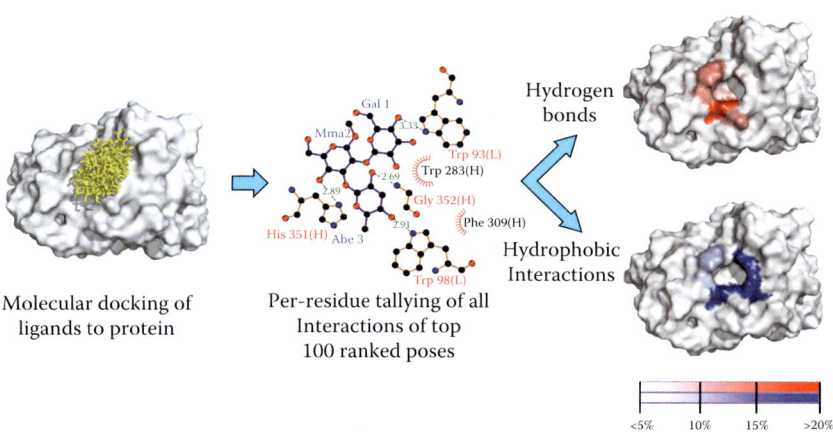

FIGURE 5.1 Illustration of the site-mapping process.

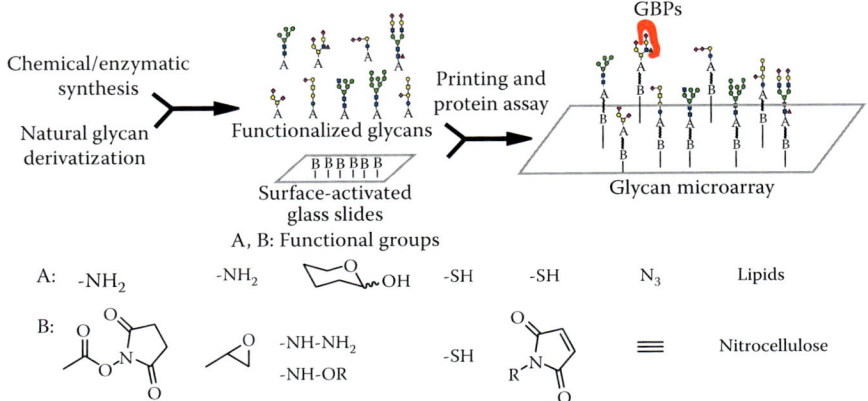

FIGURE 7.1 The concept of glycan microarrays and the chemistry involved in preparing glycan microarrays by covalent and noncovalent attachments.

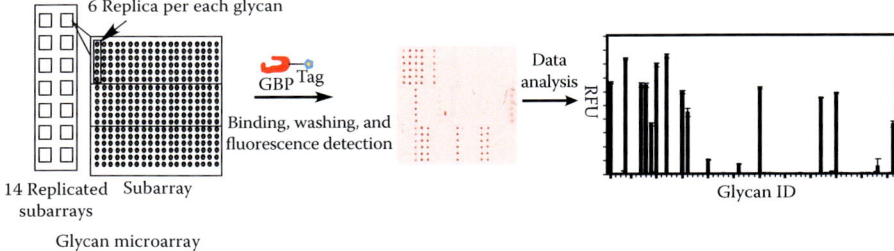

FIGURE 7.2 Schematic of a glycan microarray experiment. A glycan microarray printed as 14 identical subarrays is used as an example.

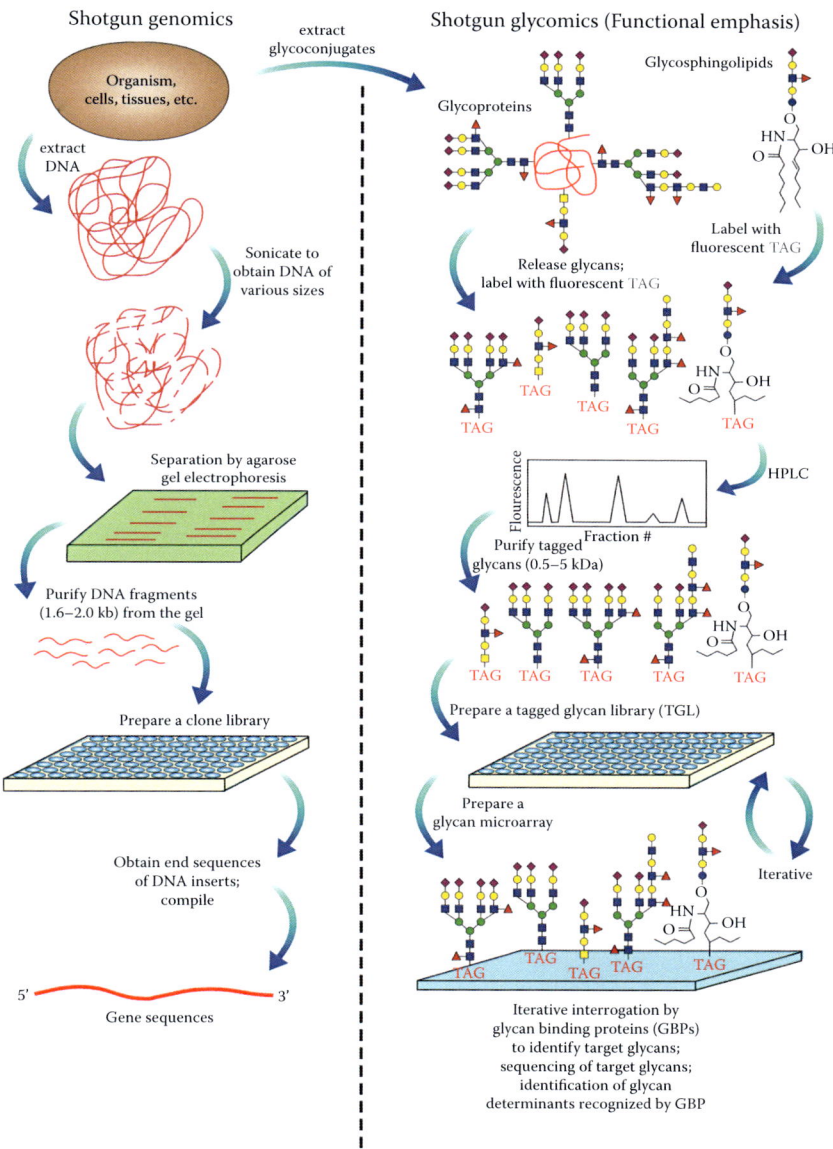

FIGURE 7.5 Strategy of shotgun glycomics.

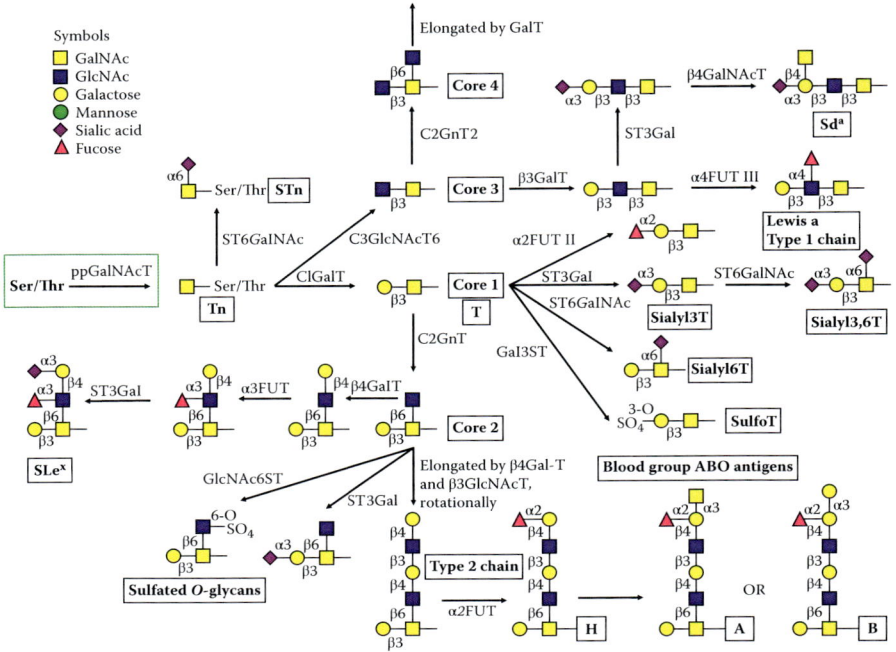

FIGURE 8.1 Biosynthesis pathways of cancer-associated *O*-glycan structures.

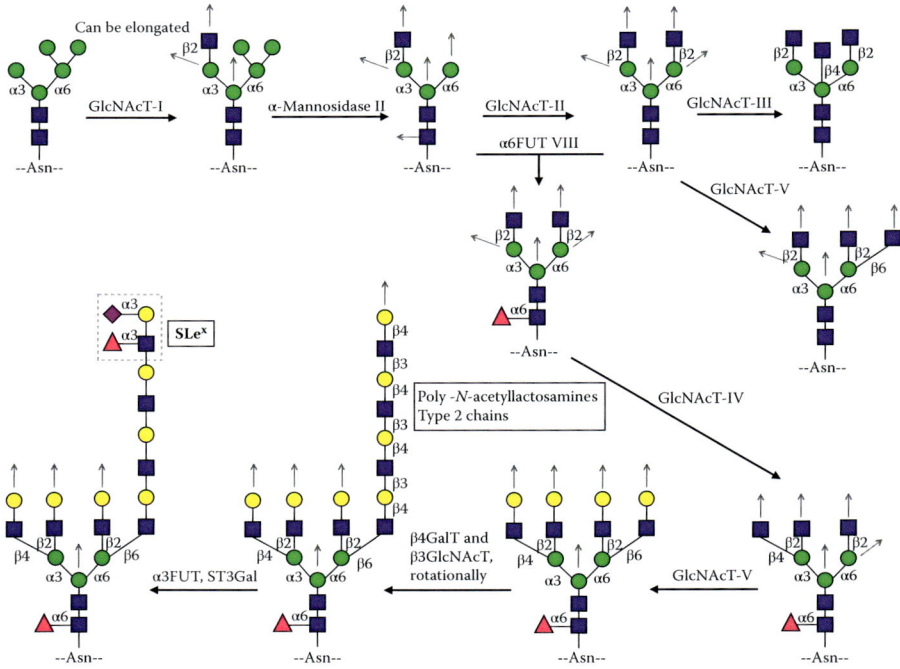

FIGURE 8.2 *N*-Glycosylation pathways in cancer cells.

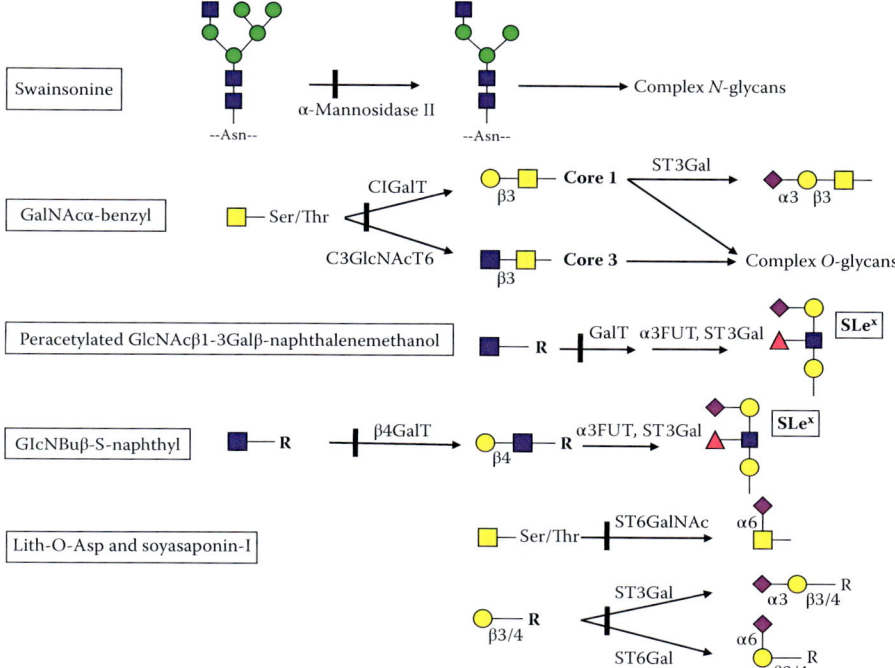

FIGURE 8.3 Strategies targeting cancer cells by glycosylation inhibitors. The symbols are explained in Figure 8.1. Black bars through reaction arrows indicate the site of inhibition.

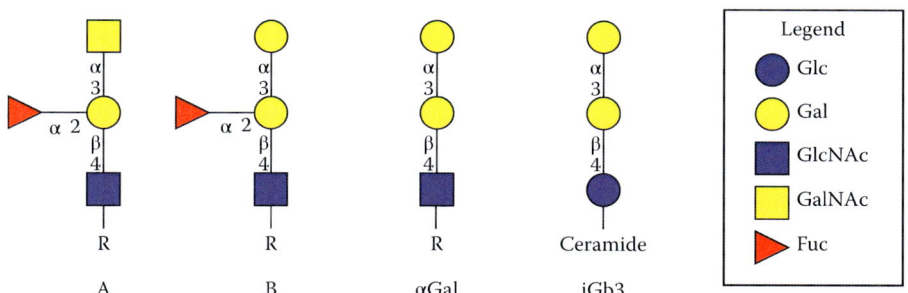

FIGURE 9.1 Diagrammatic representation of the terminal carbohydrate structures of A and B blood group antigens, αGal and iGb3.

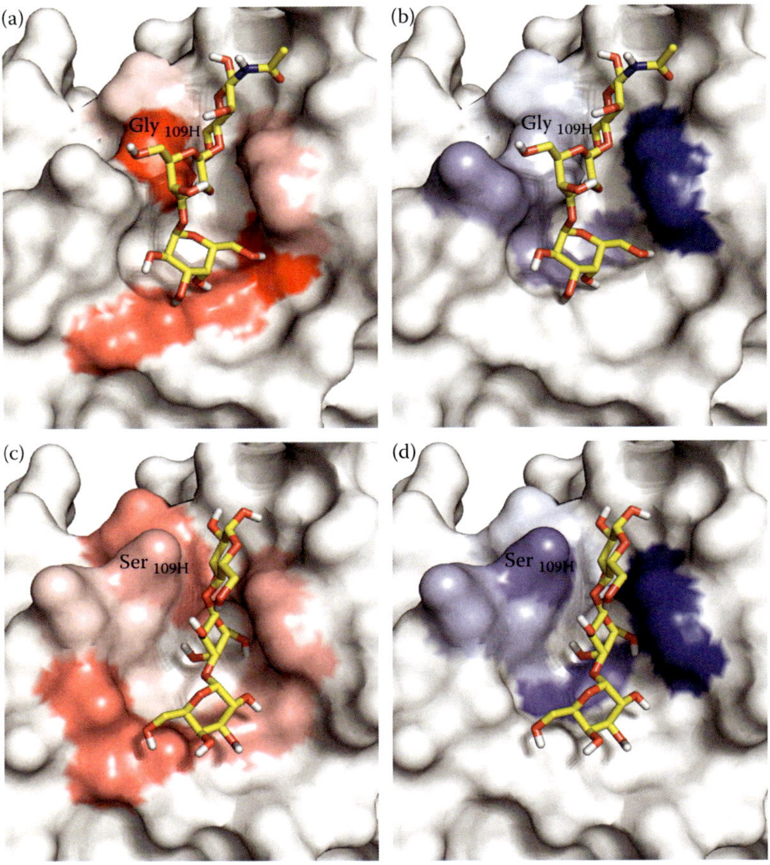

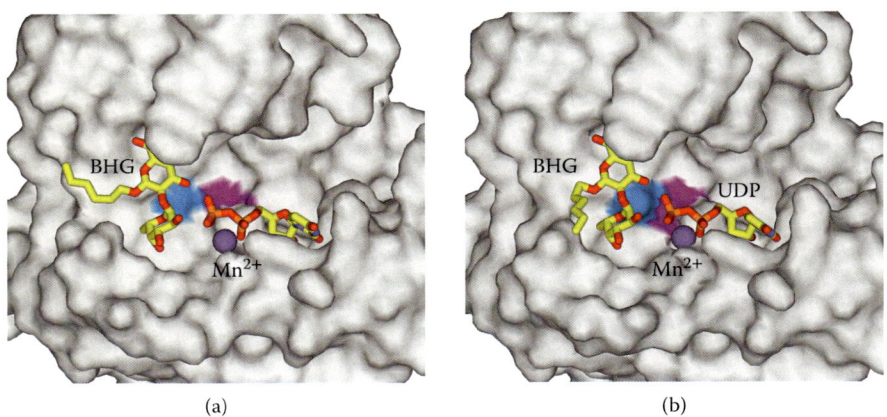

FIGURE 9.2 Interaction-based site maps and predicted αGal carbohydrate-binding modes for two murine mAbs.

FIGURE 9.3 Comparison of the crystal structures of A and B transferases.

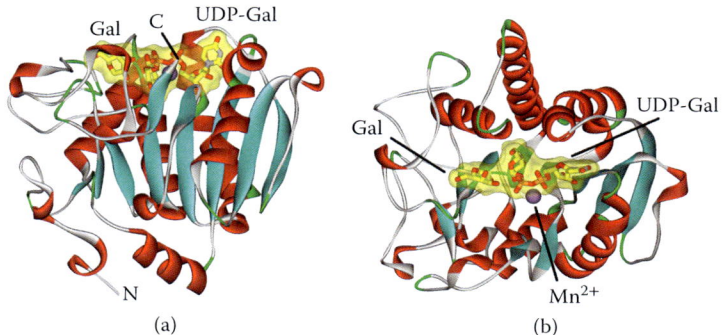

FIGURE 9.4 Crystal structure of bovine αGT, a family 6 glycosyltransferase.

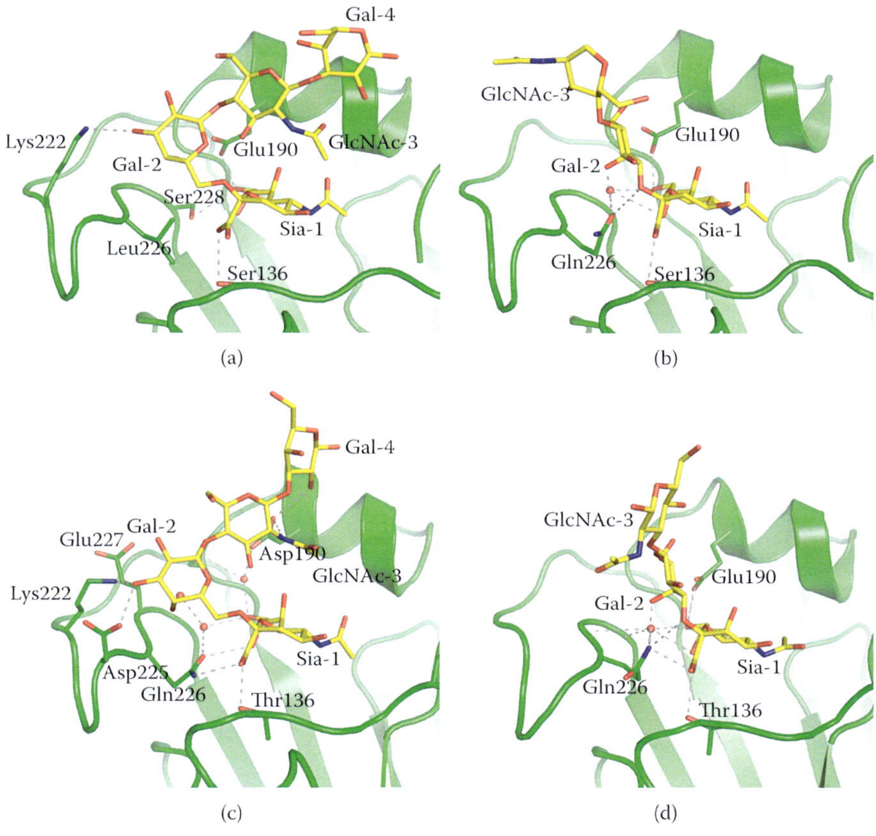

FIGURE 10.1 Receptor recognition in influenza A virus HA envelope protein.

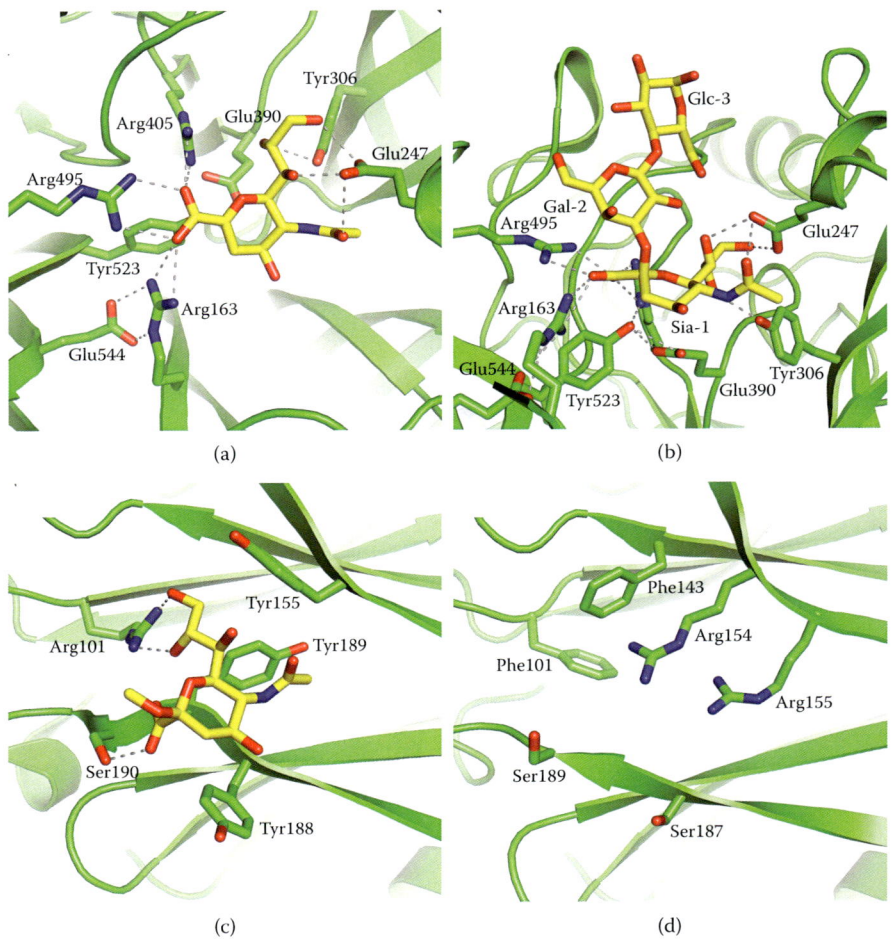

FIGURE 10.2 Receptor recognition in paramyxovirus and rotavirus.

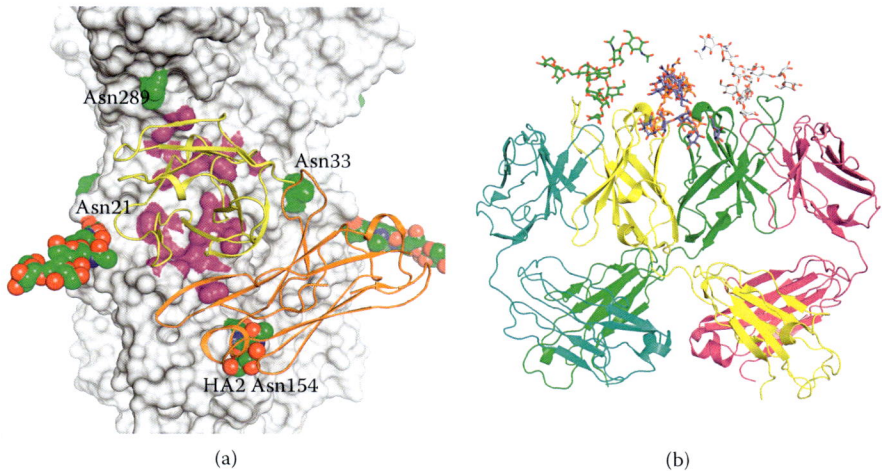

(a)   (b)

FIGURE 10.3   Broadly neutralizing antibody recognition of influenza A and HIV-1.

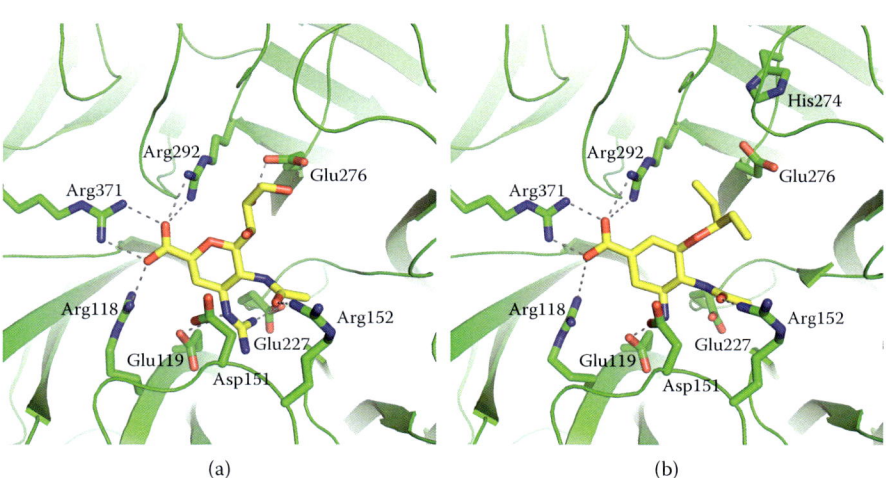

(a)   (b)

FIGURE 10.4   Influenza A virus sialidase NA as an effective target for antiviral inhibitors.

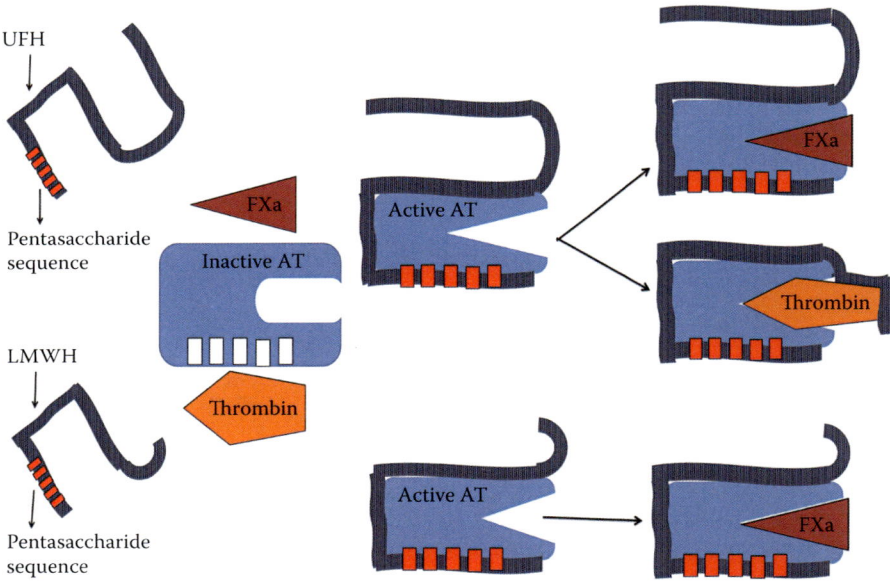

FIGURE 11.2 A cartoon illustrating differences in mechanism of action of UFH and LMWH.

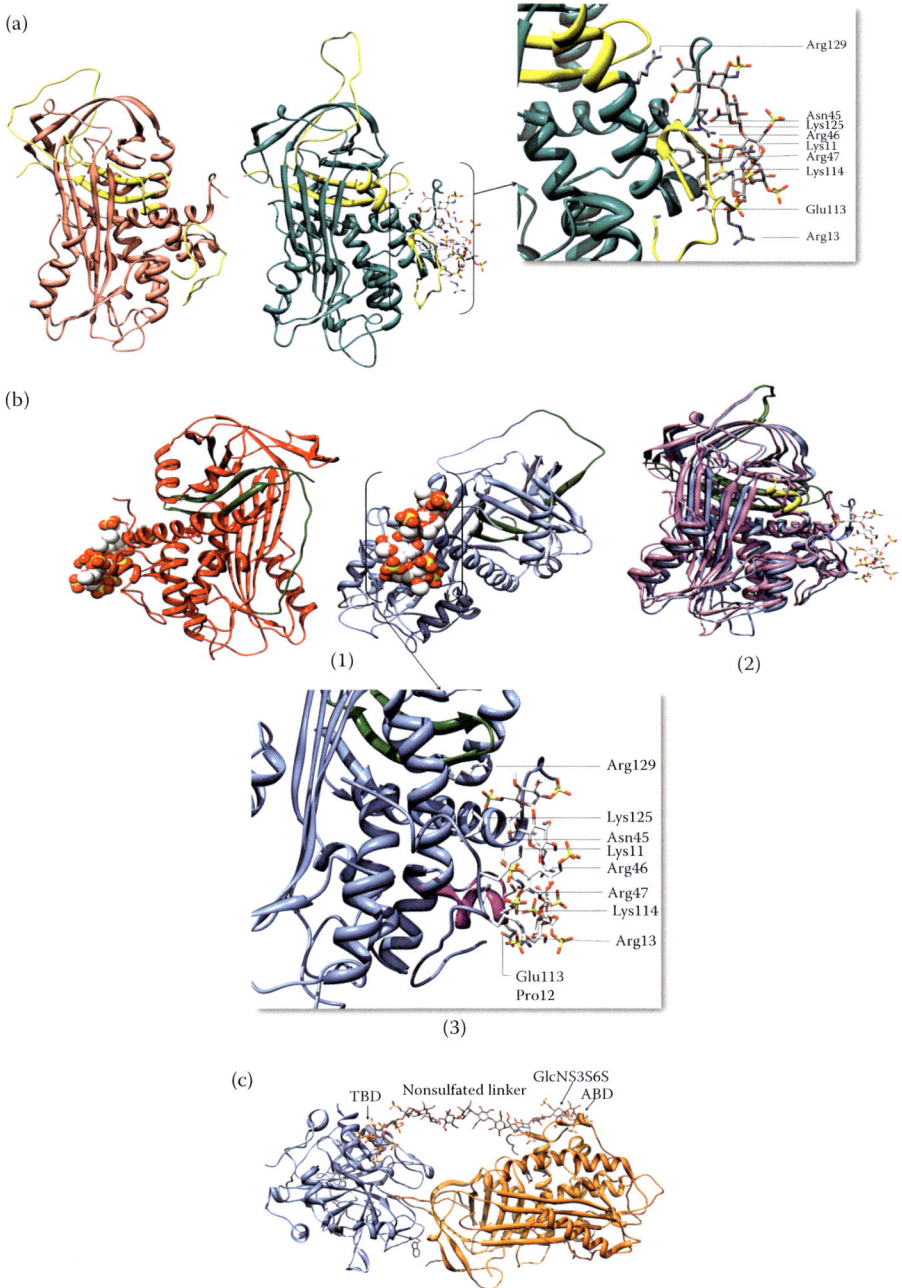

FIGURE 11.4 (a) Binding interactions of heparin with AT are shown. (b) Mechanism of action of idraparinux with AT is shown. (c) Crystal structure of the ternary complex of AT, thrombin, and SR123781A is shown.

Normal heparan sulfate structure

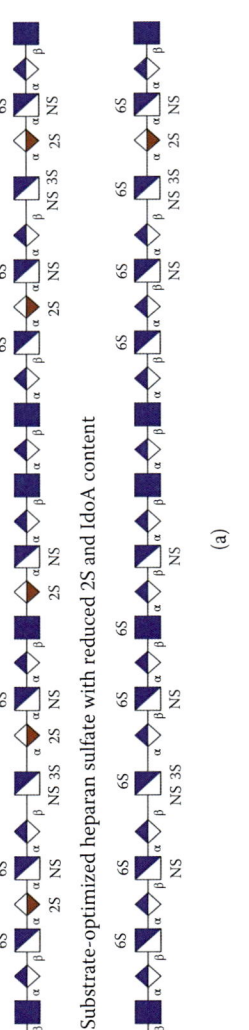

Substrate-optimized heparan sulfate with reduced 2S and IdoA content

(a)

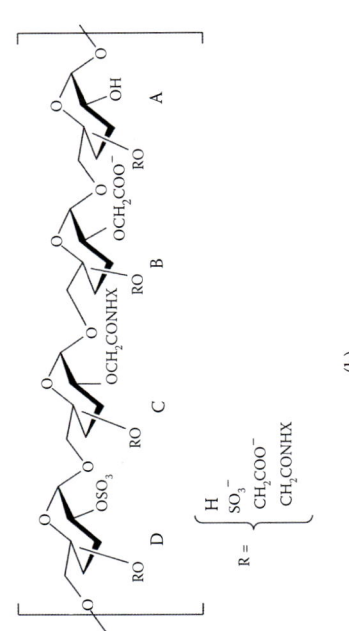

(b)

FIGURE 11.6 Novel GAG mimetics.

10 Structural Glycobiology
Applications in Viral Infection

Rui Xu and Ian A. Wilson

CONTENTS

10.1 INTRODUCTION

Protein glycosylation is an important form of posttranslational modification that plays vital roles in various cellular events. The attachment of carbohydrates on secreted and membrane-bound proteins is necessary for proper protein folding and assembly. Glycosylation can also improve protein stability and modulate protein functions. Specific recognition of glycans on cell surface proteins is an efficient way of mediating cell–cell interactions in multicellular organisms, as well as interactions between different organisms.

For eukaryotes, the biosynthesis of glycoconjugates is confined to the endoplasmic reticulum (ER) and Golgi compartments. The attached glycans display amazing structural diversity compared to proteins and nucleic acids. So far, at least 41 different glycoprotein linkages have been identified, involving 13 different monosaccharides and 8 different amino acids (Spiro 2002). The covalently linked monosaccharide is further decorated with other glycan residues through different glycosidic bonds (Varki et al. 2009). The large number of possible combinations results in glycoconjugates of significant macroheterogeneity and microheterogeneity (Jenkins et al. 1996). The type and pattern of attached oligosaccharides are influenced by their location and the environment within the protein structure as well as by host factors including, but not limited to, species, tissue distribution, and developmental stage of the host.

Over evolution, many pathogens have explored the subtle differences in glycan patterns to mount infections in precise species-specific or even tissue-specific manner. Glycan-binding proteins that recognize specific glycans or glycosidic linkages are used by microbes for adhesion and eventual invasion (Karlsson 2001; Sharon 1996; Imberty and Varrot 2008; Olofsson and Bergstrom 2005; Taube et al. 2010). In this chapter, we discuss three paradigm systems in which viral proteins use different sets of protein–glycan interactions to recognize sialic acid in cell surface glycans as receptors for viral attachment: (1) The most extensively studied influenza virus hemagglutinin (HA), (2) the paramyxovirus hemagglutinin–neuraminidase (HN) protein mediating both receptor engagement and viral fusion, and (3) the VP8* domain from nonenveloped rotaviruses.

Not content with utilizing host glycans as a gateway for viral entry, viruses also hijack the host glycosylation machinery to modify their own surface proteins. Decoration with host-like oligosaccharides is sometimes crucial for protein folding and processing; most importantly, this glycan shield can disguise a viral particle from host immune surveillance (Vigerust and Shepherd 2007). Two prominent examples are influenza A virus and human immunodeficiency virus (HIV-1), both of which are of great importance to human health. As a counterattack, the immune system attempts to generate antibodies that target functionally important regions of viral proteins that are not fully covered by the glycan shield (Kwong and Wilson 2009). These regions include the CD4-binding site of HIV-1 gp120 (Zhou et al. 2007, 2010) and the stem region of the influenza virus HA, as evidenced by several recently discovered antibodies (Ekiert et al. 2009, 2011; Sui et al. 2009; Kashyap et al. 2008; Throsby et al. 2008; Corti et al. 2011). In a more extreme case, antibody 2G12 targets the densely packed glycan shield itself and binds to the unusual cluster of high-mannose glycans atop HIV-1 gp120 (Calarese et al. 2003). Other highly potent, broadly neutralizing antibodies that recognize glycans on HIV-1 gp120 have also been recently discovered (Walker et al. 2011; Pejchal et al. 2011; McLellan et al. 2011; Walker et al. 2009).

Protein–glycan interactions in the viral life cycle provide a practical avenue for the development of antiviral agents (von Itzstein 2008). Tamiflu has become a household name because of its role in treating infections by influenza A viruses from seasonal flu to "avian flu" and the more recent (2009) "swine flu" pandemic. Its active ingredient, oseltamivir carboxylate, is an inhibitor of the viral glycan-destroying enzyme neuraminidase (NA). We summarize here the progress made on designing inhibitors for the influenza A virus as an example of antiviral drug design at the viral protein–glycan interface.

10.2 GLYCANS AS RECEPTORS FOR VIRAL INVASION

Glycans that constitute receptors for enveloped and nonenveloped viruses mostly belong to three categories: (1) glycans containing sialic acid, (2) sulfated glycosaminoglycans (GAGs), and (3) histo-blood group glycans (Olofsson and Bergstrom 2005).

Sialylated glycoconjugates are often used by viruses as receptors for entry because of their wide distribution on the surface of epithelia. Among the many variants of sialic acid, the one affecting human health the most is N-acetylneuraminic acid or sialic acid. Most sialic acid receptors are terminal sugars that are attached to a penultimate galactose through either an α2,3- or an α2,6-linkage. Viruses engaging these receptors include influenza viruses, parainfluenza viruses (PIVs), adenovirus serotype 37, and so on (Imberty and Varrot 2008, Olofsson and Bergstrom 2005). Although this occurs less frequently, an internal sialic acid in the glycan can also serve as a receptor for some viruses, including simian virus 40 and NA-insensitive strains of human rotaviruses (Imberty and Varrot 2008, Taube et al. 2010).

The GAGs are long, unbranched polysaccharides formed by linear repetitions of disaccharide motifs. They are frequently sulfated and, thus, highly negatively charged. Among the GAG family, heparan is most widely implicated as a viral receptor (Olofsson and Bergstrom 2005). The attachment of many human pathogens, including hepatitis C and dengue viruses, is mediated by heparan sulfate. However, definitive GAG glycoepitopes for viral binding have not been well characterized.

Human histo-blood group is the least used group for viral infections in humans and includes the A, B, and H antigens from individuals of blood types A, B, and O, respectively. They are complex fucosylated oligosaccharides present on the surfaces of red blood cells, the gut, and respiratory epithelia. Only a few representatives from norovirus, lagovirus, and parvovirus are known to utilize these neutral glycans as receptors (Olofsson and Bergstrom 2005). Differential expression of histo-blood group glycans in the human population correlates to susceptibility and resistance to viral infection (Lindesmith et al. 2003). In recent years, the structural basis of recognition of blood group oligosaccharides has been determined for two strains of noroviruses (Choi et al. 2008; Bu et al. 2008; Cao et al. 2007).

Among the three groups of glycoepitopes, receptor binding of the sialic acid moiety has been well documented structurally for many viral surface proteins (Imberty and Varrot 2008). The glycan recognition generally involves sialic acid bound to a shallow depression on the surface of the receptor-binding domain of the viral protein. The apparently low binding affinity is compensated by multivalent attachment between virus and host cell. Here, we will discuss in detail the structural mechanism of sialoside recognition in three well-characterized viral systems.

10.2.1 Influenza A Virus Hemagglutinin (HA)

Influenza viruses are single-stranded, negative-sense RNA viruses belonging to the *Orthomyxoviridae* family. Three genera of influenza viruses are currently known: influenza A, influenza B, and influenza C. Although all three recognize sialoside receptors on respiratory tract and cause contagious infections in humans, influenza A is of the most interest to human health and is the only type that has caused flu pandemics.

Influenza A virus displays two major surface proteins on the viral envelope: (1) The HA and (2) the NA. The HA is the protein responsible for viral attachment. On attachment and internalization through endocytosis, HA also mediates the membrane fusion between viral envelope and host endosomal membrane under stimulus from an acidic pH (Skehel and Wiley 2000). Because of its abundant presence on the viral envelope, HA is also the target for neutralizing antibodies. Complementary to HA's receptor-binding ability, the NA is a sialidase that removes sialic acid receptors, which enhances the release of progeny viruses (Colman and Ward 1985).

Based on its serotype, HAs of influenza A virus are classified into 16 subtypes. Whereas all subtypes exist in avian species, particularly aquatic birds, only a few have infected humans and only three have caused widespread human infections. In the past 100 years, only H1, H2, and H3 subtypes have been involved in four human pandemics: (1) 1918 (H1), (2) 1957 (H2), (3) 1968 (H3), and (4) 2009 (H1). In addition, isolated outbreaks of avian-origin H5, H7, and H9 influenzas were reported during the last decade, which stirred up considerable concern about the potential emergence of a "bird flu" pandemic. It is still largely a mystery why some subtypes, but not others, have been able to adapt sufficiently for infection and transmission in humans. One species barrier that has been well studied and documented is the receptor-binding specificity of HA. Human adaptation of the current H5N1 viruses is likely limited by the ability of their HA to acquire glycan-binding specificity for human receptors (Shinya et al. 2006; Stevens et al. 2006c, 2008).

Receptor-binding specificity of HA is reflected by its binding preference for cell receptors of different oligosaccharide structures, most importantly the linkage between sialic acid and the penultimate galactose. In the early 1980s, it was shown that influenza viruses isolated from different species exhibit differential specificity for sialoglycans with terminal NeuAcα(2-3)Gal and NeuAcα(2-6)Gal linkages, providing an underlying molecular basis for receptor-binding differences (Rogers and Paulson 1983). The correlation between receptor specificity and species of origin was supported by extensive surveys of human and animal isolates in later years (Stevens et al. 2006a; Rogers and D'Souza 1989; Gambaryan et al. 1997, 2005). Human-adapted viruses prefer receptors of α2,6-linkage, whereas avian viruses specifically recognize α2,3-linked sialoside glycans. Such binding preferences are determined by the different distribution patterns of these glycan linkages in humans and birds. The α2,3-linked sialic acids are expressed on epithelial cells in the intestine and respiratory tract of birds. In humans, the epithelial cells of the upper respiratory tract contain primarily α2,6-linked glycans, whereas the lower respiratory tract and lung display mainly α2,3-linked sialosides (Matrosovich et al. 2004, Shinya et al. 2006). Infection to the upper respiratory tract readily enables human influenza viruses to spread among the population via sneezing and coughing (Shinya et al. 2006). A lack of binding for α2,3-linked glycans also prevents human viruses from being trapped by mucins in the respiratory tract (Couceiro et al. 1993).

Glycan receptor binding has significant implications on the host range, tissue tropism, pathogenesis, and transmission of the influenza A virus. Tremendous progress has been made in recent years toward the characterization of HA–glycan interactions from two different aspects: From the protein side, structural determination of HAs of various subtypes, along with their receptor complexes, has elucidated the structural basis of receptor-binding specificity (Gamblin and Skehel 2010). On the glycan side, analysis of epithelial cells has revealed the distribution of glycan structures and linkages in various host species (Guo et al. 2007; Bateman et al. 2010; Ito et al. 2000; Chou et al. 2002). Identification of novel glycans, combined with microarray technology (see Chapter 7), has resulted in glycan microarrays that allow rapid assessment of viral receptor specificity against a large panel of glycan epitopes (Stevens et al. 2006b).

Influenza HA of the H3 subtype was the first viral envelope protein to have its atomic structure determined (Wilson et al. 1981). Subsequently, glycan recognition in H3 HA was studied in crystal structures of HA in complex with various sialylated glycan analogs (Eisen et al. 1997; Sauter et al. 1992; Watowich et al. 1994; Weis et al. 1988; Ha et al. 2003). Since then, all major HA subtypes, of both human origin and avian origin, have been analyzed structurally in their "apo" form and in receptor complexes. Structural analyses have been carried out for the HAs from H1 (Gamblin et al. 2004; Stevens et al. 2004; Xu et al. 2010a; Lin et al. 2009; Xu et al. 2012), H2 (Xu et al. 2010b; Liu et al. 2009), H5 (Ha et al. 2001, 2002; Stevens et al. 2006c), H7 (Yang et al. 2010; Russell et al. 2004), and H9 (Ha et al. 2001, 2002).

The crystal structures suggest a conserved architecture for all subtypes of HA. Each HA spike is a trimeric complex comprising identical HA protomers. The receptor-binding site is located in a shallow pocket on the membrane-distal globular head of HA. The binding site is surrounded by three major structural elements: The (1) 130 loop, (2) 190 helix, and (3) 220 loop (Figure 10.1). The bottom of the

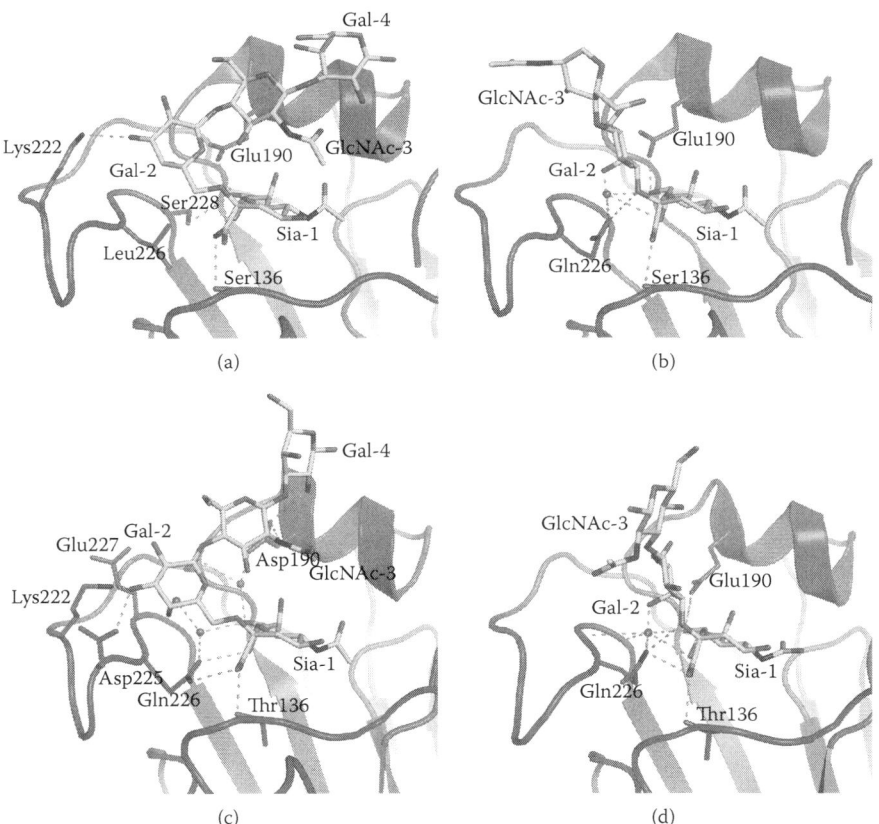

FIGURE 10.1   (See color insert.) Receptor recognition in influenza A virus HA envelope protein. (a) Human H2 HA in complex with human receptor analog (PDB: 2WR7), (b) avian H2 HA in complex with avian receptor (PDB: 2WR3), (c) human H1 HA in complex with human receptor analog (PDB: 3UBE), and (d) avian H1 HA in complex with avian receptor (PDB: 3HTP).

depression is formed by strictly conserved aromatic residues including Tyr98, Trp153, and His183. Across subtypes, molecular recognition for the terminal sialic acid is well conserved. The pyranose ring of sialic acid sits near the rim of the 130 loop, with the 2-carboxylate, the 5-acetamido, and the 8- and 9-hydroxyl groups pointing into the binding pocket. Sialic acid is bound by hydrophobic interactions and hydrogen bonding with well-conserved amino acids on the HA. The carboxylate group forms hydrogen bonds with the side-chain hydroxyl group of residue 136 (All HA residues are numbered as for the H3 subtype) and the amide of residue 137. The 8- and 9-hydroxyl groups are part of a hydrogen bonding network along with the side chains of Tyr98 and His183. Another conserved hydrogen bond is formed between the 5-acetamido nitrogen and the carbonyl oxygen of residue 135. The hydrophobic methyl group of the 5-acetamido moiety makes van der Waals contact with the indole ring of Trp153. Other interactions are unique to certain subtypes or certain HA strains. Gln226, when present, is hydrogen bonded to the carboxylate and the 8-hydroxyl of sialic acid (Figure 10.1b through d). In human H2 and H3 HA, it is replaced by Leu226 (Figure 10.1a). It is noted that Glu190 of H2 HA forms a hydrogen bond with the 9-hydroxyl group (Figure 10.1a and b). In human H1 HA, a similar hydrogen bond is mediated through a water molecule coordinated by Asp190 (Figure 10.1c).

Preferential binding toward glycans of different linkages is achieved by specific interactions beyond the sialic acid. In H1, H2, and H3 HAs, shift of receptor specificity from α2,3- to α2,6-linkage requires as few as two amino acid changes at or near the 220 loop: (1) E190D/G225D in H1 (Stevens et al. 2006a; Matrosovich et al. 2000; Tumpey et al. 2007), and (2) Q226L/G228S in H2 and H3 (Matrosovich et al. 2000; Connor et al. 1994). The substitutions remove hydrogen bonds that confer specificity to α2,3-linked glycans and add favorable interactions between HA and α2,6-sialosides (Ha et al. 2003; Liu et al. 2009; Xu et al. 2010b, 2012).

The α2,3-linked receptor analogs adopt an "extended conformation" in the crystal structures of avian HAs (Lin et al. 2009, 2009; Ha et al. 2003, 2001). The receptor analogs exit the binding pocket over the 220 loop, with the Gal-2 sugar ring contacting Gln226 (Figure 10.1b and d). Hydrogen bonds are formed between the Gln226 side chain and the O3 and O4 atoms of Gal-2. This pair of hydrogen bonds is present in all avian HAs and is the only HA–glycan interaction beyond the terminal sialic acid.

Two different strategies have been used successfully in natural viruses to switch HA-binding specificity. In H2 and H3, the polar-to-hydrophobic mutation of Q226L directly removes the hydrogen bonding interactions that are specific to α2,3-linked glycans. The combination of Q226L/G228S substitutions also expands the receptor-binding site by about 0.5 Å (Ha et al. 2003; Xu et al. 2010b). The slightly shifted 220 loop now orients the hydrophobic Leu226 to enable contact with the C6 atom of Gal-2 in the receptor analogs. Thus, the hydrophobicity switch at residue 226 directly exploits the linkage difference between receptors, replacing polar interactions for avian receptor binding with hydrophobic interactions for human receptor binding. In contrast, Gln226 is retained in human H1 HAs, but it moves away from the binding site and reduces interaction with receptor analogs (Xu et al. 2012). This loss of contacts to avian receptors is manifested by the reshaping of the 220 loop that is promoted by mutations E190D and G225D. It is noted that Glu190 is part of a hydrogen bonding network that likely contributes to the shape of loop 220 and leads to improved avian receptor binding (Gamblin et al. 2004). The G225D substitution has an even larger effect on the 220 loop conformation, given that the dihedral angles of Gly225 ($\varphi = 139.1°$; $\psi = -69.7°$) in avian H1 HAs are energetically allowed only for glycine residues (Lovell et al. 2003; Lin et al. 2009). As a result, the Gln226 side chain in human H1 HA sits relatively deep in the binding pocket and is too distant from Gal-2 of avian receptors to make hydrogen bonds (Xu et al. 2012). In addition, the E190D/G225D substitutions also introduce new polar interactions with α2,6-linked glycans. These newly formed interactions are unique to human receptors because of the difference between the overall shapes of α2,3- and α2,6-linked glycans. In the HA complexes, human receptor analogs form a bent or folded-back conformation. The Gal-2 skirts over the 220 loop, while the third sugar ring abruptly turns toward the 190 helix to make an exit for the rest of the sugars above the receptor-binding site. The Gal-2, through its O3 and O4 atoms, forms extensive hydrogen bonds with Asp225 and the neighboring Lys222. The Asp190 forms hydrogen bonds with the 2-acetamido nitrogen of GlcNAc-3 and the 2-hydroxyl group of Gal-4 (Figure 10.1c).

Based on structural analysis, receptor-binding specificity is determined by the presence or absence of a few favorable interactions beyond the terminal sialic acid. These subtle differences in receptor binding correlate with only small differences in intrinsic affinity for the receptor glycans. Human H3 HA has K_d (dissociation constant) values of 2.1 and 3.2 mM for α2,6- and α2,3-linked glycans, respectively, as revealed by direct nuclear magnetic resonance (NMR) affinity measurements (Sauter et al. 1989). An avian variant with Gln at position 226 exhibited corresponding values of 5.9 and 2.9 mM. Such small differences in absolute affinity are amplified by multivalent binding between virus and host cells and are translated into distinguishable binding properties.

In recent years, it has become increasingly clear that binding specificity of HA depends not only on the glycosidic linkage, but also on the rest of the oligosaccharide structure. Fucosylation, sulfation, branched glycans, and glycan length can greatly influence receptor binding. Glycomic profiling of respiratory epithelial cells is underway to understand the diversity and relative abundance of glycans distributed in humans and animals (Bateman et al. 2010; Lo-Guidice et al. 1994). The new knowledge about the range of glycan structures on tissues related to influenza infection goes hand in hand with the emerging data from glycan microarray analyses (Stevens et al. 2006b). Glycan microarray technology enables the coupling of up to hundreds of different glycan structures on a single chip and allows the high-throughput study of glycan binding in a matter of hours.

The glycan microarray developed by the Consortium for Functional Glycomics (CFG) (Blixt et al. 2004) is widely utilized for the analysis of influenza virus–receptor specificity. Assay protocols have been optimized for both recombinant HA samples and inactivated whole viruses. Human HAs, including those from the aforementioned four influenza pandemics, show clearly restricted preference to α2,6-linked sialosides (Stevens et al. 2006a; Xu et al. 2010b, 2012; Yang et al. 2010). Avian virus HAs exhibit specific recognition for α2,3-linked sialosides (Stevens et al. 2006a,

2006c; Xu et al. 2010b). The fine specificity of HAs has been revealed by such glycan array analyses. Human H1 HA binds only to a limited group of linear α2,6-linked glycans compared with H2 and H3 HAs. Fucosylated α2,3-linked sialosides are recognized by human rather than avian H2 HAs. The biological relevance of these binding preferences is likely related to viral adaptation and tissue tropism.

Another application of glycan assay in influenza virus research is in the rapid surveillance of H5N1 virus strains and evolution. The H5N1 influenza virus is of avian origin and has infected human beings in isolated cases in the past decade. The highly pathogenic H5N1 virus aroused global concern due to its high mortality in humans and, hence, its potential for causing a devastating human pandemic. Avian H5 showed classical specificity for α2,3-linked sialosides on the glycan microarray (Stevens et al. 2006c). H5 HA does not readily switch to α2,6-linkage specificity through a simple set of mutations that had occurred in H1, H2 and H3 (Stevens et al. 2006c, 2008; Maines et al. 2011). A recent report suggests that as few as four amino acid changes could change the binding specificity of H5 HA and allow an H5 containing virus confer respiratory droplet transmission in ferrets (Imai et al. 2012), although three of them have been strictly conserved in field strains since 2003. Instant assessment of receptor specificity for newly isolated H5N1 viruses is important in order to evaluate whether any human adaptation has occurred and, hence, whether they are more likely to be transmitted in humans.

10.2.2 PARAMYXOVIRUS HN

The *Paramyxoviridae* family includes many pathogens that affect human health, such as measles virus, human PIV (hPIV), and respiratory syncytial virus. The attachment proteins from the family share a β propeller–like fold, which enables binding to glycans and protein receptors of host cells. Based on their receptor ligands, these attachment proteins belong to three distinctive classes: (1) The HN, (2) H, and (3) attachment glycoprotein G (Bowden et al. 2010). The H and G glycoproteins use cell surface proteins as receptors and are thus out of the scope of this chapter. The HN glycoprotein utilizes terminal sialic acid for attachment and has sialidase activity, as well as the ability to bind glycan receptors. The crystal structures of HN receptor binding domains have been determined for three viruses: (1) Newcastle disease virus (NDV) (Crennell et al. 2000; Zaitsev et al. 2004; Yuan et al. 2011), (2) hPIV type 3 (hPIV3) (Lawrence et al. 2004), and (3) PIV type 5 (PIV5) (Yuan et al. 2005).

The HN glycoprotein is a tetramer of six-bladed β-propeller, similar to influenza virus NA. Each blade consists of four antiparallel β-strands. The receptor-binding site is located at the center of the β-propeller. When cocrystallized with sialic acid, only 2-deoxy-2,3-didehydro-D-N-acetylneuraminic acid (Neu5Ac2en or DANA) was found in the receptor-binding site (Crennell et al. 2000; Lawrence et al. 2004; Yuan et al. 2005). This unsaturated sialic acid derivative is a sialosyl cation transition-state analog, which confirms sialidase activity in the binding site of HN. The Neu5Ac2en binding is coordinated by a group of residues that are highly conserved in sialidases (Figure 10.2a). The carboxylate group of Neu5Ac2en interacts with a conserved arginine triad, that is, Arg163, Arg405, and Arg495 (numbering as in PIV5). The interaction provided by Arg495 is bidentate, with salt bridges formed from Arg guanadinium to both carboxylate oxygen atoms of the sugar. It

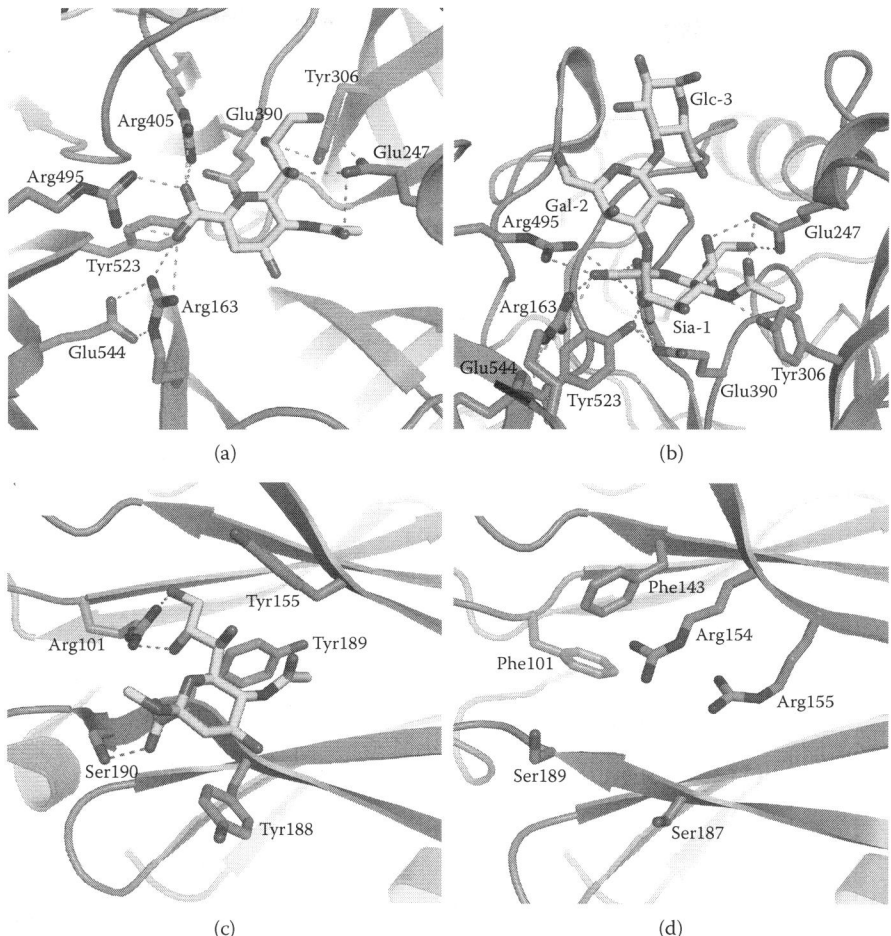

FIGURE 10.2   **(See color insert.)** Receptor recognition in paramyxovirus and rotavirus. (a) the PIV 5 HN in complex with Neu5Ac2en (PDB: 1Z50); (b) PIV 5 HN in complex with α2,3-sialyllactose (PDB: 1Z4X); (c) binding of sialic acid in the VP8* fragment of rhesus rotavirus (PDB: 1KQR); and (d) the equivalent region in human rotavirus strain Wa, which is substantially different from the rhesus rotavirus. The electron density for glycan ligand is not present in the crystal structure (PDB: 2DWR).

is mentioned that Arg163 and Arg405 approach from either side of the carboxylate in nearly symmetric directions. On the other end of the binding site, Glu247 forms hydrogen bonds with the 7- and 9-hydroxyls of the ligand. The O8 of Neu5Ac2en points into the binding pocket and is hydrogen bonded with Tyr306. It is noted that Glu390 and Tyr523 line the floor of the binding pocket.

The HN structure in complex with α2,3-sialyllactose was obtained by ligand soaking under pH conditions for attenuated enzymatic activity (Yuan et al. 2005). The terminal sialic acid is deformed into a boat-like conformation (Figure 10.2b), which is unlike the chair conformation observed in influenza HA complexes. As a result, the glycosidic bond between Gal-2 and Sia-1 is oriented in the axial position, whereas the 2-carboxylate group is positioned nearly equatorially toward the arginine triad. The lactose moiety provides only limited surface contact and no specific interactions with HN (Yuan et al. 2005).

Human paramyxovirus displays distinctive receptor-binding specificity from human influenza virus. An earlier report suggests that hPIV type 1 (hPIV1) and hPIV3 recognize α2,3-linked sialic acid with hPIV3 also showing some affinity for α2,6-linked sialic acid (Suzuki et al. 2001). A later study using the glycan microarray revealed (Amonsen et al. 2007) that hPIV1 and hPIV3 both recognize a glycan epitope on Neu5Acα(2-3)Galβ(1-4)GlcNAc. The hPIV3 shows more restricted binding than hPIV1 and generally binds only long oligosaccharides. Binding to both viruses is also detected for other α2,3-linked sialosides that are sulfated or fucosylated. No noticeable binding for α2,6-linked sialic acids was observed in the glycan microarray experiment.

The crystal structure of NDV HN uncovered a second sialic acid binding site at the dimer interface of HN (Zaitsev et al. 2004). Although the existence of the second binding site has been proposed for PIV (Bousse et al. 2004; Porotto et al. 2006, 2007), it was not observed in the crystal structures of hPIV3 and PIV5 (Lawrence et al. 2004; Yuan et al. 2005). Further studies are necessary to resolve this discrepancy.

Besides viral attachment, receptor binding in HN is responsible for inducing conformational changes that lead to F protein–mediated membrane fusion. Here again, conflicting models have been proposed based on existing HN structures. The NDV HN shows considerable conformational changes on receptor binding (Zaitsev et al. 2004). The structural changes in HN protomer are correlated with changes in the dimer interface, leading to the formation of the proposed second binding site and F protein activation (Zaitsev et al. 2004; Crennell et al. 2000). However, no ligand-induced conformational changes were observed in the HN of hPIV3 and PIV5 (Lawrence et al. 2004; Yuan et al. 2005). Rather, it was proposed that engagement of glycan receptors triggers HN tetramer dissociation, which then leads to F protein activation (Yuan et al. 2005). This difference is mirrored by recent studies on measles virus H protein (Hashiguchi et al. 2011; Navaratnarajah et al. 2011). Thus, the mechanism by which receptor binding induces membrane fusion in paramyxoviruses still remains somewhat enigmatic.

10.2.3 Rotavirus VP8*

Rotaviruses are nonenveloped viruses that cause severe diarrheal illness in children worldwide (Parashar et al. 2006). The viruses target intestinal cells through sialoside glycan receptors (Banda et al. 2009). Cell attachment is mediated by VP4 spikes on the outer shell of the virus (Lopez and Arias 2006). Trypsin cleavage of VP4 is a necessary priming step for viral entry and it yields two cleavage fragments: (1) VP5* and (2) VP8*. The VP8* is the glycan-binding domain and has been studied structurally.

Rotaviruses can be classified into two groups: (1) "NA-sensitive" and (2) "NA-insensitive" rotaviruses (Banda et al. 2009). The NA-sensitive strains recognize terminal sialic acid as a receptor, whereas NA-insensitive strains, utilize internal sialic acid as the receptor (Haselhorst et al. 2009). Pretreatment of host cells with sialidase decreases the infectivity of NA-sensitive strains, but not NA-insensitive strains.

Crystal structures of VP8* show a galectin-like β-sandwich fold in both NA-sensitive strains (Dormitzer et al. 2002b; Blanchard et al. 2007; Kraschnefski et al. 2009) and NA-insensitive strains (Monnier et al. 2006; Blanchard et al. 2007). In NA-sensitive strains, the receptor-binding site is located in a shallow groove between the two β-sheets (Figure 10.2c). It is noted that Tyr189 forms the floor of the binding pocket. Only a few hydrogen bonds anchor the protein–glycan interaction. The carboxylate group of sialic acid forms hydrogen bonds with the hydroxyl group, as well as the main-chain amide of Ser190. The glycerol side chain is stabilized by Arg101, with two hydrogen bonds to the 8- and 9-hydroxyls of the sugar. Another conserved hydrogen bond is found between the 5-acetamido nitrogen and the main-chain carbonyl oxygen of Tyr188. The Tyr188 and Tyr155 provide van der Waals contacts with the sialic acid from opposite rims of the binding site. Alanine mutations of Arg101 and Ser190 result in reduced glycan binding (Kraschnefski et al. 2009). NMR spectroscopy experiments demonstrate that the rhesus rotavirus VP8* core binds α-anomeric N-acetylneuraminic acid with a K_d of 1.2 mM without preference for α2,3- or α2,6-linked glycans (Dormitzer et al. 2002a).

The residues important for sialic acid binding are well conserved among NA-sensitive strains but not in NA-insensitive strains (Dormitzer et al. 2002b). Although similar in overall fold, the putative glycan-binding site in NA-insensitive strains shows a significantly different architecture (Figure 10.2d) (Monnier et al. 2006; Blanchard et al. 2007). Most notably, Arg101 is replaced by an aromatic Phe101. Electron density for glycans was not found at the putative binding site. Instead, it was proposed that VP8* of NA-insensitive strains recognizes internal sialic acid through an alternate site that is equivalent to the carbohydrate-binding groove of galectins (Blanchard et al. 2007).

10.3   VIRUS GLYCOSYLATION

Viruses co-opt the host translational machinery for the synthesis of viral proteins. Naturally, they also take advantage of the host glycosylation apparatus to modify viral surface proteins. The addition of N-linked and O-linked oligosaccharides on viral surface proteins serves at least two important functions: (1) Facilitating protein folding and stability and (2) preventing antibody recognition (Vigerust and Shepherd 2007). We summarize here how glycosylation improves immune evasion for two viruses affecting human health: (1) Influenza A virus and (2) HIV-1. As no viral defense can be perfect, we also discuss recent advances in the identification and structural elucidation of antibodies that have successfully overcome the barrier provided by the glycan shield and also by extreme antigenic diversity in these viruses.

10.3.1   PROTECTION OF VIRUSES BY GLYCANS FROM IMMUNE RECOGNITION

The surface proteins of influenza A HA and NA are both heavily glycosylated. The HA is the target of neutralizing antibodies and generally has 5–11 potential N-glycosylation sites in each protomer (Vigerust and Shepherd 2007). The stem region is highly glycosylated with around 5 glycosylation sites in all strains and subtypes (Das et al. 2010; Klenk et al. 2002). The distribution of glycosylation sites on the membrane-distal globular domain is more variable and evolves during viral circulation in humans. Glycosylation helps viruses to evade immune responses in two ways: (1) The added "self"-like carbohydrate prevents antibody generation against the protein surface underneath and, thus, modifies the antigenicity of HA. The difference in the location of antigenic regions in HAs from different strains and subtypes, such as H1 and H3, is largely dictated by the different locations and extent of the glycosylation sites (Caton et al. 1982; Wiley et al. 1981); (2) A newly acquired glycosylation site in the antibody-binding region of HA disrupts antibody–HA interaction and is commonly used by influenza virus to escape from antibody neutralization (Daniels et al. 1983; Tsuchiya et al. 2002). The H1 HA, since being introduced into humans in 1918, has added up to three N-glycosylation sites in the Sa antigenic site (Xu et al. 2010a; Wei et al. 2010). The tendency to add oligosaccharides for shielding antigenic sites is even greater in H3 HA, which has accumulated four more N-glycosylation sites in the globular domain since the 1968 H3N2 pandemic (Abe et al. 2004).

The HIV-1 virus uses the same strategy of glycan shielding but to an even greater extent. The HIV-1 envelope protein gp120 is one of the most heavily glycosylated proteins known with 18–33 potential N-glycosylation sites (Korber et al. 2001). These oligosaccharides limit the neutralizing antibody response and shield the viral protein gp120 from immune recognition (Reitter et al. 1998). Similar to influenza virus HA, glycosylation sites are added to gp120 over the course of infection to evade the evolving host neutralizing antibody responses (Sagar et al. 2006). Nevertheless, in an interesting twist, the extreme glycosylation in gp120 compared with human proteins has rendered these self-sugar clusters immunogenic, as exemplified by several human antibodies that can specifically recognize high-mannose sugars on the surface of gp120 (see Section 10.3.2).

10.3.2 Host Antibody Response to Viral Glycoproteins

Although effective in escaping antibody recognition, adding glycosylation sites sometimes comes with a penalty for viral fitness. Excessive glycosylation may interfere with proper protein folding. Some parts of viral proteins, especially those important for host interactions, cannot be masked by oligosaccharides. Glycosylation near the receptor-binding site of influenza HA can decrease the receptor-binding avidity of HA (Abe et al. 2004; Tsuchiya et al. 2002). During infection in humans in the late 1957–1968, the H2N2 pandemic virus failed to add new N-glycosylation sites to its globular head domain of HA. The detrimental effect of adding glycosylation sites to H2 HA may then have hampered its escape from human immunity (Tsuchiya et al. 2002). Such structural or functional limitations may prevent viral proteins from being hyperglycosylated and leave holes in the glycan shield so that antibody can penetrate to the protein surface underneath. Antibodies can then gain access to the highly conserved receptor-binding sites as in influenza HA or HIV-1 gp120 and neutralize the virus (Barbey-Martin et al. 2002; Fleury et al. 1998; Zhou et al. 2007; Whittle et al. 2011; Wu et al. 2011).

Another interesting example of antibodies breaching the glycan shield is the recently identified stem region antibodies of influenza HA (Okuno et al. 1993; Sui et al.

2009; Kashyap et al. 2008; Throsby et al. 2008; Corti et al. 2011). The stem region of HA is highly conserved in overall structure and sequences across subtypes and groups. It is also heavily glycosylated and less accessible compared to the membrane-distal globular head domain; thus, it is not usually targeted by antibodies. Instead, host immune attention is directed to the often less glycosylated membrane-distal globular domain, especially to the highly variable and more accessible loops where escape mutations have less effect on HA function and thus arise relatively easily. The stem region antibodies, however, target highly conserved epitopes that are not fully shielded by oligosaccharides in the stem region (Ekiert et al. 2009, 2011; Sui et al. 2009; Corti et al. 2011). These first identified human antibodies to this region show broad neutralization across subtypes in the group 1 family (10 of 16 HA subtypes) (Throsby et al. 2008; Sui et al. 2009; Kashyap et al. 2008; Ekiert et al. 2011; Corti et al. 2011), and it is harder for viruses to escape from these antibodies (Sui et al. 2009; Ekiert et al. 2009). In the complex structures of antibody CR6261 (Ekiert et al. 2009; Figure 10.3a), the antibody-binding footprint is relatively small (680 Å^2) and is located between N-linked oligosaccharides on the HA. The HA–antibody interaction is dominated by the heavy chain of the antibody. The strategy of heavy chain–only recognition is useful for targeting small and recessed antigenic epitopes, as also evidenced by HIV antibody b12 (Zhou et al. 2007; Kwong and Wilson 2009). Such an arrangement

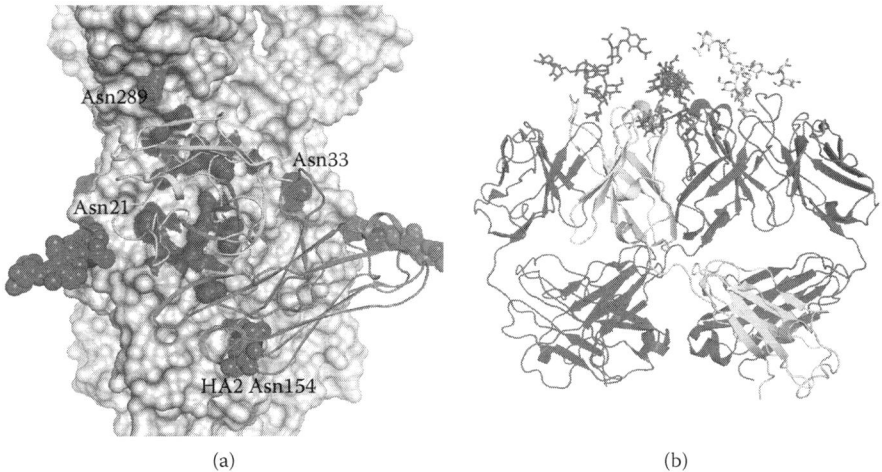

(a)   (b)

FIGURE 10.3 (**See color insert.**) Broadly neutralizing antibody recognition of influenza A and HIV-1. (a) The footprint for antibody CR6261 binding (colored purple) is confined by four N-glycosylation sites within the stem region of H1 HA (strain: A/South Carolina/1/1918) (PDB: 3GBN). The N-glycans built in the crystal structure (HA1 Asn 21 and HA2 Asn 154) are shown in a space-filling model. Other potential N-glycosylation sites (HA1 Asn33 and HA1 Asn289) are colored in green. For clarity, only the variable domains of CR6261 are shown here (heavy chain in yellow; light chain in orange). (b) The HIV-1 neutralizing antibody 2G12 in complex with the oligosaccharide Man$_9$GlcNAc$_2$ (PDB: 1OP5). The Fab V$_H$/V$_H'$ domain exchange (heavy chains colored in green and yellow) is important for glycan binding in 2G12. The Man$_9$GlcNAc$_2$ moieties are bound to the primary combining sites as well as the novel V$_H$/V$_H'$ interface that is formed in the domain exchange.

appears to be necessary for CR6261 and related V_H1-69 antibodies to avoid the HA stem glycans. One conserved glycosylation site, HA2 Asn154, is located between HA and the light chain of CR6261, eliminating possible direct light chain contacts to the HA protein residues (Ekiert et al. 2009). Another antibody was recently identified that recognizes group 2 influenza A viruses (i.e., the remaining 6 of the 16 HA subtypes) and recognizes a distinct epitope on group 2 viruses (Ekiert et al. 2011) where a glycosylation site potentially restricts antibody access on group 1 viruses. A further recent advance is the identification of other antibodies that are able to accommodate the different glycosylation in the group 1 and group 2 viruses and, hence, neutralize almost all influenza A viruses (Corti et al. 2011). In this case, the flexible sugars that might have impeded recognition are rotated out of the way and enable the antibody to specifically recognize the highly conserved protein residues below.

In the HIV virus, the glycan shield can present itself as an immunogenic epitope for antibody binding. Unlike most viral glycoproteins, HIV-1 gp120 is largely glycosylated with clusters of high-mannose oligosaccharides. The incomplete processing of glycans in gp120 is likely due to the high density of oligosaccharides that impede access by mannosidases in the Golgi (Kwong and Wilson 2009). This unique glycan composition of gp120 may play a role in viral infection (Geijtenbeek et al. 2000); it also enables the virus to be considered "foreign" and, therefore, to be placed under the surveillance of the host immune system. Antibody 2G12 differentiates these glycan differences on the glycan shield and neutralizes HIV-1 through the binding to an epitope formed by high-mannose N-glycans attached to residues 295, 332, 392 and so on (Sanders et al. 2002). In the crystal structures of 2G12 in complex with glycan oligosaccharides (Calarese et al. 2003), 2G12 adopts a unique "V_H domain–swapped dimer" architecture. Domain exchange is necessary for binding the HIV-1 glycan shield (Doores et al. 2010) as the dimer of two Fabs forms a multivalent binding surface that includes two conventional antigen-combining sites and two potential noncanonical binding sites at the newly formed V_H–$V_{H'}$ dimer interface (Calarese et al. 2003). In the primary combining site, glycan recognition is dominated by recognition of the terminal Manα(1-2)Man disaccharide. Binding specificity is achieved by high shape complementarity and a total of 12 hydrogen bonds between the Manα(1-2)Man disaccharide and 2G12 (Calarese et al. 2003). High-affinity binding, however, is possible only with multivalent binding as interaction with a single oligosaccharide is normally in the micromolar or lesser range, but it is increased to the nanomolar range by using the primary and secondary binding sites in the domain-swapped dimer. Recently, another two groups of broadly neutralizing antibodies that target the glycan shield of HIV-1 were identified (Walker et al. 2011; Walker et al. 2009). Crystal structures of two of these antibodies suggest a different mechanism for recognition (Pejchal et al. 2011; McLellan et al. 2011). These antibodies both display two glycan-binding sites in the Fab without utilizing the swapping dimer architecture in 2G12. Besides binding to well-conserved high-mannose glycans, antibody PGT128 also recognizes a protein segment under the glycan shield in the V3 loop region of gp120, potentially improving binding affinity as well as specificity (Pejchal et al. 2011). A similar motif of antibody binding to two glycans and a β-strand segment was observed for antibody PG9 binding to the V1V2 region of HIV-1 gp120, where, in this case, the β-strand comes from the V2 region and Man_5 is specifically recognized in PG9 compared to Man_{8-9} in PGT128/PGT127 for one of the glycans, whereas the other glycan interaction seems less restricted by glycan type and position (Walker et al. 2009; McLellan et al. 2011).

10.4 PROTEIN–GLYCAN INTERFACES AS DRUG TARGETS

The role of glycan recognition in the life cycle of many viruses has stimulated much interest in drug design at the protein–glycan interface (von Itzstein 2008). For influenza A virus, both HA and NA are potential targets for the discovery and development of viral inhibitors (Matrosovich and Klenk 2003; von Itzstein 2007).

The development of HA inhibitors was inspired by the identification of sialic acid as a viral receptor (Paulson 1985) and later by x-ray structural determination of HA in complex with sialic acid (Weis et al. 1988) in the 1980s. Initial studies explored the possibility of designing sialic acid analogs with high-affinity binding to the receptor-binding site, often with the aid of structure-guided optimization (Sauter et al. 1989; Sauter et al. 1992; Watowich et al. 1994; Toogood et al. 1991). However, low-affinity binding to the shallow receptor-binding site proved to be an insurmountable challenge for monovalent sialic acid analog ligands (Matrosovich 1989). The balance between potency and specificity also handicaps the design of sialic acid analogs, as tight binding to all the known HA subtypes, or at least subtypes of interest to human health, such as the pandemic H1, H2, and H3 and other human-infecting zoonotic viruses, H5, H7, and H9, is the desired outcome. Substitutions on the designed inhibitors that improve affinity likely also enhance only interactions with nonconserved residues at the receptor-binding site. As a result, derivatives of Neu5Ac often have variable effects on different viral strains (Matrosovich et al. 1991, 1992, 1993). Later approaches have used polyvalent presentation of sialic acid analogs, mimicking the multivalent binding between virus and host cell. Methods of presentation include linear polymers, synthetic dendrimers, sialic acid–containing liposomes (Carlescu et al. 2009) and, more recently, nanostructures (Papp et al. 2011). Sialylglycopolymer is the best studied among the group and shows greatly improved inhibitory activity over monovalent ligands (Mammen et al. 1996). Notwithstanding, a polyvalent inhibitor against influenza virus has not yet been approved for clinical trials, as concerns of toxicity and drug delivery remain to be solved. In recent years, sialic acid–mimic peptides have been selected from phage display as potential HA inhibitors and show improved binding affinities over sialic acid (Matsubara et al. 2010).

The development of NA inhibitors has been much more fruitful. As for HA, inhibitor design was also aided by high-resolution NA crystal structures. The crystal structure of NA revealed a tetramer of β-propellers (Varghese et al. 1983; Colman et al. 1983), which represents a structural fold conserved in sialidases, as described in Section 10.2.2 for PIV HN. In the active site, Neu5Ac is oriented by interaction with an arginine triad, and ligand binding is facilitated by significant contacts with five other highly conserved residues. Later, higher-resolution structures suggested the glycan moiety that is trapped at the active site is Neu5Ac2en, an unsaturated Neu5Ac derivative and a known low-affinity inhibitor of influenza virus (Varghese et al. 1992; Burmeister et al. 1993). The moiety Neu5Ac2en displays a distorted boat conformation that proved to be a valuable starting point for the design of potent inhibitors.

Under the guidance of structural analysis, the 4-hydroxyl group of Neu5Ac2en was replaced by a large basic functional group for optimal formation of salt bridges with the conserved Glu119 and Glu227 (von Itzstein et al. 1993). The resultant 4-deoxy-4-guanidino-Neu5Ac2en is a highly effective inhibitor of influenza NA (Woods et al. 1993) and was later marketed as zanamivir (Relenza) by Glaxo (now GlaxoSmithKline, London, United Kingdom). In the crystal structure of NA from the 1918 H1N1 pandemic in complex with zanamivir (Xu et al. 2008), the Glu119 carboxylate adopts a different orientation from the unliganded structure to accommodate the 4-guanidino group of zanamivir (Figure 10.4a). Electrostatic interactions are, indeed, evident between zanamivir and the two conserved glutamate residues as intended. Ligand binding also induces conformational changes in the 150 loop for more extensive interaction and brings another acidic residue, Asp151, within the hydrogen bonding distance of the guanidinyl group of zanamivir (Xu et al. 2008).

In later years, continuing efforts to improve the bioavailability of Neu5Ac2en derivatives led to the development of oseltamivir (Tamiflu) (Kim et al. 1997, 1998). The active metabolite of oseltamivir is a cyclohexene derivative that mimics the structure of Neu5Ac2en. An amine group replaces the 4-hydroxyl group to create charge interactions with Glu119 and Glu227. The glycerol moiety of Neu5Ac2en is replaced by a pentyl ether side chain to improve its lipophilicity. X-ray crystallographic studies suggest that oseltamivir engages most of the conserved interaction seen in the NA complexes with Neu5Ac2en and zanamivir (Figure 10.4b). The Glu276 reorients away from the ligand to make room for the bulky hydrophobic pentyloxy group (Kim et al. 1997, 1998; Russell et al. 2006). In the oseltamivir-resistant mutant, substitution of H274Y presents a bulkier tyrosine residue that pushes the carboxyl group of Glu276 into the binding site and reduces inhibition by oseltamivir (Collins et al. 2008).

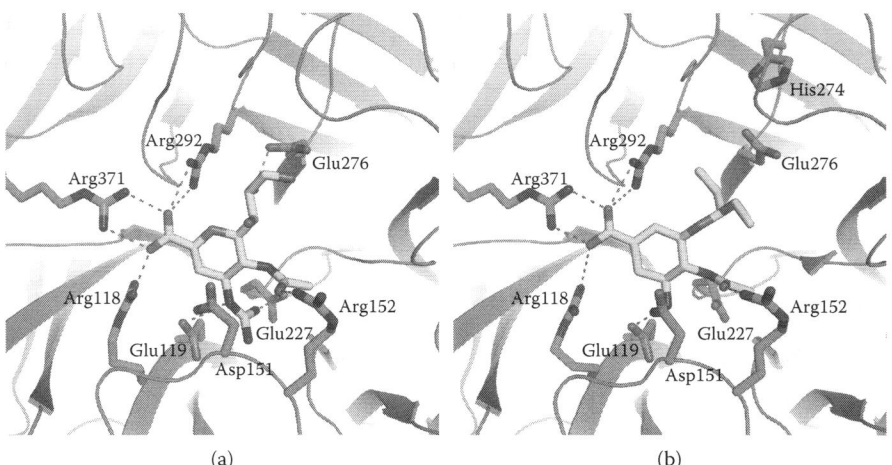

(a)   (b)

FIGURE 10.4   **(See color insert.)** Influenza A virus sialidase NA as an effective target for antiviral inhibitors. (a) The N1 NA (viral strain: A/Brevig Mission/1/1918) in complex with zanamivir (PDB: 3B7E) and (b) the N1 NA from an avian flu virus in complex with oseltamivir (PDB: 2HU4).

The successful story of drug design in influenza NA has inspired similar approaches in other viral sialidases. Inhibitors of hPIV have been designed based on crystal structures of NDV HN and these efforts have led to the micromolar inhibition of HN from hPIV3 (Alymova et al. 2004; Tindal et al. 2007).

ACKNOWLEDGMENTS

This study was supported in part by the National Institute of Allergy and Infectious Diseases (NIAID; Bethesda, Maryland) grant AI058113 (to Ian A. Wilson) and the Skaggs Institute for Chemical Biology (La Jolla, California). This is publication #21401 from the Scripps Research Institute (La Jolla, California).

REFERENCES

Abe, Y., Takashita, E., Sugawara, K. et al. 2004. Effect of the addition of oligosaccharides on the biological activities and antigenicity of influenza A/H3N2 virus hemagglutinin. *J Virol,* 78, 9605–9611.

Alymova, I. V., Taylor, G., Takimoto, T. et al. 2004. Efficacy of novel hemagglutinin-neuraminidase inhibitors BCX 2798 and BCX 2855 against human parainfluenza viruses in vitro and in vivo. *Antimicrob Agents Chemother,* 48, 1495–1502.

Amonsen, M., Smith, D. F., Cummings, R. D., and Air, G. M. 2007. Human parainfluenza viruses hPIV1 and hPIV3 bind oligosaccharides with α2-3-linked sialic acids that are distinct from those bound by H5 avian influenza virus hemagglutinin. *J Virol,* 81, 8341–8345.

Banda, K., Kang, G., and Varki, A. 2009. 'Sialidase sensitivity' of rotaviruses revisited. *Nat Chem Biol,* 5, 71–72.

Barbey-Martin, C., Gigant, B., Bizebard, T. et al. 2002. An antibody that prevents the hemagglutinin low pH fusogenic transition. *Virology,* 294, 70–74.

Bateman, A. C., Karamanska, R., Busch, M. G. et al. 2010. Glycan analysis and influenza A virus infection of primary swine respiratory epithelial cells: The importance of NeuAcα2-6 glycans. *J Biol Chem,* 285, 34016–34026.

Blanchard, H., Yu, X., Coulson, B. S., and von Itzstein, M. 2007. Insight into host cell carbohydrate-recognition by human and porcine rotavirus from crystal structures of the virion spike associated carbohydrate-binding domain (VP8*). *J Mol Biol,* 367, 1215–1226.

Blixt, O., Head, S., Mondala, T. et al. 2004. Printed covalent glycan array for ligand profiling of diverse glycan binding proteins. *Proc Natl Acad Sci USA,* 101, 17033–17038.

Bousse, T. L., Taylor, G., Krishnamurthy, S. et al. 2004. Biological significance of the second receptor binding site of Newcastle disease virus hemagglutinin-neuraminidase protein. *J Virol,* 78, 13351–13355.

Bowden, T. A., Crispin, M., Jones, E. Y., and Stuart, D. I. 2010. Shared paramyxoviral glycoprotein architecture is adapted for diverse attachment strategies. *Biochem Soc Trans,* 38, 1349–1355.

Bu, W., Mamedova, A., Tan, M. et al. 2008. Structural basis for the receptor binding specificity of Norwalk virus. *J Virol,* 82, 5340–5347.

Burmeister, W. P., Henrissat, B., Bosso, C., Cusack, S., and Ruigrok, R. W. 1993. Influenza B virus neuraminidase can synthesize its own inhibitor. *Structure,* 1, 19–26.

Calarese, D. A., Scanlan, C. N., Zwick, M. B. et al. 2003. Antibody domain exchange is an immunological solution to carbohydrate cluster recognition. *Science,* 300, 2065–2071.

Cao, S., Lou, Z., Tan, M. et al. 2007. Structural basis for the recognition of blood group trisaccharides by norovirus. *J Virol,* 81, 5949–5957.

Carlescu, I., Scutaru, D., Popa, M., and Uglea, C. V. 2009. Synthetic sialic-acid-containing polyvalent antiviral inhibitors. *Med Chem Res,* 18, 477–494.

Caton, A. J., Brownlee, G. G., Yewdell, J. W., and Gerhard, W. 1982. The antigenic structure of the influenza virus A/PR/8/34 hemagglutinin (H1 subtype). *Cell,* 31, 417–427.

Choi, J. M., Hutson, A. M., Estes, M. K., and Prasad, B. V. 2008. Atomic resolution structural characterization of recognition of histo-blood group antigens by Norwalk virus. *Proc Natl Acad Sci USA,* 105, 9175–9180.

Chou, H. H., Hayakawa, T., Diaz, S. et al. 2002. Inactivation of CMP-*N*-acetylneuraminic acid hydroxylase occurred prior to brain expansion during human evolution. *Proc Natl Acad Sci USA,* 99, 11736–11741.

Collins, P. J., Haire, L. F., Lin, Y. P. et al. 2008. Crystal structures of oseltamivir-resistant influenza virus neuraminidase mutants. *Nature,* 453, 1258–1261.

Colman, P. M., Varghese, J. N., and Laver, W. G. 1983. Structure of the catalytic and antigenic sites in influenza virus neuraminidase. *Nature,* 303, 41–44.

Colman, P. M. and Ward, C. W. 1985. Structure and diversity of influenza virus neuraminidase. *Curr Top Microbiol Immunol,* 114, 177–255.

Connor, R. J., Kawaoka, Y., Webster, R. G., and Paulson, J. C. 1994. Receptor specificity in human, avian, and equine H2 and H3 influenza virus isolates. *Virology,* 205, 17–23.

Corti, D., Voss, J., Gamblin, S. J. et al. 2011. A Neutralizing Antibody Selected from Plasma Cells That Binds to Group 1 and Group 2 Influenza A Hemagglutinins. *Science,* 333, 850–856

Couceiro, J. N., Paulson, J. C., and Baum, L. G. 1993. Influenza virus strains selectively recognize sialyloligosaccharides on human respiratory epithelium; the role of the host cell in selection of hemagglutinin receptor specificity. *Virus Res,* 29, 155–165.

Crennell, S., Takimoto, T., Portner, A., and Taylor, G. 2000. Crystal structure of the multifunctional paramyxovirus hemagglutinin-neuraminidase. *Nat Struct Biol,* 7, 1068–1074.

Daniels, R. S., Douglas, A. R., Skehel, J. J., and Wiley, D. C. 1983. Analyses of the antigenicity of influenza haemagglutinin at the pH optimum for virus-mediated membrane fusion. *J Gen Virol,* 64, 1657–1662.

Das, S. R., Puigbo, P., Hensley, S. E. et al. 2010. Glycosylation focuses sequence variation in the influenza A virus H1 hemagglutinin globular domain. *PLoS Pathog,* 6, e1001211.

Doores, K. J., Fulton, Z., Huber, M., Wilson, I. A., and Burton, D. R. 2010. Antibody 2G12 recognizes di-mannose equivalently in domain- and nondomain-exchanged forms but only binds the HIV-1 glycan shield if domain exchanged. *J Virol,* 84, 10690–10699.

Dormitzer, P. R., Sun, Z. Y., Blixt, O. et al. 2002a. Specificity and affinity of sialic acid binding by the rhesus rotavirus VP8* core. *J Virol,* 76, 10512–10517.

Dormitzer, P. R., Sun, Z. Y., Wagner, G., and Harrison, S. C. 2002b. The rhesus rotavirus VP4 sialic acid binding domain has a galectin fold with a novel carbohydrate binding site. *EMBO J,* 21, 885–897.

Eisen, M. B., Sabesan, S., Skehel, J. J., and Wiley, D. C. 1997. Binding of the influenza A virus to cell-surface receptors: Structures of five hemagglutinin-sialyloligosaccharide complexes determined by x-ray crystallography. *Virology,* 232, 19–31.

Ekiert, D. C., Bhabha, G., Elsliger, M. A. et al. 2009. Antibody recognition of a highly conserved influenza virus epitope. *Science,* 324, 246–251.

Ekiert, D. C., Friesen, R. H., Bhabha, G. et al. 2011. A highly conserved neutralizing epitope on group 2 influenza A viruses. *Science,* 333, 843–850.

Fleury, D., Wharton, S. A., Skehel, J. J., Knossow, M., and Bizebard, T. 1998. Antigen distortion allows influenza virus to escape neutralization. *Nat Struct Biol,* 5, 119–123.

Gambaryan, A. S., Karasin, A. I., Tuzikov, A. B. et al. 2005. Receptor-binding properties of swine influenza viruses isolated and propagated in MDCK cells. *Virus Res,* 114, 15–22.

Gambaryan, A. S., Tuzikov, A. B., Piskarev, V. E. et al. 1997. Specification of receptor-binding phenotypes of influenza virus isolates from different hosts using synthetic sialylglycopolymers: Non-egg-adapted human H1 and H3 influenza A and influenza B viruses share a common high binding affinity for 6'-sialyl(N-acetyllactosamine). *Virology,* 232, 345–350.

Gamblin, S., Haire, L., Russell, R. et al. 2004. The structure and receptor binding properties of the 1918 influenza hemagglutinin. *Science,* 303, 1838–1842.

Gamblin, S. J. and Skehel, J. J. 2010. Influenza hemagglutinin and neuraminidase membrane glycoproteins. *J Biol Chem,* 285, 28403–28409.

Geijtenbeek, T. B., Kwon, D. S., Torensma, R. et al. 2000. DC-SIGN, a dendritic cell-specific HIV-1-binding protein that enhances trans-infection of T cells. *Cell,* 100, 587–597.

Guo, C. T., Takahashi, N., Yagi, H. et al. 2007. The quail and chicken intestine have sialyl-galactose sugar chains responsible for the binding of influenza A viruses to human type receptors. *Glycobiology,* 17, 713–724.

Ha, Y., Stevens, D., Skehel, J., and Wiley, D. 2001. X-ray structures of H5 avian and H9 swine influenza virus hemagglutinins bound to avian and human receptor analogs. *Proc Natl Acad Sci USA,* 98, 11181–11186.

Ha, Y., Stevens, D., Skehel, J., and Wiley, D. 2002. H5 avian and H9 swine influenza virus haemagglutinin structures: Possible origin of influenza subtypes. *EMBO J,* 21, 865–875.

Ha, Y., Stevens, D., Skehel, J., and Wiley, D. 2003. X-ray structure of the hemagglutinin of a potential H3 avian progenitor of the 1968 Hong Kong pandemic influenza virus. *Virology,* 309, 209–218.

Haselhorst, T., Fleming, F. E., Dyason, J. C. et al. 2009. Sialic acid dependence in rotavirus host cell invasion. *Nat Chem Biol,* 5, 91–93.

Hashiguchi, T., Ose, T., Kubota, M. et al. 2011. Structure of the measles virus hemagglutinin bound to its cellular receptor SLAM. *Nat Struct Mol Biol,* 18, 135–141.

Imai, M., Watanabe, T., Hatta, M., et al. 2012. Experimental adaptation of an influenza H5 HA confers respiratory droplet transmission to a reassortant H5 HA/H1N1 virus in ferrets. *Nature*, doi:10.1038/nature10831

Imberty, A. and Varrot, A. 2008. Microbial recognition of human cell surface glycoconjugates. *Curr Opin Struct Biol,* 18, 567–576.

Ito, T., Suzuki, Y., Suzuki, T. et al. 2000. Recognition of N-glycolylneuraminic acid linked to galactose by the α2,3-linkage is associated with intestinal replication of influenza A virus in ducks. *J Virol,* 74, 9300–9305.

Jenkins, N., Parekh, R. B., and James, D. C. 1996. Getting the glycosylation right: Implications for the biotechnology industry. *Nat Biotechnol,* 14, 975–981.

Karlsson, K. A. 2001. Pathogen-host protein-carbohydrate interactions as the basis of important infections. *Adv Exp Med Biol,* 491, 431–443.

Kashyap, A. K., Steel, J., Oner, A. F. et al. 2008. Combinatorial antibody libraries from survivors of the Turkish H5N1 avian influenza outbreak reveal virus neutralization strategies. *Proc Natl Acad Sci USA,* 105, 5986–5991.

Kim, C. U., Lew, W., Williams, M. A. et al. 1997. Influenza neuraminidase inhibitors possessing a novel hydrophobic interaction in the enzyme active site: Design, synthesis, and structural analysis of carbocyclic sialic acid analogues with potent anti-influenza activity. *J Am Chem Soc,* 119, 681–690.

Kim, C. U., Lew, W., Williams, M. A. et al. 1998. Structure-activity relationship studies of novel carbocyclic influenza neuraminidase inhibitors. *J Med Chem,* 41, 2451–2460.

Klenk, H. D., Wagner, R., Heuer, D., and Wolff, T. 2002. Importance of hemagglutinin glycosylation for the biological functions of influenza virus. *Virus Res,* 82, 73–75.

Korber, B., Gaschen, B., Yusim, K. et al. 2001. Evolutionary and immunological implications of contemporary HIV-1 variation. *Br Med Bull,* 58, 19–42.

Kraschnefski, M. J., Bugarcic, A., Fleming, F. E. et al. 2009. Effects on sialic acid recognition of amino acid mutations in the carbohydrate-binding cleft of the rotavirus spike protein. *Glycobiology,* 19, 194–200.

Kwong, P. D. and Wilson, I. A. 2009. HIV-1 and influenza antibodies: Seeing antigens in new ways. *Nat Immunol,* 10, 573–578.

Lawrence, M. C., Borg, N. A., Streltsov, V. A. et al. 2004. Structure of the haemagglutinin-neuraminidase from human parainfluenza virus type III. *J Mol Biol,* 335, 1343–1357.

Lin, T., Wang, G., Li, A. et al. 2009. The hemagglutinin structure of an avian H1N1 influenza A virus. *Virology,* 392, 73–81.

Lindesmith, L., Moe, C., Marionneau, S. et al. 2003. Human susceptibility and resistance to Norwalk virus infection. *Nat Med,* 9, 548–553.

Liu, J., Stevens, D. J., Haire, L. F. et al. 2009. Structures of receptor complexes formed by hemagglutinins from the Asian Influenza pandemic of 1957. *Proc Natl Acad Sci USA,* 106, 17175–17180.

Lo-Guidice, J. M., Wieruszeski, J. M., Lemoine, J. et al. 1994. Sialylation and sulfation of the carbohydrate chains in respiratory mucins from a patient with cystic fibrosis. *J Biol Chem,* 269, 18794–18813.

Lopez, S. and Arias, C. F. 2006. Early steps in rotavirus cell entry. *Curr Top Microbiol Immunol,* 309, 39–66.

Lovell, S. C., Davis, I. W., Arendall, W. B. 3rd, et al. 2003. Structure validation by Cα geometry: Φ, Ψ and Cβ deviation. *Proteins,* 50, 437–450.

Maines, T. R., Chen, L. M., Van Hoeven, N. et al. 2011. Effect of receptor binding domain mutations on receptor binding and transmissibility of avian influenza H5N1 viruses. *Virology,* 413, 139–147.

Mammen, M., Helmerson, K., Kishore, R. et al. 1996. Optically controlled collisions of biological objects to evaluate potent polyvalent inhibitors of virus-cell adhesion. *Chem Biol,* 3, 757–763.

Matrosovich, M. N. 1989. Towards the development of antimicrobial drugs acting by inhibition of pathogen attachment to host cells: A need for polyvalency. *FEBS Lett,* 252, 1–4.

Matrosovich, M. N., Gambaryan, A. S., and Chumakov, M. P. 1992. Influenza viruses differ in recognition of 4-*O*-acetyl substitution of sialic acid receptor determinant. *Virology,* 188, 854–858.

Matrosovich, M. N., Gambaryan, A. S., Reizin, F. N., and Chumakov, M. P. 1991. Recognition by human A and B influenza viruses of 8- and 7-carbon analogues of sialic acid modified in the polyhydroxyl side chain. *Virology,* 182, 879–882.

Matrosovich, M. N., Gambaryan, A. S., Tuzikov, A. B. et al. 1993. Probing of the receptor-binding sites of the H1 and H3 influenza A and influenza B virus hemagglutinins by synthetic and natural sialosides. *Virology,* 196, 111–121.

Matrosovich, M. and Klenk, H. D. 2003. Natural and synthetic sialic acid-containing inhibitors of influenza virus receptor binding. *Rev Med Virol,* 13, 85–97.

Matrosovich, M. N., Matrosovich, T. Y., Gray, T., Roberts, N. A., and Klenk, H. D. 2004. Human and avian influenza viruses target different cell types in cultures of human airway epithelium. *Proc Natl Acad Sci USA,* 101, 4620–4624.

Matrosovich, M., Tuzikov, A., Bovin, N. et al. 2000. Early alterations of the receptor-binding properties of H1, H2, and H3 avian influenza virus hemagglutinins after their introduction into mammals. *J Virol,* 74, 8502–8512.

Matsubara, T., Onishi, A., Saito, T. et al. 2010. Sialic acid-mimic peptides as hemagglutinin inhibitors for anti-influenza therapy. *J Med Chem,* 53, 4441–4449.

McLellan, J. S., Pancera, M., Carrico, C. et al. 2011. Structure of HIV-1 gp120 V1V2 domain with broadly neutralizing antibody PG9. *Nature,* 480, 336–343.

Monnier, N., Higo-Moriguchi, K., Sun, Z. Y. et al. 2006. High-resolution molecular and anti-gen structure of the VP8* core of a sialic acid-independent human rotavirus strain. *J Virol,* 80, 1513–1523.

Navaratnarajah, C. K., Oezguen, N., Rupp, L. et al. 2011. The heads of the measles virus attachment protein move to transmit the fusion-triggering signal. *Nat Struct Mol Biol,* 18, 128–134.

Okuno, Y., Isegawa, Y., Sasao, F., and Ueda, S. 1993. A common neutralizing epitope con-served between the hemagglutinins of influenza A virus H1 and H2 strains. *J Virol,* 67, 2552–2558.

Olofsson, S. and Bergstrom, T. 2005. Glycoconjugate glycans as viral receptors. *Ann Med,* 37, 154–172.

Papp, I., Sieben, C., Sisson, A. L. et al. 2011. Inhibition of influenza virus activity by multiva-lent glycoarchitectures with matched sizes. *Chembiochem,* 12, 887–895.

Parashar, U. D., Gibson, C. J., Bresse, J. S., and Glass, R. I. 2006. Rotavirus and severe child-hood diarrhea. *Emerg Infect Dis,* 12, 304–306.

Paulson, J. C. 1985. Interactions of animal viruses with cell surface receptors. In *The Receptors,* ed. M. P. Conn, pp. 131–219, Orlando, FL: Academic Press.

Pejchal, R., Doores, K. J., Walker, L. M. et al. 2011. A potent and broad neutralizing antibody recognizes and penetrates the HIV glycan shield. *Science,* 334, 1097–1103.

Porotto, M., Fornabaio, M., Greengard, O. et al. 2006. Paramyxovirus receptor-binding mol-ecules: Engagement of one site on the hemagglutinin-neuraminidase protein modulates activity at the second site. *J Virol,* 80, 1204–1213.

Porotto, M., Fornabaio, M., Kellogg, G. E., and Moscona, A. 2007. A second receptor binding site on human parainfluenza virus type 3 hemagglutinin-neuraminidase contributes to activation of the fusion mechanism. *J Virol,* 81, 3216–3228.

Reitter, J. N., Means, R. E., and Desrosiers, R. C. 1998. A role for carbohydrates in immune evasion in AIDS. *Nat Med,* 4, 679–684.

Rogers, G. N. and D'souza, B. L. 1989. Receptor binding properties of human and animal H1 influenza virus isolates. *Virology,* 173, 317–322.

Rogers, G. N. and Paulson, J. C. 1983. Receptor determinants of human and animal influenza virus isolates: Differences in receptor specificity of the H3 hemagglutinin based on spe-cies of origin. *Virology,* 127, 361–373.

Russell, R. J., Gamblin, S. J., Haire, L. F. et al. 2004. H1 and H7 influenza haemagglutinin struc-tures extend a structural classification of haemagglutinin subtypes. *Virology,* 325, 287–296.

Russell, R. J., Haire, L. F., Stevens, D. J. et al. 2006. The structure of H5N1 avian influenza neuraminidase suggests new opportunities for drug design. *Nature,* 443, 45–49.

Sagar, M., Wu, X., Lee, S., and Overbaugh, J. 2006. Human immunodeficiency virus type 1 V1-V2 envelope loop sequences expand and add glycosylation sites over the course of infection, and these modifications affect antibody neutralization sensitivity. *J Virol,* 80, 9586–9598.

Sanders, R. W., Venturi, M., Schiffner, L. et al. 2002. The mannose-dependent epitope for neu-tralizing antibody 2G12 on human immunodeficiency virus type 1 glycoprotein gp120. *J Virol,* 76, 7293–7305.

Sauter, N. K., Bednarski, M. D., Wurzburg, B. A. et al. 1989. Hemagglutinins from two influ-enza virus variants bind to sialic acid derivatives with millimolar dissociation constants: A 500-MHz proton nuclear magnetic resonance study. *Biochemistry,* 28, 8388–8396.

Sauter, N. K., Hanson, J. E., Glick, G. D. et al. 1992. Binding of influenza virus hemagglutinin to analogs of its cell-surface receptor, sialic acid: Analysis by proton nuclear magnetic resonance spectroscopy and x-ray crystallography. *Biochemistry,* 31, 9609–9621.

Sharon, N. 1996. Carbohydrate-lectin interactions in infectious disease. *Adv Exp Med Biol,* 408, 1–8.

Shinya, K., Ebina, M., Yamada, S. et al. 2006. Avian flu: Influenza virus receptors in the human airway. *Nature,* 440, 435–436.

Skehel, J. J. and Wiley, D. C. 2000. Receptor binding and membrane fusion in virus entry: The influenza hemagglutinin. *Annu Rev Biochem,* 69, 531–569.

Spiro, R. G. 2002. Protein glycosylation: Nature, distribution, enzymatic formation, and disease implications of glycopeptide bonds. *Glycobiology,* 12, 43R–56R.

Stevens, J., Blixt, O., Chen, L. M. et al. 2008. Recent avian H5N1 viruses exhibit increased propensity for acquiring human receptor specificity. *J Mol Biol,* 381, 1382–1394.

Stevens, J., Blixt, O., Glaser, L. et al. 2006a. Glycan microarray analysis of the hemagglutinins from modern and pandemic influenza viruses reveals different receptor specificities. *J Mol Biol,* 355, 1143–1155.

Stevens, J., Blixt, O., Paulson, J. C., and Wilson, I. A. 2006b. Glycan microarray technologies: Tools to survey host specificity of influenza viruses. *Nat Rev Microbiol,* 4, 857–864.

Stevens, J., Blixt, O., Tumpey, T. M. et al. 2006c. Structure and receptor specificity of the hemagglutinin from an H5N1 influenza virus. *Science,* 312, 404–410.

Stevens, J., Corper, A., Basler, C. et al. 2004. Structure of the uncleaved human H1 hemagglutinin from the extinct 1918 influenza virus. *Science,* 303, 1866–1870.

Sui, J., Hwang, W. C., Perez, S. et al. 2009. Structural and functional bases for broad-spectrum neutralization of avian and human influenza A viruses. *Nat Struct Mol Biol,* 16, 265–273.

Suzuki, T., Portner, A., Scroggs, R. A. et al. 2001. Receptor specificities of human respiroviruses. *J Virol,* 75, 4604–4613.

Taube, S., Jiang, M., and Wobus, C. E. 2010. Glycosphingolipids as receptors for non-enveloped viruses. *Viruses,* 2, 1011–1049.

Throsby, M., Van Den Brink, E., Jongeneelen, M. et al. 2008. Heterosubtypic neutralizing monoclonal antibodies cross-protective against H5N1 and H1N1 recovered from human IgM+ memory B cells. *PLoS One,* 3, e3942.

Tindal, D. J., Dyason, J. C., Thomson, R. J. et al. 2007. Synthesis and evaluation of 4-O-alkylated 2-deoxy-2,3-didehydro-N-acetylneuraminic acid derivatives as inhibitors of human parainfluenza virus type-3 sialidase activity. *Bioorg Med Chem Lett,* 17, 1655–1658.

Toogood, P. L., Galliker, P. K., Glick, G. D., and Knowles, J. R. 1991. Monovalent sialosides that bind tightly to influenza A virus. *J Med Chem,* 34, 3138–3140.

Tsuchiya, E., Sugawara, K., Hongo, S. et al. 2002. Effect of addition of new oligosaccharide chains to the globular head of influenza A/H2N2 virus haemagglutinin on the intracellular transport and biological activities of the molecule. *J Gen Virol,* 83, 1137–1146.

Tumpey, T. M., Maines, T. R., Van Hoeven, N. et al. 2007. A two-amino acid change in the hemagglutinin of the 1918 influenza virus abolishes transmission. *Science,* 315, 655–659.

Varghese, J. N., Laver, W. G., and Colman, P. M. 1983. Structure of the influenza virus glycoprotein antigen neuraminidase at 2.9 Å resolution. *Nature,* 303, 35–40.

Varghese, J. N., Mckimm-Breschkin, J. L., Caldwell, J. B., Kortt, A. A., and Colman, P. M. 1992. The structure of the complex between influenza virus neuraminidase and sialic acid, the viral receptor. *Proteins,* 14, 327–332.

Varki, A., Cummings, R. D., Esko, J. D. et al. 2009. *Essentials of glycobiology,* Cold Spring Harbor (NY): Cold Spring Harbor Laboratory Press.

Vigerust, D. J. and Shepherd, V. L. 2007. Virus glycosylation: Role in virulence and immune interactions. *Trends Microbiol,* 15, 211–218.

von Itzstein, M. 2007. The war against influenza: Discovery and development of sialidase inhibitors. *Nat Rev Drug Discov,* 6, 967–974.

von Itzstein, M. 2008. Disease-associated carbohydrate-recognising proteins and structure-based inhibitor design. *Curr Opin Struct Biol,* 18, 558–566.

von Itzstein, M., Wu, W. Y., Kok, G. B. et al. 1993. Rational design of potent sialidase-based inhibitors of influenza virus replication. *Nature,* 363, 418–423.

Walker, L. M., Huber, M., Doores, K. J. et al. 2011. Broad neutralization coverage of HIV by multiple highly potent antibodies. *Nature,* 477, 466–470.

Walker, L. M., Phogat, S. K., Chan-Hui, P. Y. et al. 2009. Broad and potent neutralizing antibodies from an African donor reveal a new HIV-1 vaccine target. *Science,* 326, 285–289.

Watowich, S. J., Skehel, J. J., and Wiley, D. C. 1994. Crystal structures of influenza virus hemagglutinin in complex with high-affinity receptor analogs. *Structure,* 2, 719–731.

Wei, C. J., Boyington, J. C., Dai, K. et al. 2010. Cross-neutralization of 1918 and 2009 influenza viruses: Role of glycans in viral evolution and vaccine design. *Sci Transl Med,* 2, 24ra21.

Weis, W., Brown, J. H., Cusack, S. et al. 1988. Structure of the influenza virus haemagglutinin complexed with its receptor, sialic acid. *Nature,* 333, 426–431.

Whittle, J. R., Zhang, R., Khurana, S. et al. 2011. Broadly neutralizing human antibody that recognizes the receptor-binding pocket of influenza virus hemagglutinin. *Proc Natl Acad Sci USA.* 108, 14216–14221.

Wiley, D. C., Wilson, I. A., and Skehel, J. J. 1981. Structural identification of the antibody-binding sites of Hong Kong influenza haemagglutinin and their involvement in antigenic variation. *Nature,* 289, 373–378.

Wilson, I. A., Skehel, J. J., and Wiley, D. C. 1981. Structure of the haemagglutinin membrane glycoprotein of influenza virus at 3 Å resolution. *Nature,* 289, 366–373.

Woods, J. M., Bethell, R. C., Coates, J. A. et al. 1993. 4-Guanidino-2,4-dideoxy-2,3-dehydro-*N*-acetylneuraminic acid is a highly effective inhibitor both of the sialidase (neuraminidase) and of growth of a wide range of influenza A and B viruses in vitro. *Antimicrob Agents Chemother,* 37, 1473–1479.

Wu, X., Zhou, T., Zhu, J. et al. 2011. Focused evolution of HIV-1 neutralizing antibodies revealed by structures and deep sequencing. *Science,* 333, 1593–1602.

Xu, R., Ekiert, D. C., Krause, J. C. et al. 2010a. Structural basis of preexisting immunity to the 2009 H1N1 pandemic influenza virus. *Science,* 328, 357–360.

Xu, R., Mcbride, R., Nycholat, C. M., Paulson, J. C., and Wilson, I. A. 2012. Structural characterization of the receptor specificity of the hemagglutinin from the 2009 influenza pandemic. *J Virol,* 86, 982–990.

Xu, R., Mcbride, R., Paulson, J. C., Basler, C. F., and Wilson, I. A. 2010b. Structure, receptor binding, and antigenicity of influenza virus hemagglutinins from the 1957 H2N2 pandemic. *J Virol,* 84, 1715–1721.

Xu, X., Zhu, X., Dwek, R. A., Stevens, J., and Wilson, I. A. 2008. Structural characterization of the 1918 influenza virus H1N1 neuraminidase. *J Virol,* 82, 10493–10501.

Yang, H., Chen, L. M., Carney, P. J., Donis, R. O., and Stevens, J. 2010. Structures of receptor complexes of a North American H7N2 influenza hemagglutinin with a loop deletion in the receptor binding site. *PLoS Pathog,* 6, e1001081.

Yuan, P., Swanson, K. A., Leser, G. P. et al. 2011. Structure of the Newcastle disease virus hemagglutinin-neuraminidase (HN) ectodomain reveals a four-helix bundle stalk. *Proc Natl Acad Sci USA,* 108, 14920–14925.

Yuan, P., Thompson, T. B., Wurzburg, B. A. et al. 2005. Structural studies of the parainfluenza virus 5 hemagglutinin-neuraminidase tetramer in complex with its receptor, sialyllactose. *Structure,* 13, 803–815.

Zaitsev, V., von Itzstein, M., Groves, D. et al. 2004. Second sialic acid binding site in Newcastle disease virus hemagglutinin-neuraminidase: Implications for fusion. *J Virol,* 78, 3733–3741.

Zhou, T., Georgiev, I., Wu, X. et al. 2010. Structural basis for broad and potent neutralization of HIV-1 by antibody VRC01. *Science,* 329, 811–817.

Zhou, T., Xu, L., Dey, B. et al. 2007. Structural definition of a conserved neutralization epitope on HIV-1 gp120. *Nature,* 445, 732–737.

11 Developing Drugs from Sugars
Current State of Glycosaminoglycan-Derived Therapeutics

Neha S. Gandhi and Ricardo L. Mancera

CONTENTS

11.1 INTRODUCTION

Sulfated glycosaminoglycans (GAGs) such as heparin and heparan sulfate (HS) are linear glycans consisting of 1,4-linked uronic acid and glucosamine (Figure 11.1) and encompassing varying degrees of sulfation (Gandhi and Mancera 2008). The HS and

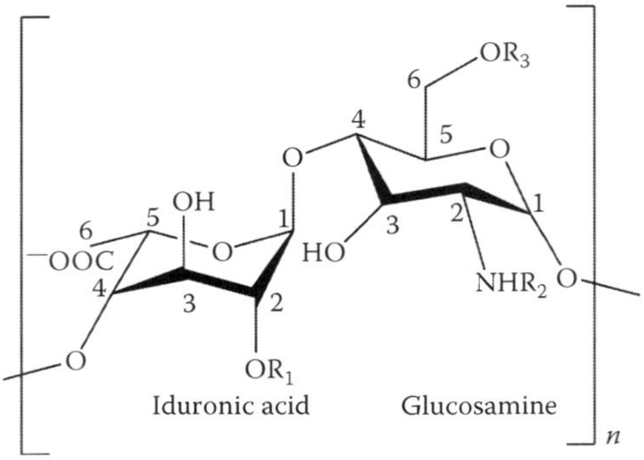

FIGURE 11.1 Chemical structure of the general repeating disaccharide unit in heparin and heparan sulfate, where R_1 = H or SO_3^-, R_2 = H/$COCH_3$ or SO_3^-, and R_3 = H or SO_3^-. The heparin disaccharide comprises a 2-O-sulfated L-iduronic acid linked to a 6-O-sulfated, N-sulfated D-glucosamine by $\alpha 1,4$-glycosidic linkage. The α-L-IdoA can be replaced by its C-5 epimer, β-D-GlcA.

heparin, a minor ubiquitous form of HS, are polydisperse, polyanionic polysaccharides in which the negative charges arise from the presence of sulfate and carboxylate groups. These molecules found inside the cell and in the extracellular matrix (ECM) have manifold regulatory functions and are critically relevant to many disease processes, such as inflammation (Parish 2006), neurodegeneration (Díaz-Nido et al. 2002), angiogenesis (Iozzo and San Antonio 2001), cardiovascular disorders (Rosenberg et al. 1997), cancer (Yip et al. 2006), and infectious diseases (Rostand and Esko 1997).

Thus, GAGs represent novel targets for the development of therapeutic agents for treatment of numerous diseases (Brown et al. 2007; Fuster and Esko 2005; Shriver et al. 2004), with drugs already being developed for the treatment of metabolic disorders, cancer, and infection. The pharmacological and therapeutic value of HS/heparin and their mimetics resides in their ability to bind and cause immobilization and/or activation of a variety of proteins, such as enzymes, enzyme inhibitors, growth factors, ECM proteins, cytokines, adhesion molecules, and receptor proteins (Gesslbauer and Kungl 2006; Volpi 2006). To date, many of these agents have proved useful as anticoagulant and antithrombin (AT) agents and in the inhibition of carbohydrate–lectin interactions (Osborn et al. 2004).

Heparin, which is isolated from mammalian tissues, is the largest selling pharmaceutical by weight in the world and has been used in the clinic to prevent blood clotting for over 60 years. Unfractionated heparin (UFH) remains the anticoagulant of choice during cardiopulmonary bypass as its anticoagulant activity can be quickly and completely neutralized by protamine. The UFH modulates the contact activation pathway by inactivating factor XIa and, to a lesser extent, factor XIIa through an AT-III-dependent mechanism (Figure 11.2) (Olson et al. 2004). Nevertheless, the use of UFH has some major disadvantages (Krishnaswamy et al. 2010), such as the potential for pathogen contamination due to its sourcing from animal tissues (Liu, Zhang et al. 2009) and a variable dose–response curve (Hirsh et al. 2007). Long-term use of UFH is associated with change in bone mineral density, as it directly affects

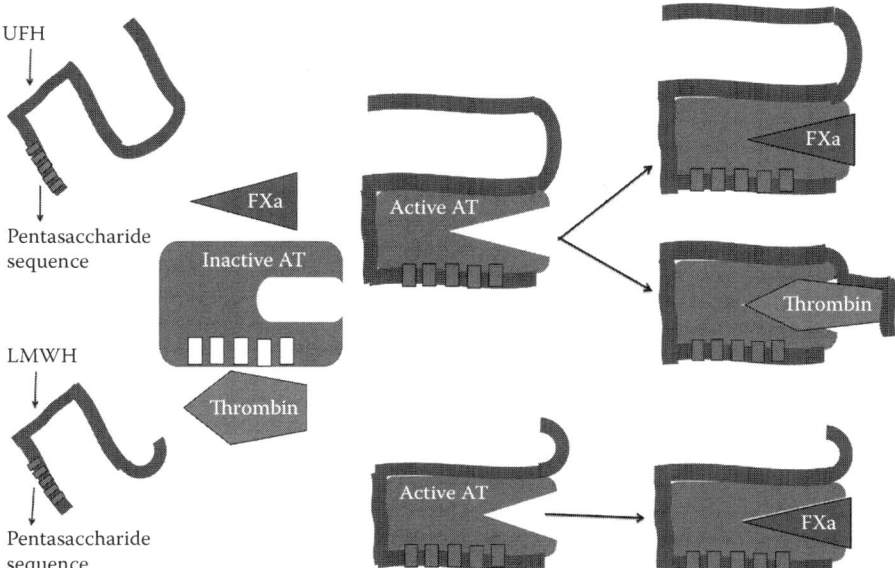

FIGURE 11.2 **(See color insert.)** A cartoon illustrating differences in mechanism of action of UFH and LMWH. Activated AT III degrades thrombin and FXa and binding of UFH to AT through the pentasaccharide sequence increases the catalytic action of AT. Soon after, the UFH–AT complex binds to and sequesters both FXa and thrombin, thereby preventing the cascade of events that normally allows blood to clot. The LMWH, which also consists of a pentasaccharide sequence, binds to AT and more selectively increases degradation of FXa but not thrombin and consequently prevents blood clotting.

the activity of bone cells (osteoclasts/osteoblasts) as well as indirectly influencing intercellular signaling and mineral metabolism that leads to heparin-induced osteo-porosis (fractures of vertebral bodies or femoral neck) (Matzsch et al. 1990; Monreal et al. 1990). Administration of UFH can lead to devastating heparin-induced throm-bocytopenia (HIT) (Arepally and Ortel 2010). These side effects are attributed to the structural diversity (Hirsh 1991) of UFH attributed to its large chain length and the presence of numerous negatively charged groups, resulting in various potential binding sites along the heparin chain, which can bind numerous proteins such as AT, von Willebrand's factor (vWF), growth factors, platelets, leucocytes, vascular endo-thelial cells, and other biological components. The variable anticoagulant response is caused by AT-independent binding of UFH to plasma proteins and to proteins released from platelets.

Low-molecular-weight heparins (LMWHs) have been produced by fragmenta-tion of heparin to avoid the aforementioned problems with UHF, in particular the occurrence of HIT. This disease originates from immunoglobulin G (IgG)-type antibodies that activate platelets. The activation of platelets occurs on formation of the antigenic complex of platelet factor 4 (PF4) and UHF, followed by binding of the tail of the antibody to the platelet FcγIIa receptor (Kelton et al. 1994). The PF4 is a chemo-attractant cytokine released from the α-granules of activated platelets involved in neutralizing heparin on the endothelial surface of blood vessels, thereby inhibiting local AT activity and promoting coagulation (Hermodson et al. 1977; Nath et al. 1975). The LMWHs are less prone to interact with PF4 and have thus been investigated in various clinical trials (Warkentin et al. 2000), offering simplified parenteral anticoagulant therapy for the treatment of pulmonary embolism (PE) and deep vein thrombosis (DVT). The LMWHs have an average molecular weight of less than 8000 Da. The small size of LMWHs also allows them to target antifactor Xa (aXa) and reduces their ability to inhibit thrombin (anti-FIIa activity; see Figure 11.2) (Olson et al. 2004). Various LMWHs, such as Lovenox® (enoxaparin)/ Sanofi-Aventis Laval, Quebec, Fragmin® (dalteparin)/Pfizer and Eisai Woodcliff Lake, New Jersey, Sandoparin® (certoparin)/Novartis Nürnberg, Bavaria, Germany, Innohep® (tinzaparin)/LEO Ballerup, Denmark, and Fraxiparin® (nadroparin)/ GSK and Sanofi-Synthélabo Canada, Markham, Ontario, have been manufactured using various methods to depolymerize heparin (Fareed and Walenga 2007). The LMWHs have a more predictable dose response and a longer half-life than UFH. The LMWH is cleared principally by the renal route. In contrast, UFH is not cleared by the kidneys and, hence, has potential benefits and higher safety than LMWH in patients with renal insufficiency. Recently, several clinical studies have reinforced the significance of thromboprophylaxis with LMWH in cancer patients (Robert 2010). Unlike UFH-induced osteoporosis in pregnancy, adequate data are not available to assess the change in bone density and resorption associated with the use of different doses of LMWH in different patient groups (Lefkou et al. 2010; Le Templier and Rodger 2008).

Heparin and LMWH are recommended for prolonged use in women with antiphospholipid (aPL) syndrome (APS) or other blood-clotting disorders (thrombophilia disorders) that are linked to recurrent pregnancy loss (RPL). In thrombophilia disorders, heparin is used to thin the blood in the developing placenta, eventually preventing miscarriage, although the use of heparin and LMWH may increase the risk of bone loss (Rajgopal et al. 2008). The combination of heparin or LMWH with aspirin is superior to aspirin or heparin monotherapy in achieving more live births in patients with positive aPL antibodies and RPL (Clark et al. 2010; Mak et al. 2010); however, a recent study found that neither heparin nor low-dose aspirin improved birth rates compared to placebo among women with unexplained RPL (Kaandorp et al. 2010). Furthermore, in similar studies heparin was shown to be beneficial for women having RPL but showing negative tests for aPL antibodies, postulating that recurrent miscarriages could be due to some unrecognized blood-clotting disorder (Di Nisio et al. 2005; Kaandorp et al. 2009). Therefore, the potential use of heparin/LMWH during conception in women with thrombophilia remains controversial (Ricci et al. 2010).

Medium-molecular-weight heparins (MMWHs) were designed to inactivate free and fibrin-bound thrombin, as well as activated factor Xa (FXa). The GH9001 (Leo Pharmaceuticals), a mixture of MMWH and low-molecular-weight dermatan sulfate (LMWDS), showed bioavailability and half-life similar to those of LMWH; however, its development was halted following the results of a study that showed an increase in liver enzymes in subjects given repeated doses of the drug (Weitz et al. 2003). Danaparoid or heparinoid (Orgaran®; Org 10172) is an anticoagulant consisting of GAG mixtures, such as HS, dermatan sulfate (DS), and chondroitin sulfate (CS), rendering it chemically distinct from LMWH and MMWH (de Pont et al. 2007; Meuleman 1992). It is used in patients who develop HIT during high-risk surgeries (Chong and Magnani 2007). This compound has a lower affinity for AT, significant anti-FXa (aXa) activity, and a lesser amount (approximately 20%) of cross-reactivity with heparin-induced antibodies (Magnani and Gallus 2006). These properties are attributed to its composition, which is devoid of heparin fragments. As a consequence, its use is also recommended as an alternative anticoagulant in pregnancies having high risk for thrombosis and heparin intolerance (Magnani 2010; Schindewolf et al. 2007). High-molecular-weight heparins (e.g., Astenose; Glycomed, Inc., Hastings, New York) were also developed to treat restenosis following balloon angioplasty (Ian et al. 1995). It had negligible anticoagulant activity and significantly enhanced potency to inhibit specific vWF/platelet interactions (Sobel et al. 1996; Sternbergh et al. 1995); however, the molecule was withdrawn following toxicity results in preclinical trials.

Anticoagulants have been widely used for the prophylaxis of surgical thrombosis over many years. Clinicians have a wide therapeutic arsenal of anticoagulants to choose from, including UFH, LMWH, warfarin, synthetic heparin derivatives (such as fondaparinux, idraparinux, and idrabiotaparinux), direct inhibitors of FXa (such as rivaroxaban and apixaban), and direct inhibitors of thrombin (such as dabigatran) (Eikelboom and Weitz 2010; Samama and Gerotziafas 2010). The mechanism of action of these anticoagulants is extensively reviewed by several researchers (Bauer 2002; Klement and Rak 2006). The AT–glycan interaction has been successfully targeted by UFH and LMWHs for many decades. The discovery of the mechanism of binding of heparin to AT has increased interest in the development of small, structurally defined heparin mimetics with AT activity but with reduced side effects (De Kort et al. 2005).

The structural diversity of heparin–/HS–protein interactions is enormous (Coombe and Kett 2005; Gama and Hsieh-Wilson 2005; Kreuger et al. 2006). These interactions and mechanisms are being investigated at the molecular and functional levels with the use of new experimental techniques, such as gel mobility shift assays, filter-binding assays, isothermal titration calorimetry (ITC), surface-plasmon resonance (SPR), carbohydrate microarrays, and chemical synthesis (Fugedi 2003; Powell et al. 2004; Rek et al. 2009). Several approaches are available to develop GAG-based agents to target the HS-dependent binding of a protein ligand to its cell-surface receptor (Lindahl 2007). The GAG mimetics can activate (agonists) or inactivate (antagonists) cell-surface receptors by directly binding to the receptors, form ternary complexes with a protein ligand and receptor, or interfere with the binding of endogenous heparin. Typical examples are heparin mimetics binding to AT to increase its ability to inhibit serine proteases and GAG inhibitors of heparanase, the endo-β-D-glucuronidase that plays a role in metastasis and angiogenesis. In malaria, an oligosaccharide lacking an AT-binding site is capable of blocking the interaction between the plasmodium-induced protein PfEMP1 and HS on erythrocyte or vascular endothelial surfaces (Vogt et al. 2006). Low-molecular-weight compounds, such as guanidinylated neomycin and peptidic foldamers, can bind specifically to HS and block HS–protein interactions (Choi et al. 2005; Elson-Schwab et al. 2007). Another therapeutic approach involves the binding of antiangiogenic polypeptides, such as endostatin and histidine-rich glycoprotein (GP), to HS, interfering with the binding of endogenous protein ligands and thereby inhibiting receptor signaling (Kreuger et al. 2002; Vanwildemeersch et al. 2006). Metabolic and enzyme inhibitors are known to alter the biosynthesis of HS (Brown et al. 2007). Compounds such as selenate; sodium chlorate, an inhibitor of the universal sulfate donor 3'-phospho-adenyl-5'-phosphosulfate (PAPS) (Baeuerle and Huttner 1986); and β-D-xylosides are used to block HS biosynthesis in cells and tissues (Fritz et al. 1994). The GAG–protein interactions have also been explored by the use of soluble GAGs or GAG mimetics, such as sucrose octasulfate, suramin, pentosan polysulfate, and dextran sulfates, which presumably occupy the GAG-binding sites in these proteins (Manetti et al. 2000; Zhu et al. 1993). Polypeptides containing clusters of positively charged amino acids can bind to the negatively charged sulfate and carboxyl groups, as is the case in protamine (Portmann and Holden 1949) and lactoferrin (Hekman 1971). Synthetic peptides containing lysine and arginine (Morad et al. 1984; Schick et al. 2004; Wang and Rabenstein 2006) also antagonize protein interactions with HS/heparin.

This chapter reviews recent advances in the development of GAG (heparin/HS)-based inhibitors that have been or are likely to be introduced into current clinical practice both as anticoagulants and as non-anticoagulants and helps one to understand their mechanism of action.

11.2   HEPARIN/HS MIMETICS AS ANTICOAGULANTS

For decades, antithrombotic and anticoagulants based on heparin/HS have been the mainstay of polytherapy for venous thrombosis, cardiovascular disorders, thrombotic and ischemic strokes, cancer, and neurodegenerative diseases (Alban 2008). The unique pentasaccharide (sometimes referred to as AGA*IA or DEFGH) GlcNAc/NS6S->GlcA->GlcNS3S6S->IdoA2S->GlcNS6S (where Glc is glucosamine, IdoA is iduronic acid, and GlcA is glucoronic acid, all of which are either sulfated or acetylated) is responsible for the blood anticoagulant activity of heparin. Notably, presence of the rare 3-O-sulfate group at position F in the pentasaccharide, in addition to the more common N- and 6-O-sulfate groups of GlcN residues, is responsible for the strong and very specific interactions of the pentasaccharide with the AT-binding domain (ABD) (Atha et al. 1985). The structural requirements for the binding of heparin (1 in Figure 11.3a) to AT, as shown in Figure 11.3, were determined by analyzing homology models based on the crystal structure of cleaved α_1-antitrypsin (Grootenhuis and Van Boeckel 1991; Huber and Carrell 1989; van Boeckel et al. 1991) and crystal structures of cleaved and intact AT (PDB codes 1ATT and 1ATH) (Mourey et al. 1990; Schreuder et al. 1994) and by determining the structure–activity relationships of a series of oligosaccharides that had various combinations of sulfate and carboxylate groups (Boeckel and Petitou 1993; Petitou and Boeckel 2004). The hydrophobic interactions between the heparin pentasaccharide and AT were determined to increase the binding affinity (Jairajpuri et al. 2003). Charged groups, as depicted in Figure 11.3, were found to either be absolutely essential for the activation of AT (highlighted in the solid boxes) or be required to increase the biological activity (in the dotted boxes). On binding to this specific sequence, AT undergoes a

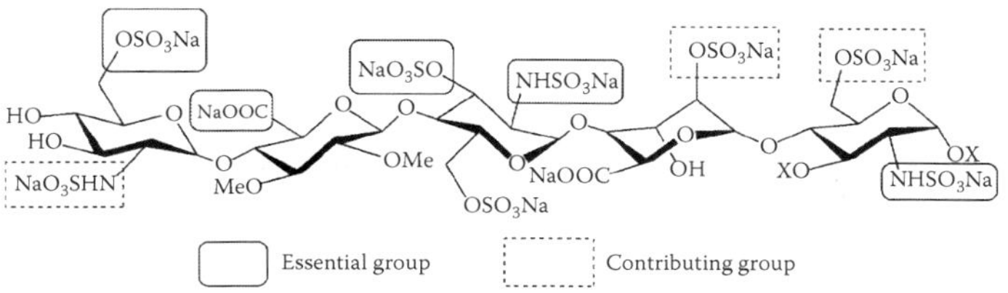

FIGURE 11.3 Heparin/HS mimetics as anticoagulants. (a) Chemical structures of heparin pentasaccharide derivatives. (1) The AT-binding pentasaccharide motif. The functional groups highlighted are essential for AT activation. (2) Structure of fondaparinux. (3) Structure of idraparinux. (4) Structure of idrabiotaparinux, the biotinylated analog of idraparinux. (b) Chemical structures of synthetic HS mimetics. (5) The M118 is a low-molecular-weight anticoagulant rationally designed specifically to treat acute coronary syndromes. It has an averaged molar mass between 5500 and 9000 Da and a polydispersity of approximately 1.0. The chemical structure of M118 consists of $C_{12m}H_{14m+1}O_{10m}N_mNa_mR_{3m-1}R1_m$, $C_{12m}H_{14m+2}O_{10m+1}N_mNa_mR_{3m}R1_m$, where n is equal to average number of disaccharide repeats, $m = 1 + n$, R is H or SO_3Na, and R_1 is SO_3Na or $COCH_3$. (6) Structure of the neutralized anticoagulant EP217609, a biotinylated form of EP42675. The biotin and lysine moieties are highlighted by dotted rectangles. These molecules are the first representatives of a new class of synthetic, parenteral anticoagulants with a dual mechanism of action combining the properties of an indirect FXa inhibitor and a direct thrombin inhibitor.

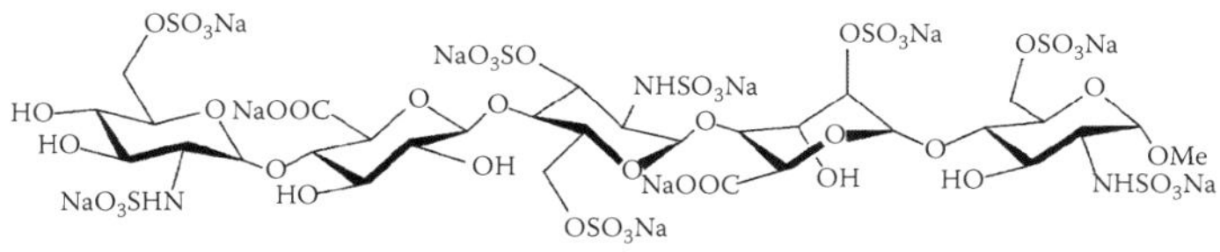

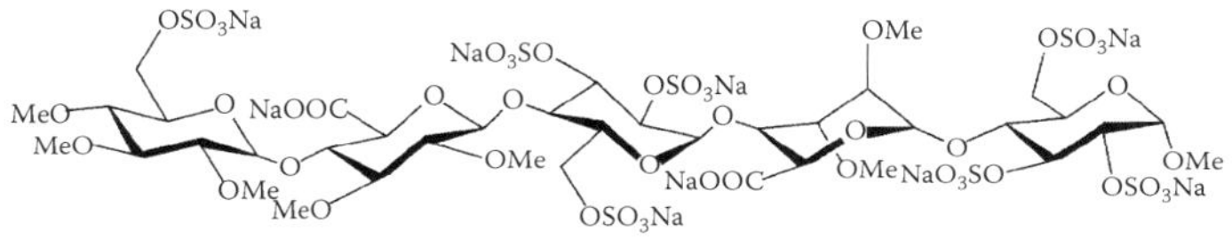

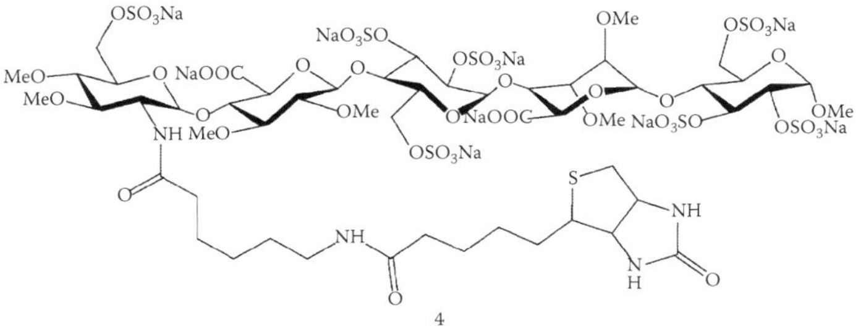

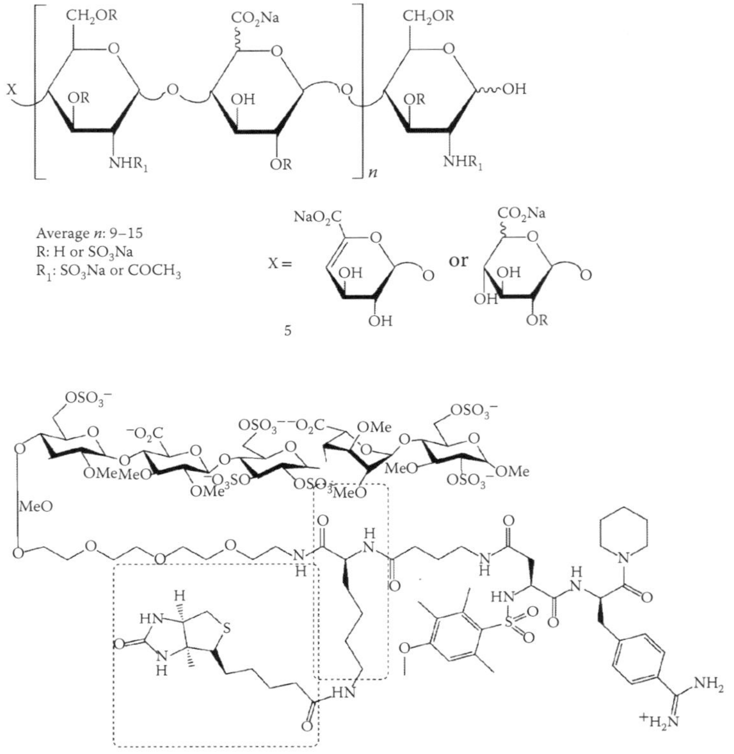

FIGURE 11.3 (*Continued*) Heparin/HS mimetics as anticoagulants.

conformational change that allows the inhibitory protein to be recognized by FXa, which is then inhibited. However, the inhibition of thrombin requires a much larger oligosaccharide containing the ABD (not necessarily the AT-specific pentasaccharide) and a thrombin-binding domain (Oosta et al. 1981). Several recent review articles describe the structure–activity relationships and the mechanism of action of heparin-mimetic anticoagulants (Boeckel and Petitou 1993; De Kort et al. 2005; Petitou and Boeckel 2004; Petitou et al. 2009).

11.2.1 Anticoagulant Fondaparinux

Pharmaceutical companies such as Organon and Sanofi-Aventis have been working on the development of oligosaccharides closely related to the natural heparin sequence that can bind AT and thereby cause anticoagulation efficiently without the need for frequent administration compared to full-length heparin. A synthetic analog of heparin (2 in Figure 11.3a) with a methyl group at the anomeric center was selected as a development candidate based on the observation that such derivatives are easier to synthesize and avoid nonspecific binding to plasma proteins (Boeckel and Petitou 1993; Petitou et al. 1987). Fondaparinux (Arixtra®, SR90107, Org31540) binds specifically with a dissociation constant K_d of ~40 nM to plasma AT in a 1:1 stoichiometric relationship (Herbert et al. 1997). The crystal structure of the complex of Arixtra with AT (PDB: 2GD4) allows elucidation of the mechanism of the pentasaccharide-induced conformational change in AT, which is required for efficient "allosteric inhibition" of the coagulation factors IXa and Xa (Figure 11.4), whereas thrombin inhibition is accelerated by ~1000-fold solely through a "bridging mechanism" (Johnson, Li et al. 2006). Determination of several crystal structures of wild-type and mutant AT in different states (native and inhibitory) bound to fondaparinux also revealed the presence of a native-like hinge region (Figure 11.4)

(a)

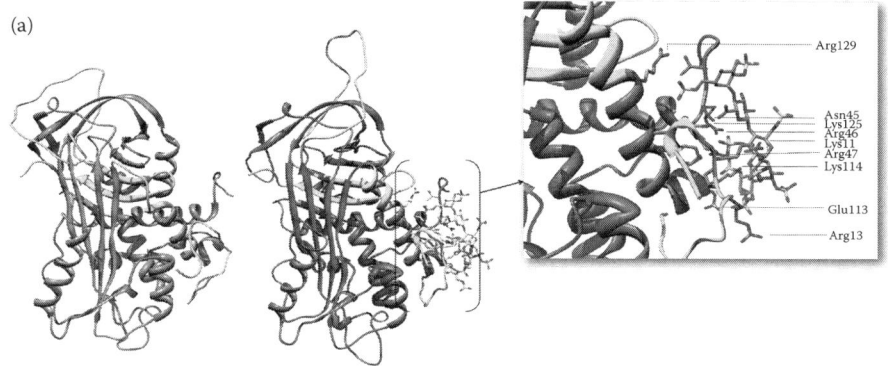

FIGURE 11.4 **(See color insert.)** (a) Binding interactions of heparin with AT are shown. The structures show ribbon diagrams of native monomeric AT in orange and fondaparinux-activated AT (PDB: 2GD4) in cyan. The chains of FXa are not shown for clarity. The pentasaccharide-activation of AT is seen to involve closing of the β-sheets A, an extension of D-helix (yellow) and an expulsion of residues of the reactive site loop (yellow). The specific AT–fondaparinux (shown in sticks) interactions within 4 Å are indicated. The pentasaccharide interactions involve Arg46, Arg47, Lys114, Lys125, and Arg129. (b) Mechanism of action of idraparinux with AT is shown. (1) Dimer of latent AT (red) and inhibitory AT (blue) each complexed with a pentasaccharide analog of idraparinux (represented as spheres; PDB: 1AZX) are shown. The reactive site loop in each molecule is shown in green ribbons. (2) Superimposition of structures of inhibitory AT (magenta; PDB: 1E05) and pentasaccharide analog of idraparinux complexed with inhibitory AT (blue; PDB: 1E03). The heparin-induced conformational change in the reactive site loop and induced helical segment is shown in green and yellow, respectively. The pentasaccharide is represented as sticks. (3) Interactions of the pentasaccharide analog of idraparinux with inhibitory AT with the lowermost induced helical segment shown in magenta and the reactive site loop in green. Only residues within 3.5 Å forming heparin-binding site are shown. The conformation and interactions of the idraparinux pentasaccharide with AT are similar to those of fondaparinux and involve interactions with Arg129 and Lys125 in the D-helix, Arg46 and Arg47 in the A-helix, Lys114 and Glu113 in the P-helix, and Lys11 and Arg13 in a cleft formed by the *N*-terminus. (c) Crystal structure of the ternary complex of AT, thrombin, and SR123781A (Li et al. 2004) is shown. The proteins are shown in a ribbon representation (thrombin in blue and AT in orange) and the 16-mer heparin is shown as sticks.

(b)

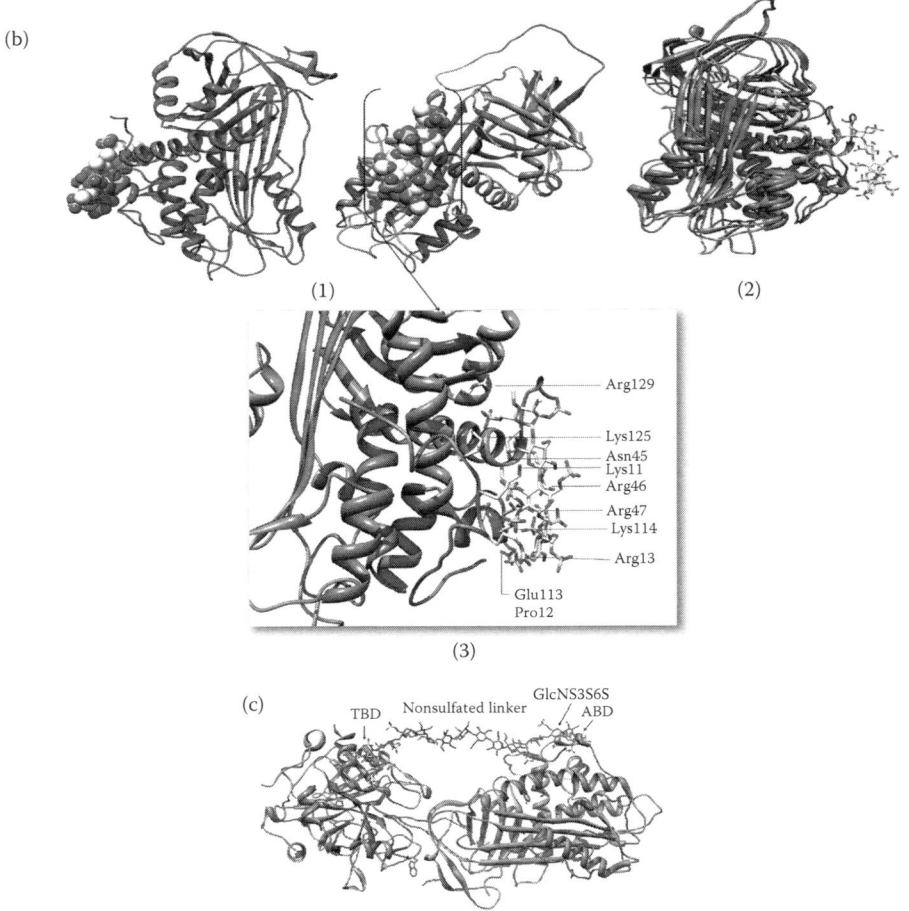

FIGURE 11.4 **(See color insert.)** (*Continued*) (a) Binding interactions of heparin with AT are shown. (b) Mechanism of action of idraparinux with AT is shown. (c) Crystal structure of the ternary complex of AT, thrombin, and SR123781A is shown.

and suggested the existence of a three-step induced-fit binding mechanism (Johnson, Langdown et al. 2006; Langdown et al. 2009). The crystal structure of the synthetic pentasaccharide–thrombospondin-1 (TSPN-1) complex has been resolved (PDB: 1ZA4), indicating that residues Arg29, Arg42, and Arg77 in an extensive positively charged patch at the bottom of the domain specifically associate with the sulfate groups of heparin; however, the formation of these interactions is dependent on the length of the oligosaccharide (Tan et al. 2006).

In 2001, following successful clinical development programs in the 1990s, GlaxoSmithKline (London, United Kingdom) registered fondaparinux as a new anti-thrombotic drug under the name Arixtra after approval from the U.S. Food and Drug Administration (FDA) and the European Committee for Proprietary Medical Products (CPMP) (Turpie 2004). Fondaparinux was recently approved for patients with acute coronary syndrome (ACS) after successful completion of phase II and III clinical trials (Sharma et al. 2010). The phase II studies PENTALYSE (Coussement et al. 2001) and PENTUA (Simoons et al. 2004) showed a promising benefit–risk ratio in patients with ACS. Fondaparinux was found to be associated with less bleeding and improved survival in non-ST elevation ACS and ST elevation myocardial infarction (MI), according to the OASIS-5 and OASIS-6 phase III clinical trials (Schiele 2010; Wienbergen and Zeymer 2007). Fondaparinux is not recommended for primary percutaneous coronary intervention (PCI) or as the sole anticoagulant for PCI (Mehta et al. 2007; Wienbergen and Zeymer 2007). The once-daily dose of 2.5 mg was chosen for phase III studies on the basis of a number of dose-ranging clinical trials (Coussement et al. 2001; Mehta et al. 2005; Simoons et al. 2004). Although the intrinsic bleeding risk is minimal, recombinant factor VIIa (fVIIa) reverses the anticoagulant effect in patients treated with fondaparinux who develop bleeding complications (Lisman et al. 2003). A registry study showed that the routine use of fondaparinux in patients with ACS increases considerably and that the mortality rate is lower compared to the use of LMWH or UFH (Schiele et al. 2010).

There is increased demand in the market for an oral formulation of fondaparinux as an alternative to subcutaneous delivery, which is the current mode of administration. A recent study reported a small-intestine-targeted delivery system that increases the oral bioavailability of fondaparinux based on thiolated polycarbophil (PCP-Cys) and glutathione (GSH) combined with sodium decanoate (Vetter et al. 2010). Endotis Pharma (Romainville, France; EP371510) and Emisphere (Cedar Knolls, New Jersey) are also developing oral formulations of fondaparinux.

11.2.2 ANTICOAGULANT IDRAPARINUX

Idraparinux (SanOrg34006, SR34006; 3 in Figure 11.3a) has a structure and mechanism of interaction similar to those of fondaparinux, but it can be synthesized more easily due to the presence of a "pseudo" alternating sequence that can be readily prepared from glucose (Westerduin et al. 1994). The presence of methyl and O-sulfates (Herbert et al. 1998) in idraparinux (K_d of 1 nM) results in stronger binding to AT than fondaparinux (K_d of 50 nM) (Hjelm and Schedin-Weiss 2007); this also imparts superior aXa activity to idraparinux compared to fondaparinux (1600 vs. 700 U μg^{-1}) (Desai et al. 1998a, 1998b; Herbert et al. 1998). Crystal structures of the complexes of a pentasaccharide analog of idraparinux (modified by the addition of a sulfate at the H unit) (Basten et al. 1992) with a dimer consisting of inhibitory, latent, and intermediate α-AT (Jin et al. 1997; Johnson and Huntington 2003; McCoy et al. 2003; PDB: 1AZX, 1E03, 1NQ9) and with β-AT (McCoy et al. 2003; PDB: 1E04) reveal pentasaccharide-induced conformational changes in AT involving the reactive site loop (Figure 11.4). This mechanism further leads to the activation of AT, the allosteric inhibition of the principal coagulation factors IXa and Xa, and the nonallosteric inhibition of thrombin.

Idraparinux has been indicated for the treatment and secondary prevention of venous thromboembolism (VTE) and the prevention of thromboembolic events associated with atrial fibrillation (AF) in various clinical trials (Prandoni et al. 2008). Unlike fondaparinux, it is not metabolized, does not have an antidote to reverse the anticoagulant effect, and is excreted unchanged through the kidneys (Faaij et al. 1999). In addition, it has a longer half-life (80–120 hours) in the blood-stream, thus enabling once-weekly administration (Ma and Fareed 2004). The phase II dose-ranging PERSIST study established that subcutaneous once-weekly administration of 2.5 mg of idraparinux was just as effective as existing treatments in VTE prevention such as warfarin, with potential advantages in terms of both safety and patient convenience (Buller et al. 2004; The PERSIST Investigators 2002).

A subcutaneous once-weekly dose of 2.5 mg of idraparinux was administered in the three comparative Van Gogh phase III trials, focusing on the long-term treatment of patients with confirmed DVT or PE (Buller et al. 2007a, 2007b). The Van Gogh DVT trials (Van Gogh PE, Van Gogh DVT, and Van Gogh extension) compared idraparinux with standard oral anticoagulant therapy (heparin or LMWH followed by a vitamin K antagonist) to demonstrate that idraparinux is at least as effective as the standard agents but superior in safety. Unfortunately, the results of the Van Gogh PE and DVT trials showed that the rate of symptomatic recurrence was significantly higher with idraparinux than with conventional therapy. In addition, the elimination half-life of 60 days led to a prolonged anticoagulant effect after completion of the therapy (Harenberg, Jörg et al. 2008). The Van Gogh extension trial enrolled patients completing 6 months of treatment in the PE and DVT trials to receive once-weekly either idraparinux or placebo for an additional period of 6 months. Overall, the results of the Van Gogh trials showed that idraparinux is effective as standard therapy for DVT but not for PE, and for longer-term prophylaxis, although the increased risk of clinically relevant bleeding outweighs its benefits (Harenberg, Vukojevic et al. 2008; Harenberg et al. 2009).

The AMADEUS phase III trial was designed to compare the efficacy and safety of idraparinux with vitamin K antagonists for the long-term prevention of thromboembolic events associated with AF (The AMADEUS Investigators et al. 2008). The results of the trial showed that idraparinux is at least as effective as oral, dose-adjusted anticoagulant therapy; however, the trial was stopped because of excessive bleeding with idraparinux and a few documented ischemic events in patients. The results of a pharmacokinetic (PK) study that evaluated creatinine clearance in patients enrolled for the AMADEUS phase III and van Gogh clinical trials led to the recommendation that a lower dose of idraparinux should be used in patients with renal impairment (Veyrat-Follet et al. 2009).

11.2.3 Anticoagulant Idrabiotaparinux

Idrabiotaparinux (biotinylated idraparinux, SSR126517, SSR126517E; 4 in Figure 11.3a) is a novel, long-acting, synthetic anticoagulant, which has a chemical structure similar to its predecessor idraparinux but contains a biotin segment (Figure 11.3). It is indicated for the treatment of DVT and/or PE and for the secondary prevention of VT3 (Harenberg 2009). The linkage of biotin at position 2 of the nonreducing end of glucose in the pentasaccharide and an optimal 6C-length of the spacer arm prevent the interaction of the molecule with either AT or FXa in vitro (Savi et al. 2008). Idrabiotaparinux has a similar PK profile to idraparinux, allowing once-weekly subcutaneous administration with the added safety feature of easy reversibility using avidin as antidote. Avidin is a tetrameric protein derived from the egg white of many species with each monomer having a biotin molecule attached to it. As a consequence, injection of avidin can trigger the immediate elimination of biotinylated idrabiotaparinux from the blood-stream of animals and humans, resulting in the complete neutralization of the anti-thrombotic (aXa) activity (Paty et al. 2010; Savi et al. 2008).

11.2.4 ANTICOAGULANT SR123781

The SR123781 (SR123781A) is a short-acting hexadecasaccharide analog of hepa-rin with N-sulfate groups replaced by O-sulfates, and alkylated hydroxyl groups in the ABD. It has tailor-made FXa- and thrombin-inhibitory activities combined with more selectivity in its mode of action. The likely molecular interactions of this hexa-decasaccharide (Figure 11.4) have been determined from the x-ray crystal structure of the ternary complex of AT–thrombin–HS (Li et al. 2004). This hexadecasaccha-ride consists of an ABD (S12–S16) at the reducing end of the nonsulfated linker, a nonsulfated linker region (S5–S11), and a TBD (S1–S4) at the nonreducing end of the linker. The highly sulfated glucose units allow nonspecific binding to throm-bin, which is dependent primarily on the overall charge density of the GAG frag-ment rather than on the precise sequence of the variously substituted sugar residues. The nonsulfated linker region (S6–S11) does not interact with any protein residues; rather, it facilitates the formation of the ternary complex, giving rise to increased AT activity with minimal interaction with PF4.

The SR123781 was found to have more potent activity than heparin and fondaparinux in different experimental models of arterial and venous thrombosis and showed high affinity for human AT (K_d of 58 nM) (Herbert et al. 2001). Ex vivo SR123781 displayed prolonged aXa and AT activity after intravenous and subcutane-ous administration to rats, rabbits, and baboons, and inhibited thrombus formation. It also exhibited a favorable antithrombotic/bleeding ratio as compared with standard heparin, which encourages its use in the treatment and prevention of various throm-botic diseases (Herbert et al. 2001). The DRIVE phase IIb study used SR123781 for the prevention of VTE events in patients undergoing total hip replacement and confirmed a correlation between dose and clinical response, with good efficacy and safety. The compound was undergoing its phase IIb trial (SHINE) for the prevention of major cardiovascular events in patients with non-ST elevated acute coronary syn-drome. However, Sanofi-Aventis discontinued the development of SR123781 follow-ing the success of their heparin-mimetic AVE 5026 (Xu-song and Bing-ren 2009).

11.2.5 ANTICOAGULANT AVE5026

The AVE5026 (Sanofi-Aventis) is a polydisperse mixture (polydispersity of approxi-mately 1.0) of oligomeric ultra-LMWH fragments (molecular weights of 2000–3000 Da) prepared by partial and controlled chemoselective depolymerization of porcine UFH; it is currently in clinical development phase for the prevention of VTE (Viskov et al. 2009). The AVE5026 binds to and activates AT, which may result in the inhibition of activated factor Xa and, to a minimal extent, factor IIa (thrombin). It exhibits an even higher ratio of aXa to anti-factor IIa activity (>30:1) and also shows dose-dependent antithrombotic activity in a rat microvascular thrombosis disease model, suggesting that this agent may inhibit tumor growth by regulating angiogenesis and apoptosis (Hoppensteadt et al. 2008a, 2008b). The half-life of AVE5026 is 16–20 hours, enabling once-daily subcutaneous administration. The AVE5026, similar to fondaparinux, is excreted renally and its anticoagulant effects are not neutralized by protamine sulfate.

The efficacy and safety of the administration of AVE5026 in conjunction with enoxaparin as an antithrombotic were evaluated in patients undergoing total knee replacement surgery. Once-daily subcutaneous doses of 5, 10, 20, 40, or 60 mg of AVE5026 and 40 mg of enoxaparin were injected to patients undergoing knee arthroplasty in a phase II (TREK) dose ranging study. The study demonstrated high statistically significant dose-dependent response with AVE5026 for the prevention of VTE (Lassen et al. 2009). In this study, a 20 mg dose of AVE5026 showed superior efficacy and a good safety profile for confirmed adjudicated VTE and, hence, this dose will be evaluated in an extensive phase III trial in patients undergoing hip, knee, or abdominal surgery and in cancer patients receiving chemotherapy.

11.2.6 ANTICOAGULANT M118

Momenta Pharmaceuticals (Cambridge, Massachusetts) developed M118, which is presented in Figure 11.3 (5 in the figure). It is a next-generation anticoagulant engineered from UFH using a specific enzymatic depolymerization process (Kishimoto et al. 2009). The M118 can effectively prevent thrombosis in diseased arteries in an animal model of vascular plaque formation and attenuate the formation of platelet–neutrophil conjugates (Chakrabarti et al. 2009). Preclinical and phase I studies have shown that M118 combines the positive attributes of both UFH and LMWH, such as potent activity against factors Xa and IIa (Fier et al. 2007; Volovyk et al. 2009), predictable pharmacokinetics after both intravenous and subcutaneous administration, ability to be monitored by the use of point-of-care coagulation assays, reversibility with protamine sulfate (Draganov et al. 2009), and lack of drug–drug interactions when coadministered with aspirin and clopidogrel or with GP IIb/IIIa inhibitors such as eptifibatide (Fier et al. 2009). The M118 has been evaluated in a phase II clinical trial involving patients undergoing PCI (Chiara et al. 2009); results of the trial demonstrate that M118 is well tolerated, noninferior to UFH, and feasible to use as an anticoagulant in patients undergoing PCI in ACS. These results form the basis for further investigation of M118 in treatment of ischemic heart disease (Rao et al. 2010).

11.2.7 ANTICOAGULANTS EP42675 AND EP217609

The anticoagulant EP42675 is the first representative of a new class of synthetic parenteral anticoagulants. It inhibits FXa in the presence of AT and inhibits both fibrin-bound and fluid-phase thrombin due to the presence of a direct thrombin-inhibiting moiety (Bal Dit Sollier et al. 2009a). This dual mechanism is attributed to its structure, consisting of fondaparinux (AT-binding pentasaccharide and an indirect FXa inhibitor) coupled to a peptidomimetic α-NAPAP analog (a direct inhibitor of the active site of both free and clot-bound thrombin) (Bal Dit Sollier et al. 2009a). This dual mechanism imparts a favorable PK profile to EP42675 that ensures prolonged anticoagulant coverage with improved control over its therapeutic window and no cross-reaction with PF4 antibodies. The EP42675 has successfully completed phase I trials, in which it did not show any clinically relevant negative effects in 100 healthy subjects enrolled in the study. The PK profile of this molecule showed low intra- and inter-subject variability (Gueret et al. 2009a). Furthermore, a pharmacodynamic (PD) interaction study of EP42675 with a combination of oral aspirin and clopidogrel did not show any relevant drug–drug interactions (Gueret et al. 2009b). As a result, it is expected that EP42675 will be used either intravenously or subcutaneously on a fixed dose regimen, without the need for routine anticoagulation monitoring (Bal Dit Sollier et al. 2009b).

Endotis has also developed neutralized anticoagulant EP217609, the biotinylated form of EP42675. In EP42675, the biotin entity is covalently linked to the spacer between the pentasaccharide portion and the direct thrombin inhibitor portion of the molecule (Figure 11.3), with the molecule having a dual mechanism of action similar to that of EP42675 (De Kort and Van Boeckel 2010). The presence of the biotin moiety (6 in Figure 11.3b) allows it to be neutralized on administration of avidin. The PK/PD profiles of EP217609 and EP42675 in animal models were found to be similar, with half-lives of approximately 3 hours (Petitou et al. 2009). Furthermore, both EP217609 and its specific antidote avidin showed no evidence of clotting in the canine model of extracorporeal circulation (ECC), making it a promising candidate for ECC in cardiac surgery and PCI in human subjects with high risk of bleeding (Fromes et al. 2010). The phase I study assessed neutralization of EP217609 by avidin in 24 healthy subjects receiving single, ascending doses. In this study, administration of EP217609 showed a dose-dependent increase in standard and specific coagulation tests followed by a decrease in thrombin generation (Gueret et al. 2010).

11.3 NON-ANTICOAGULANT HEPARIN/HS MIMETICS

Heparin and LMWH were shown to attenuate cancer progression and metastasis in several mouse models of carcinomas and melanomas, in which the anticoagulant activity of heparin was of no consequence (Casu et al. 2008; Kragh et al. 2005; Yoshitomi et al. 2004). Therefore, attention has recently been focused on the non-anti-coagulant properties of heparin, which is known to inhibit inflammation (Lassen et al. 2009; Young 2008) and the metastatic spread of tumor cells (Borsig 2010; Yip et al. 2006).

11.3.1 PI-88

Heparanase is an endoglycosidase enzyme that cleaves HS and plays vital roles in inflammation, tumor cell invasion, metastasis, and angiogenesis (McKenzie 2007). Sulfated sugar molecules such as cyclitols and glycol-split derivatives have been reported to inhibit heparanase and suppress tumor metastasis (Miao et al. 2006). The PI-88 (muparfostat; $6\text{-}O\text{-}PO_3H_2\text{-}Man\alpha(1\text{-}3)Man\alpha(1\text{-}3)Man\alpha(1\text{-}3)Man\alpha(1\text{-}2)Man$; a in Figure 11.5) (Wall et al. 2001) is one such selective inhibitor of heparanase–heparin interactions derived from chemical sulfonation of the oligosaccharide phosphate fraction of phosphomannan produced by the yeast *Pichia holstii* NRRL Y-2448 (Ferro et al. 2001). The PI-88 inhibits heparanase and competes with HS for the binding of growth factors, such as fibroblast growth factor (FGF)-1, FGF-2, and vascular endothelial growth factor (VEGF), involved in angiogenesis, thereby preventing HS

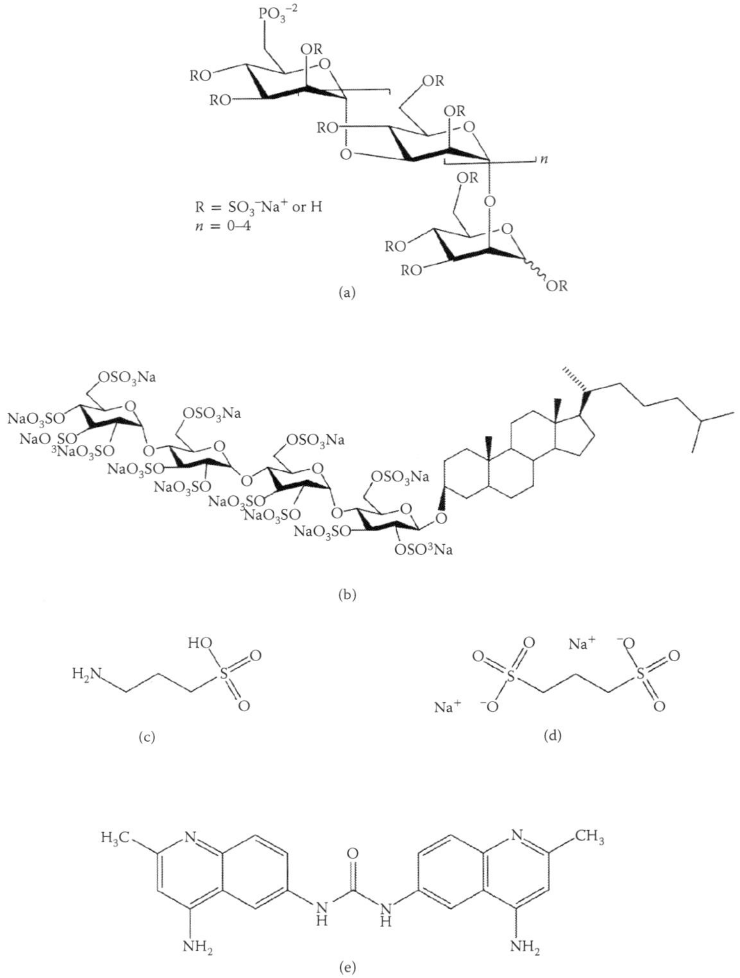

FIGURE 11.5 Non-anti-coagulant heparin/HS mimetics. (a) Chemical structure of PI-88, a potent antiangiogenic, antitumor, and antimetastatic agent that inhibits heparanase activity and competes with heparan sulfate binding of peptide growth factors involved in angiogenesis, such as FGF and VEGF. It is a highly sulfonated oligosaccharide with an average sulfation of three sulfates per mannose residue and a molecular weight of 2100–2585 Da. (b) Chemical structure of PG545, a potent inhibitor of angiogenesis and metastasis. (c) Tramiprosate is the first GAG mimetic to be investigated as an anti-Alzheimer drug in advanced clinical trials. Tramiprosate was shown to slow or arrest the progression of AD by binding preferentially to soluble Aβ (amyloid) peptide and maintaining Aβ in a nonfibrillar form. (d) Chemical structure of eprodisate sodium, a structural analog of tramiprosate. Eprodisate binds to the GAG-binding site and interferes with heparan sulfate interactions on serum AA and consequently arrest amyloidosis. (e) Surfen, an aminoquinoline derivative, is a small-molecule heparan sulfate proteoglycan antagonist. It binds to various sulfated GAGs and thereby neutralizes the ability of various heparins to activate AT and inactivate FXa. In addition, surfen possesses anti-inflammatory, antibacterial, and antiviral properties.

degradation (Parish et al. 1999). Apart from its anticipated anticoagulant effects, PI-88 was well tolerated in in vivo models of mouse-derived tumors, demonstrating potent antiangiogenic and antimetastatic effects (Joyce et al. 2005). Besides its use in halting cancer progression, PI-88 has proved its potential in treating inflammatory diseases, thrombosis, and viral infections (Ferro and Don 2003; Lee et al. 2006).

Promising results from phase I/II trials have been reported with PI-88, both as a single agent and in combination with chemotherapy in treating a variety of tumor types, including melanoma and hepatocellular carcinoma (HCC; Kudchadkar et al. 2008). The first phase I trial determined the recommended dose and toxicity profile of PI-88 in patients with advanced malignant disease, in which the drug was administered subcutaneously for 4 consecutive days either bimonthly or weekly (Basche et al. 2006). The outcome of this study showed that PI-88 at a dose of 2.28 mg/kg/day for 14 days results in dose-limiting thrombocytopenia, which appeared to be immunologically mediated through the development of antiheparin PF4 complex antibodies (Rosenthal et al. 2002). Administration of PI-88 (250 mg/day subcutaneous) with fixed weekly (30 mg/m^2) docetaxel (a chemotherapeutic agent) showed improved pharmacological properties and broad antitumor activities in the second phase I trial (Chow et al. 2008). However, minor toxicity responses during the course of the therapy, such as fatigue, dysgeusia, thrombocytopenia, diarrhea, nausea, and emesis were noticed. In another similar phase I trial, the recommended dose of PI-88 was reported to be 190 mg/m^2 alone and 1000 mg/m^2 in combination with dacarbazine administered every 3 weeks (Millward et al. 2007). However, a phase I/II trial of daily PI-88 alone or with dacarbazine in patients with hormone refractory prostate cancer (HRPC) did not meet its end point due to cases of significant febrile neutropenia (Khasraw et al. 2010). Furthermore, the development of antibody-induced thrombocytopenia limited its use in some patients. The PI-88 was investigated further in patients with advanced melanoma in phase II trials: a fixed dose of 250 mg/day was given by injection for 4 consecutive days followed by 3 drug-free days in a 28 day cycle (Kudchadkar et al. 2008). Nonetheless, in patients with advanced melanoma, PI-88 demonstrated significant activity, but further investigations are warranted for its use in combination with chemotherapy. It is noted that PI-88 was evaluated for the treatment of HCC in a phase II trial (Liu, Lee et al. 2009). Followed by the encouraging results of this study, which showed low toxicity with PI-88 and a well tolerated dose of 160 mg/day, the second stage of this phase II trial was commenced. Progen Pharmaceuticals (Queensland, Australia) has halted further development of PI-88 at this moment, although an ongoing phase III trial in the setting of postoperative recurrence of HCC will continue to completion.

Progen has capitalized on the development of semisynthetic heparin mimetics (the PG500 series) as potent inhibitors of angiogenesis and metastasis (Ferro et al. 2007). These compounds, based on the PI-88 scaffold, are fully sulfated and have single-entity oligosaccharides attached to a lipophilic moiety, such as a glycone, at the reducing end of the molecule (Karoli et al. 2005). These compounds are believed to have a similar mechanism of action to PI-88 in cancer progression, through the inhibition of VEGF, FGF-1, and FGF-2, and in metastasis, through the inhibition of heparanase, thereby blocking cell signaling (Dredge et al. 2009). The PG545 (b in Figure 11.5) has been chosen as the lead candidate based on its complete target profile, incorporating aspects such as its efficacy, pharmacokinetics, toxicology, and ease of manufacture

(Bytheway et al. 2009; Hammond et al. 2009). This compound has been evaluated in preclinical trials in animal models of cancer, and it is being progressed for the filing of an investigational new drug (IND) application with the U.S. FDA.

11.3.2 Tramiprosate

Sulfated GAGs such as HS/heparin have been widely reported to promote the aggregation of amyloid β-peptide (Aβ) into the insoluble amyloid fibrils that play a pivotal role in plaque formation in Alzheimer's disease (AD) (McLaurin et al. 1999; van Horssen et al. 2003). Several molecules such as LMWH (Ariga et al. 2010), pentosan polysulfate and dextran sulfate (Leveugle et al. 1994), small-molecule anionic sulfonates or sulfates (Kisilevsky et al. 1995), and sulfonated dyes such as Congo red and thioflavin S have been proposed to be used to prevent HS-induced aggregation of Aβ. Several of the compounds that inhibit the formation of Aβ fibrils, such as phenothiazines, polyphenols, and porphyrins, can also alter Tau protein aggregation (Taniguchi et al. 2005). Tramisprosate (Alzhemed™; c in Figure 11.5) is one such GAG mimetic designed to interfere with the actions of Aβ early in the cascade of amyloidogenic events (Aisen et al. 2007; Geerts 2004) preventing the formation of Tau–actin aggregates without interfering with the binding of Tau to microtubules (Santa-Maria et al. 2007).

Tramiprosate (also referred to as 3-amino-1-propanesulfonic acid, 3-aminopropylsulfonic acid, 3-APS, homotaurine, or NC-531) was the first investigational candidate (Neurochem, Inc. [now Bellus Health], Quebec, Canada) to enter a phase III trial for treating mild-to-moderate dementia in AD (c in Figure 11.5). Tramiprosate, a modification of the amino acid taurine, binds preferentially to the soluble Aβ peptide and maintains Aβ in a random-coil/α-helical rich conformation and in nonfibrillar form, thereby inhibiting aggregation and hence plaque formation and deposition (Wright 2006). Preclinical results showed that it can cross the blood–brain barrier (BBB) with low toxicity in various animal species, particularly TgCRND8 transgenic mice, a model of early-onset brain amyloidosis, and it exhibits protective effects against Aβ-induced cytotoxicity (Gervais et al. 2007).

A phase II trial demonstrated that tramiprosate crossed the BBB and dose dependently reduced $A\beta_{42}$ in the cerebrospinal fluid of patients with mild-to-moderate AD (Aisen et al. 2006), but there were no significant differences observed in cognitive functioning after 3 months of treatment. In the U.S. phase III trial, the drug appeared safe and well tolerated in patients, who were randomly assigned to receive a placebo or 100 or 150 mg twice-daily doses of tramiprosate; but it was not found to be significantly better than placebo in improving cognitive function or dementia (Rafii and Aisen 2009). Most of the results of the trials were inconclusive, and no further reports on the drug are available except magnetic resonance imaging (MRI) results, which show less hippocampal shrinkage on treatment with tramiprosate (Gauthier et al. 2009; Saumier et al. 2009). Subsequently, the European phase III trial has been discontinued owing to lack of clinical efficacy compared to placebo. After the failure of phase III trials, Bellus Health promoted the molecule as a nutraceutical VIVIMIND™, a supplement against memory loss (Neugroschl and Sano 2010). It has also completed a phase II trial (CEREBRIL™) for the treatment of hemorrhagic stroke resulting from cerebral amyloid angiopathy (CAA) (Greenberg et al. 2006).

The NRM8499, a prodrug of tramiprosate (Bellus Health), is undergoing phase I clinical trials to investigate its safety, tolerability, and PK profile in a group of up to 84 young and elderly healthy subjects for the treatment of AD. Preclinical studies conducted in rodents showed that the prodrug increases plasma concentration and brain exposure to tramiprosate (by 1.5- to 3-fold), which may help to improve the therapeutic effect on cognitive function and other clinical outcomes in AD.

11.3.3 EPRODISATE SODIUM

Eprodisate (NC-503; 1,3-propanedisulfonic acid disodium salt; Kiacta®; Fibrillex™; d in Figure 11.5) is a low-molecular-weight negatively charged sulfonated molecule that is a structural analog of tramiprosate. It interferes with the binding of HS and serum amyloid protein A (SAA), thus targeting amyloid fibril polymerization and inhibiting amyloid deposition in tissues (Ancsin and Kisilevsky 1999; Kisilevsky et al. 1995; Revill et al. 2006). Eprodisate inhibits the development of amyloid deposits in in vivo mouse models of amyloid protein A (AA) amyloidosis (Gervais et al. 2003). Preclinical PK studies of eprodisate showed good bioavailability on oral administration in various animal models and the absence of interactions with plasma proteins. The drug is not metabolized, is excreted primarily by the kidney, and is known to have an approximate terminal half-life of 10–20 hours based on a multiple rising oral dose study. High inter-individual variability in its plasma concentrations was observed in PK analyses during phase I trials (Kisilevsky 2000). Phase II–III trials in AA amyloidosis-patients (ClinicalTrials.gov identifier NCT00035334) showed that eprodisate was well tolerated and that its adverse events profile was comparable to that of the placebo. The results of a recent trial showed that it delays the progression of AA amyloidosis-related renal disease compared with placebo (Dember et al. 2007; Manenti et al. 2008). Patients receiving eprodisate had better creatinine clearance and, consequently, they were at reduced risk of worsening of renal function; however, no effect was seen on SAA levels, proteinuria, and amyloid content of abdominal fat in these patients compared to those receiving placebo (Dember et al. 2007; Manenti et al. 2008). Despite having previously been granted orphan and fast-track status, the company Bellus Health withdrew its application for marketing authorization for eprosidate in the treatment of AA amyloidosis (SARL 2008).

Eprosidate was also evaluated in clinical trials for diabetes linked to amyloidosis. It was found to decrease glucose, cholesterol, and triglycerides in the blood of obese diabetic Zucker rats compared with antidiabetic drugs such as metformin or a sulfonylurea agent while preserving 40% more pancreatic islet cells compared with the control group, consequently having some protective effect on renal function. However, the development of eprodisate in relation to diabetes was halted after a clinical proof-of-concept trial, as the study did not meet its primary efficacy end point (Bellini 2010).

11.3.4 SURFEN

Esko (Carlton County, Minnesota) has been focused on identifying small-molecule antagonists of HS function. One such molecule identified is bis-2-methyl-4-amino-quinolyl-6-carbamide, also known as "surfen" (e in Figure 11.5). This molecule was first described in 1938 as an excipient for the production of depot insulin (Umber et al.

1938). It was known to have anti-inflammatory and antibacterial properties as it blocks the binding of C5a receptor and anthrax lethal factor, respectively (Lanza et al. 1992; Panchal et al. 2004). Fluorescence-based titrations indicated that surfen could bind to GAGs and neutralize the aXa activity of heparin–AT complexes (Schuksz et al. 2008). Addition of surfen to cultured cells was shown to block binding and signaling mediated by FGF2, which depend on the formation of ternary complexes of FGF, FGF receptors, and cell-surface HS (Schuksz et al. 2008). Furthermore, surfen blocked cell adhesion to the heparin-binding Hep-II domain of fibronectin and prevented infection by herpes simplex virus dependent on GP D binding to cell-surface HS (Schuksz et al. 2008).

Recently, surfen was shown to impair the action of a factor in semen that greatly enhances viral infection (Roan et al. 2010). In semen, proteolytic peptide fragments from prostatic acid phosphatase can form amyloid fibrils termed "semen-derived enhancer of viral infection" (SEVI). These fibrils greatly enhance HIV infectivity by increasing the attachment of virions to their target cells. Therefore, SEVI may have a significant impact on whether HIV is successfully transmitted during sexual contact. Surfen interferes with the binding of SEVI to both target cells and HIV-1 virions; but it does not cause SEVI fibrils to break up.

11.4   OTHER NOVEL GAG MIMETICS

Drug discovery programs based on heparin and its mimetics have predominantly been focused towards the antithrombotic and anticoagulant fields. In recent years, however, GAG mimetics designed to mimic the natural ligands (or conjugated to a non-carbohydrate moiety to increase the potency) have been developed to treat genetic diseases affecting the central nervous system. GAG mimetics are also being used as a tool to investigate the effects of GAG moieties on cellular behavior.

11.4.1   SUBSTRATE-OPTIMIZED GLYCANS

Mucopolysaccharidosis (MPS) is a form of lysosomal storage disease (LSD) caused by a deficiency in the lysosomal enzymes responsible for the degradation of GAGs. Inhibitors of GAG synthesis, although not tested in clinical trials, have been used for substrate reduction therapy (SRT) in LSD and as antiviral agents (Roberts et al. 2006; Tiwari et al. 2007). Zacharon Pharmaceuticals (San Diego, California) has developed small-molecule inhibitors of GAG biosynthesis for LSD based on substrate optimization therapy (SOT). The SOT offers a novel therapeutic approach by selectively modifying GAG structure (generally the sulfation pattern) without reducing the overall amount produced despite enzyme deficiency (Crawford et al. 2011). An inhibitor of the biosynthetic step involving the addition of 2-O-sulfate is a more feasible approach in MPS II patients with 2-sulfatase deficiency. The activity of such an inhibitor results in the production of glycans with less 2-O sulfation and increased 6-O sulfation (Figure 11.6a). Screening of 74,000 compounds resulted in the identification of ZP2345 as a lead compound that reduces 2-O sulfation in a dose-dependent manner in a cultured human cell model of MPS-II (Crawford et al. 2011). Ongoing studies focus on the synthesis of congeners and biological testing to improve the potency and efficacy of these inhibitors across multiple MPS classes.

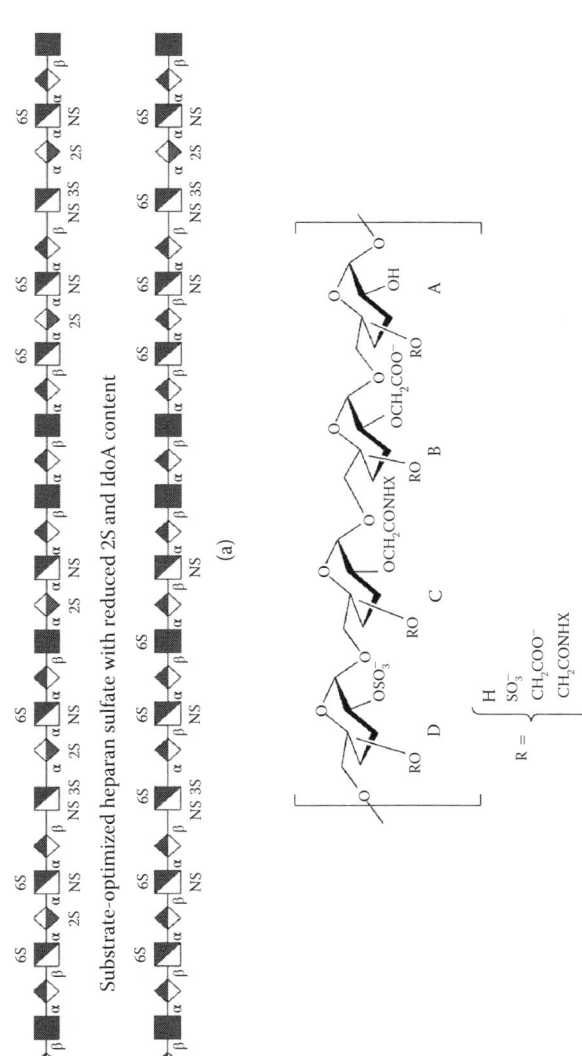

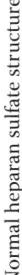

FIGURE 11.6 **(See color insert.)** Novel GAG mimetics. (a) Substrate-optimized glycan platform (e.g., HS in MPS). (b) The RGTAs with structurally defined carboxymethyl and sulfate groups enhanced tissue repair in various models, including skin, bone, and muscle healing, and prevented some of the damage caused by myocardial ischemia. The composition of OTR4120 as reported by titrimetry and ¹H NMR includes differently substituted units: A (<1%), B (=32%), C (=0%), and D (=67%). It is noted R represents the proportion of substituted groups in positions C3 and C4, arranged to define the global dextran sulfate of each group. (Modified from Crawford, B. et al., *Mol Genet Metab*, 102, S12–S13, 2011; Papy-Garcia, D. et al., *Macromolecules*, 38, 4647–4654, 2005.)

11.4.2 HEPARIN MIMETICS IN CANCER

Carbohydrate ligand-binding protein molecules such as heparanase and P- and L-selectins facilitate the metastasis of cancer cells. Exogenous heparin, LMWH, and their mimetics exercise their antimetastatic and antiangiogenic activity by specifically inhibiting selectin-mediated cell–cell interactions and the enzymatic activity of heparanase (Casu et al. 2010; Stevenson et al. 2005). Heparin-related inhibitors such as aza sugar derivatives, glycol-split derivatives, and cyclitols have been reported to inhibit heparanase by blocking its active site or its heparin/HS binding sites, or both (Vlodavsky et al. 2007). In addition, antiangiogenic and antimetastatic activities were also reported with several LMWH derivatives, including a deoxycholic acid conjugate and a periodate-oxidized heparin fragment (Lee et al. 2009; Mousa et al. 2006).

Endotis Pharma has chemically synthesized short oligosaccharides that have potent affinity for and cause selective inhibition of several growth factors and proteins (VEGF-A, FGF-2, PDGF-B, stromal-derived factor [SDF]-1α, and heparanase) involved in tumor growth and dissemination (Cabannes et al. 2009). These "glycodrugs" were identified from a library of over 100 synthetic oligosaccharides of different sizes and substitutions screened for their affinity for specific targets, efficacy on cell proliferation and migration, and in vitro endothelial tubule formation. The lead compound in the series, EP80061 (structure not disclosed), induced a very potent antimetastatic effect on a disseminated tumor model in C57Bl/6 mice (Serina et al. 2010).

The M402 (Momenta Pharmaceuticals), a novel HS mimetic containing a mixture of linear sugar chains, has been evaluated in preclinical studies for antimetastatic properties (Zhou et al. 2009). The M402 showed reduced anticoagulant activity and inhibited tumor metastasis in vitro and in vivo through modulation of multiple factors, such as P-selectin, VEGF, FGF-2, and SDF-1α (Chu et al. 2009; Zhou et al. 2009). The M402 on its own or in combination with chemotherapeutics increased survival in animal models with aggressive tumors (Chu et al. 2009; Zhou et al. 2010). Combination therapy of M402 and gemcitabine demonstrated significantly prolonged survival and reduced metastasis compared with groups treated with a monotherapy of saline solution or gemcitabine in a murine pancreatic model (Lolkema et al. 2010). A reduction in the epithelial-to-mesenchymal transition stage in cancer metastasis was observed in mice treated with M402.

11.4.3 ReGeneraTing Agents

ReGeneraTing Agents (RGTAs) are large biopolymers engineered to promote tissue repair and regeneration (Barbier-Chassefière et al. 2009) by stabilizing many heparin-binding growth factors (HBGFs), such as FGF-2 (Tardieu et al. 1992), transforming growth factor-β (TGF-β) (Meddahi, Benoit et al. 1996), and VEGF (Rouet et al. 2005). These polymers also protect proteins bound to the ECM from proteolysis. The RGTAs have been found to inhibit human leukocyte elastase (Meddahi, Lemdjabar et al. 1996), plasmin (Meddahi et al. 1995), and heparanase (Rouet et al. 2006). The RGTA derivatives are potent activators of tissue repair in various in vivo wound-healing models, such as wounds (Meddahi et al. 1994), bone defects (Albo et al. 1996), infarcted myocardium (Yamauchi et al. 2000), colic ulceration (Meddahi, Alexakis et al. 2002), and periodontitis (Escartin et al. 2003). These

RGTAs have also been shown to stimulate satellite cell growth and differentiation in primary cultures (Papy-Garcia et al. 2002). The RGTA polymers are easier to synthesize and more cost-effective than growth factors.

The RGTAs are dextran derivatives comprising varying amounts of substituted carboxymethyl, benzylamide, and sulfonate groups (Figure 11.6b). The RGTAs with an increased level of sulfation and benzylamidation show anti-prion activity by blocking the conversion of prion protein PrP^C into some abnormal forms in scrapie-infected GTI cells (Schonberger et al. 2003) and scrapie-infected and bovine spongi-form encephalopathy-infected mice (Adjou et al. 2003). However, RGTA OTR4120 (alternatively called RGD120 or RG1192), derived from a glycosidic polymer of dextran T40 and functionalized with a certain degree of substitution (ds) of sulfate residues (ds = 1.2) and carboxymethyl residues (ds = 0.5) (Barbosa et al. 2005), is known to enhance tissue repair in several animal models, including peripheral nerve injury in rats (Zuijdendorp et al. 2008), burned skin in rats (Garcia-Filipe et al. 2007), chronic skin ulcers in mice (Barbier-Chassefière et al. 2009), and cutaneous wound repair in rats (Tong et al. 2009). An nuclear magnetic resonance (NMR) study of OTR4120 showed the presence of a 15-sugar-unit repeat sequence (Martelly et al. 2010; Papy-Garcia et al. 2005). This polymeric compound resembles heparin in its structure, but it has at least 10 times less anticoagulant activity than heparin (Aamiri et al. 1995). A PK study in a muscle crush mouse model showed that OTR4120 could replace degraded HS–GAG following tissue injury and bind to the heparin-binding sites present on many ECM proteins that have been freed from occupation by their endogenous GAGs (Meddahi, Brée et al. 2002). In a recent clinical pilot study, an OTR4120 ophthalmic solution was found to improve the healing of severe corneal ulcers and dystrophy (Khammari Chebbi et al. 2008).

The OTR4131 is a compound similarly characterized by the presence of dex-tran T20, and carboxylate and sulfated groups as in OTR4120, but it differs from OTR4120 by the presence of acetate groups (ds = 0.2). The GAG mimetics OTR4120 and OTR4131 bind directly to RANTES/CCL5 and inhibit RANTES-induced hepatoma cell migration and invasion (Sutton et al. 2007). Nondifferential biolog-ical activities were observed for both RGTAs, suggesting that the acetate groups in OTR4131 are not essential for RANTES interaction. Similar observations were recorded for SDF-1, a cytokine that binds to these GAG mimetics and, consequently, has the same effect on the mobilization of progenitor cells in peripheral blood in mice (Albanese et al. 2009). However, the acetyl groups of OTR4131 mediate the high affinity for chemoattractant factors of hematopoietic stem cells. The OTR4131 has been tested in vitro and in vivo for the treatment of oral mucositis (Mangoni et al. 2009), controlling ocular surface inflammation and promoting corneal wound healing (Brignole et al. 2005).

11.4.4 Heptagonists

Protamine and low-molecular-weight protamine are antidotes used in the reversal of heparin and LMWH during bleeding complications such as those occurring in cardio-vascular surgeries. PolyMedix, Inc., Radnor, Pennsylvania has developed novel, small synthetic salicylamide derivatives called "heptagonists", which do not cause the side effects caused by protamine. Heptagonists act as universal anticoagulation-reversing agents and are active against heparin, LMWH, idraparinux, and fondaparinux (Jeske et al. 2007). One of these molecules, PMX-60056 (Jeske et al. 2009), can neutralize effectively the AT and aXa activities of LMWH. The molecule PMX-50056 has been shown to completely reverse the anticoagulant effects of heparin and normalize blood-clotting time in six human subjects in less than 10 minutes in a phase IB clinical trial (www.polymedix.com/heptagonists.php).

11.5   CONCLUSIONS

Early work by many research groups was focused on generating semisynthetic anticoagulant analogs by selectively modifying the functional groups in heparin/HS. Maintaining a specific sulfate substitution pattern or spatial arrangement of anionic groups along a saccharide chain is certainly critical to achieving the agonist-like actions of heparin, such as the binding and activation of AT by a unique heparin pentasaccharide. However, heparin/HS are complex structures endowed with a plethora of biological activities in numerous disease processes, such as angiogenesis, cancer, tissue repair, cardiovascular disease, immune system function, and microbial and viral pathogenesis, to name a few. Consequently, it is necessary to design antagonists that bind to the HS-binding site of a protein and block HS–protein interactions. There are clear differences in the structural requirements for heparin-like anticoagulants or agonists, where critical contacts between protein and anionic oligosaccharides must occur to allosterically modulate or otherwise mediate protein function, versus antagonists of HS–protein interactions, where affinity and selectivity of GAG mimetics for the HS-binding site of a protein are critical factors regardless of oligosaccharide conformation and/or oligosaccharide–protein contacts. Using structure–function relationships based on chemical synthesis and molecular modeling to more specifically modify heparin and heparin-based molecules with structurally diverse nonanionic moieties holds great potential for ultimately identifying novel, selective therapeutics and diagnostics of specific HS-mediated biological processes.

The PD and PK properties of several previous-generation GAG oligosaccharides make them inadequate for direct therapeutic application. Their anionic nature may result in significant interactions with multiple, physiologically important proteins leading to many side effects, such as low tissue permeability, short serum half-life, and poor stability. In addition, the chemical sulfation of heparin-related saccharides (greater than 40 saccharides in size) is difficult, tedious, and costly at both drug development and production scales (Bonnaffé 2011; Codée et al. 2004). Furthermore, new therapeutic applications of sulfated GAGs in the areas of treatment of infectious diseases and inflammation, and the control of cell growth in wound-healing and cancer require elimination of the anticoagulant activity of heparin oligosaccharides and engineering of appropriate PK properties and optimal oral bioavailability.

Despite these shortcomings, LMWHs and their derivatives offer advantages over oral anticoagulant drugs such as warfarin and dabigatran, as well as aXa drugs such as epixiban. These orally available small molecules are capable of crossing the placental barrier and hence cannot be used directly in pregnancy. The GAG-based agents can, however, be used as alternative anticoagulants in pregnant women with high risk for thrombosis. Furthermore, cancer patients are at an increased risk of both VTE and bleeding. The LMWHs offer a favorable balance of anticoagulation and antitumor effects, which could contribute to the improved survival of cancer patients when given for long-term prophylaxis. Detailed insight on GAG–protein interactions has predominantly been provided by recent progress in NMR spectroscopy, x-ray crystallography, and molecular modeling techniques. The chemoenzymatic synthesis that mimics the biosynthetic pathway of HS (Peterson et al. 2009) offers an alternative to the chemical synthesis of GAGs, which poses challenges for medicinal chemists. A library of sulfated polysaccharides can be synthesized by selecting the appropriate combination of biosynthetic enzymes such as glycosyl transferases, a GlcA C5-epimerase, and sulfotransferases. The synthesis of GAGs is further accelerated by the use of both microarray-based and microfluid-based platforms, seamlessly coupled with bioactivity screening (de Paz and Seeberger 2008; Laremore et al. 2009). As a consequence, a detailed understanding of the molecular basis of selective modulation, as well as new insights into the biology of interactions of GAGs with proteins, provides significant opportunities for the discovery of improved treatment mechanisms for many diseases. Many newly established companies, such as Intellihep (Liverpool, United Kingdom), Zacharon Pharmaceuticals, GlycoMimetics (Gaithersburg, United Kingdom), Endotis Pharma, PolyMedix, Progen Pharmaceuticals, and OTR3 Momenta, have ongoing discovery programs in GAG therapeutics. Nevertheless, it will probably take a long time before analog candidates whose safety and efficacy are robustly proved undergo early clinical evaluation.

REFERENCES

Aamiri, A., Mobarek, A., Carpentier, G., Barritault, D., and Gautron, J. 1995. Effect of a substituted dextran on reinnervation during regeneration of adult rat skeletal muscle. *C R Acad Sci III, Sci Vie*, 318, 1037–1044.

Adjou, K. T., Simoneau, S., Sales, N. et al. 2003. A novel generation of heparan sulfate mimetics for the treatment of prion diseases. *J Gen Virol*, 84, 2595–2603.

Aisen, P. S., Gauthier, S., Vellas, B. et al. 2007. Alzhemed: A potential treatment for Alzheimers disease. *Curr Alzheimer Res*, 4, 473–478.

Aisen, P. S., Saumier, D., Briand, R. et al. 2006. A Phase II study targeting amyloid-β with 3APS in mild-to-moderate Alzheimer disease. *Neurology*, 67, 1757–1763.

Alban, S. 2008. Natural and synthetic glycosaminoglycans. Molecular characteristics as the basis of distinct drug profiles. *Hamostaseologie*, 28, 51–61.

Albanese, P., Caruelle, D., Frescaline, G. et al. 2009. Glycosaminoglycan mimetics-induced mobilization of hematopoietic progenitors and stem cells into mouse peripheral blood: Structure/function insights. *Exp Hematol*, 37, 1072–1083.

Albo, D., Long, C., Jhala, N. et al. 1996. Modulation of cranial bone healing with a heparin-like dextran derivative. *J Craniofac Surg*, 7, 19–22.

Ancsin, J. B. and Kisilevsky, R. 1999. The heparin/heparan sulfate-binding site on apo-serum Amyloid A. *J Biol Chem*, 274, 7172–7181.

Arepally, G. M. and Ortel, T. L. 2010. Heparin-induced thrombocytopenia. *Annu Rev Med*, 61, 77–90.

Ariga, T., Miyatake, T., and Yu, R. K. 2010. Role of proteoglycans and glycosaminoglycans in the pathogenesis of Alzheimer's disease and related disorders: Amyloidogenesis and therapeutic strategies—A review. *J Neurosci Res*, 88, 2303–2315.

Atha, D. H., Lormeau, J. C., Petitou, M., Rosenberg, R. D., and Choay, J. 1985. Contribution of monosaccharide residues in heparin binding to antithrombin III. *Biochemistry*, 24, 6723–6729.

Baeuerle, P. A. and Huttner, W. B. 1986. Chlorate—a potent inhibitor of protein sulfation in intact cells. *Biochem Biophys Res Commun*, 141, 870–877.

Bal Dit Sollier, C., Neuhart, E., Krezel, C. et al. 2009a. Anticoagulant activities of EP42675—synthetic direct inhibitor and indirect factor Xa inhibitor, In *XXII Congress of the International Society of Thrombosis and Haemostasis*, Boston.

Bal Dit Sollier, C., Neuhart, E., Krezel, C. et al. 2009b. Pharmacokinetics and pharmacodynamics of EP42675—A new synthetic anticoagulant with a dual mechanism of action, In *XII Congress of the International Society of Thrombosis and Haemostasis*, Boston.

Barbier-Chassefière, V., Garcia-Filipe, S., Yue, X. L. et al. 2009. Matrix therapy in regenerative medicine, a new approach to chronic wound healing. *J Biomed Mater Res A*, 90A, 641–647.

Barbosa, I., Morin, C., Garcia, S. et al. 2005. A synthetic glycosaminoglycan mimetic (RGTA) modifies natural glycosaminoglycan species during myogenesis. *J Cell Sci*, 118, 253–264.

Basche, M., Gustafson, D. L., Holden, S. N. et al. 2006. A phase I biological and pharmacologic study of the heparanase inhibitor PI-88 in patients with advanced solid tumors. *Clin Cancer Res*, 12, 5471–5480.

Basten, J., Jauran, G., Olde-Hanter, B., Petitou, M., and van Boekel, C. A. A. 1992. Biologically active heparin-like fragments with a "non-glycosamino" glycan structure. Part 2: A tetra-*o*-methylated pentasaccharide with high affinity for antithrombin III. *Bioorg Med Chem Lett*, 2, 901–904.

Bellini, R. 2010. BELLUS Health ends NC-503 diabetes development program following results, Bellus Health Inc., Laval, Quebec, Canada.

Bauer, K. A. 2002. Selective inhibition of coagulation factors: Advances in antithrombotic therapy. *Semin Thromb Hemost*, 28, 015–024.

Boeckel, C. A. A. v. and Petitou, M. 1993. The unique antithrombin III binding domain of heparin: A lead to new synthetic antithrombotics. *Angew Chem Int Ed Engl*, 32, 1671–1690.

Bonnaffé, D. 2011. Bioactive synthetic heparan sulfate and heparin derivatives: From long fragments mimetics to chimeras. *Comptes Rendus Chimie*, 14, 59–73.

Borsig, L. 2010. Antimetastatic activities of heparins and modified heparins. Experimental evidence. *Thromb Res*, 125, S66–S71.

Brignole, F., Potron, L., Martin, C. et al. 2005. Effects of RGTA OTR4131 on Ocular Surface: In vivo Evaluation on a Rabbit Corneal Wound Healing Model and in vitro Toxicological Studies. *Invest Ophthalmol Vis Sci*, 46, E-Abstract 2166.

Brown, J. R., Crawford, B. E., and Esko, J. D. 2007. Glycan antagonists and inhibitors: A fount for drug discovery. *Crit Rev Biochem Mol Biol*, 42, 481–515.

Buller, H. R., Cohen, A. T., Davidson, B. et al. 2007a. Extended prophylaxis of venous thromboembolism with idraparinux. *N Engl J Med*, 357, 1105–1112.

Buller, H. R., Cohen, A. T., Davidson, B. et al. 2007b. Idraparinux versus standard therapy for venous thromboembolic disease. *N Engl J Med*, 357, 1094–1104.

Buller, H. R., Cohen, A. T., Lensing, A. W. A. et al. 2004. A novel long-acting synthetic factor Xa inhibitor (SanOrg34006) to replace warfarin for secondary prevention in deep vein thrombosis. A Phase II evaluation. *J Thromb Haemost*, 2, 47–53.

Bytheway, I., Hammond, E., Handley, P., and Dredge, K. 2009. The dual angiogenesis/heparanase inhibitor PG545, but not the tyrosine kinase inhibitor sorafenib, inhibits spontaneous metastasis in models of breast and lung cancer, In *AACR-NCI-EORTC International Conference: Molecular Targets and Cancer Therapeutics, Mol Cancer Ther*, Boston, MA.

Cabannes, E., Caravano, A., Motte, V. et al. 2009. Heparan Sulfate Mimetics as anticancer small-glyco drugs, In *67th Harden Conference*, Cambridge, UK.

Casu, B., Naggi, A., and Torri, G. 2010. Heparin-derived heparan sulfate mimics to modulate heparan sulfate-protein interaction in inflammation and cancer. *Matrix Biol.* in Press, Corrected Proof.

Casu, B., Vlodavsky, I., and Sanderson, R. D. 2008. Non-anticoagulant heparins and inhibition of cancer. *Pathophysiol Haemost Thromb*, 36, 195–203.

Chakrabarti, S., Beaulieu, L., Reyelt, L., Iafrati, M., and Freedman, J. 2009. M118, a novel low-molecular weight heparin with decreased polydispersity leads to enhanced anticoagulant activity and thrombotic occlusion in ApoE knockout mice. *J Thromb Thrombolysis*, 28, 394–400.

Chiara, M., Ian, F., James, R. et al. 2009. Design and rationale of the Evaluation of M118 IN pErcutaNeous Coronary intErvention (EMINENCE) trial. *Am Heart J*, 158, 726–733.

Choi, S., Clements, D. J., Pophristic, V. et al. 2005. The design and evaluation of heparin-binding foldamers. *Angew Chem Int Ed Engl*, 117, 6843–6847.

Chong, B. and Magnani, H. 2007. Danaparoid for the treatment of heparin-induced thrombocytopenia. In *Heparin-Induced Thrombocytopenia*, Warkentin, T. E. and Greinacher, A, pp. 295–318. New York: Informa Healthcare.

Chow, L., Gustafson, D., O'Bryant, C. et al. 2008. A phase I pharmacological and biological study of PI-88 and docetaxel in patients with advanced malignancies. *Cancer Chemother Pharmacol*, 63, 65–74.

Chu, C., Duffner, J., Dussault, N. et al. 2009. M-ONC 402, a novel non-anticoagulant heparin, inhibits P-Selectin function and metastatic seeding of tumor cells in mice, In 100th Annual Meeting of American Association for Cancer Research (AACR), Dever, Colorado.

Clark, P., Walker, I. D., Langhorne, P. et al. 2010. SPIN (Scottish Pregnancy Intervention) study: A multicenter, randomized controlled trial of low-molecular-weight heparin and low-dose aspirin in women with recurrent miscarriage. *Blood*, 115, 4162–4167.

Codée, J. D. C., Overkleeft, H. S., van der Marel, G. A., and van Boeckel, C. A. A. 2004. The synthesis of well-defined heparin and heparan sulfate fragments. *Drug Discov Today: Technol*, 1, 317–326.

Coombe, D. R. and Kett, W. C. 2005. Heparan sulfate-protein interactions: Therapeutic potential through structure-function insights. *Cell Mol Life Sci*, 62, 410–424.

Coussement, P. K., Bassand, J. P., Convens, C. et al. 2001. A synthetic factor-Xa inhibitor (ORG31540/SR9017A) as an adjunct to fibrinolysis in acute myocardial infarction: The Pentalyse study. *Eur Heart J*, 22, 1716–1724.

Crawford, B., Brown, J., Tolmie, K. et al. 2011. Small molecule inhibitors of Glycosaminoglycan Biosynthesis as substrate optimization therapy for the Mucopolysaccharidoses. *Mol Genet Metab*, 102, S12–S13.

De Kort, M., Buijsman, R. C., and van Boeckel, C. A. A. 2005. Synthetic heparin derivatives as new anticoagulant drugs. *Drug Discov Today*, 10, 769–779.

De Kort, M. and Van Boeckel, C. A. A. 2010. Antithrombotic dual inhibitors comprising a biotin residue, N. V. ORGANON (Oss, NL), United States.

Dember, L. M., Hawkins, P. N., Hazenberg, B. P. C. et al. 2007. Eprodisate for the Treatment of Renal Disease in AA Amyloidosis. *N Engl J Med*, 356, 2349–2360.

de Paz, J. L. and Seeberger, P. H. 2008. Deciphering the glycosaminoglycan code with the help of microarrays. *Mol Biosyst*, 4, 707–711.

de Pont, A.-C., Hofstra, J.-J., Pik, D., Meijers, J., and Schultz, M. 2007. Pharmacokinetics and pharmacodynamics of danaparoid during continuous venovenous hemofiltration: A pilot study. *Crit Care*, 11, R102.

Desai, U. R., Petitou, M., Bjork, I., and Olson, S. T. 1998a. Mechanism of heparin activation of antithrombin: Evidence for an induced-fit model of allosteric activation involving two interaction subsites. *Biochemistry*, 37, 13033–13041.

Desai, U. R., Petitou, M., Björk, I., and Olson, S. T. 1998b. Mechanism of Heparin Activation of Antithrombin: Role of individual residues of the pentasaccharide activating sequence in the recognition of native and activated states of antithrombin *J Biol Chem*, 273, 7478–7487.

Di Nisio, M., Peters, L., and Middeldorp, S. 2005. Anticoagulants for the treatment of recurrent pregnancy loss in women without antiphospholipid syndrome. *Cochrane Database Syst Rev*, CD004734.

Díaz-Nido, J., Wandosell, F., and Avila, J. 2002. Glycosaminoglycans and β-amyloid, prion and tau peptides in neurodegenerative diseases. *Peptides*, 23, 1323–1332.

Draganov, D., Wright, T., Avery, W. et al. 2009. Pharmacokinetics of M118, unfractionated heparin and enoxaparin sodium in normal and 5/6 nephrectomized uremic rats. *Toxicol Lett*, 189, S113–S113.

Dredge, K., Hammond, E., Davis, K. et al. 2009. The PG500 series: Novel heparan sulfate mimetics as potent angiogenesis and heparanase inhibitors for cancer therapy. *Invest New Drugs*, 28, 276–283.

Eikelboom, J. W. and Weitz, J. I. 2010. New anticoagulants. *Circulation*, 121, 1523–1532.

Elson-Schwab, L., Garner, O. B., Schuksz, M. et al. 2007. Guanidinylated neomycin delivers large, bioactive cargo into cells through a heparan sulfate-dependent pathway. *J Biol Chem*, 282, 13585–13591.

Escartin, Q., Lallam-Laroye, C., Baroukh, B. et al. 2003. A new approach to treat tissue destruction in periodontitis with chemically modified dextran polymers. *FASEB J*, 17, 644–651.

Faaij, R. A., Burggraaf, J., Schoemaker, R. C., van Amsterdam, R. G. M., and Cohen, A. F. 1999. A phase I single rising dose study to investigate the safety, tolerance and pharmacokinetics of subcutaneous SANORG 34006 in healthy male and female elderly volunteers. *Thromb Haemost*, 490–491.

Fareed, J. and Walenga, J. 2007. Why differentiate low molecular weight heparins for venous thromboembolism? *Thromb J*, 5, 8.

Ferro, V. and Don, R. 2003. The development of the novel angiogenesis inhibitor PI-88 as an anticancer drug. *Australas Biotechnol*, 13, 38–40.

Ferro, V., Dredge, K., Liu, L. et al. 2007. PI-88 and novel heparan sulfate mimetics inhibit angiogenesis. *Semin Thromb Hemost*, 33, 557–568.

Ferro, V., Fewings, K., Palermo, M. C., and Li, C. 2001. Large-scale preparation of the oligosaccharide phosphate fraction of Pichia holstii NRRL Y-2448 phosphomannan for use in the manufacture of PI-88. *Carbohydr Res*, 332, 183–189.

Fier, I. D., Brandquist, C. M., Abu-Rashid, M. et al. 2009. Lack of pharmacokinetic and pharmacodynamic interactions between M118, a novel low-molecular-weight-heparin and Eptifibatide in healthy subjects. *J Clin Pharmacol*, 49, 73.

Fier, I., Nedelman, M. A., Qi, Y. W., Guerrero, J. L., and Venkataraman, G. 2007. A novel, rationally engineered heparin (M118) prevents thrombosis more effectively than unfractionated heparin in a canine model of deep arterial injury. *J Am Coll Cardiol*, 49, 379A–380A.

Fritz, T. A., Lugemwa, F. N., Sarkar, A. K., and Esko, J. D. 1994. Biosynthesis of heparan sulfate on beta-D-xylosides depends on aglycone structure. *J Biol Chem*, 269, 300–307.

Fromes, Y., Neuhart, E., Krezel, C. et al. 2010. EP217609. A new neutralizable anticoagulant for cardiopulmonary bypass during cardiac surgery., In The French Society of Thoracic and Cardiovascular Surgery's 63th annual meeting (Pharma, E., Ed.), Tours, France.

Fugedi, P. 2003. The potential of the molecular diversity of heparin and heparan sulfate for drug development. *Mini Rev Med Chem*, 3, 659–667.

Fuster, M. M. and Esko, J. D. 2005. The sweet and sour of cancer: Glycans as novel therapeutic targets. *Nat Rev Cancer*, 5, 526–542.

Gama, C. I. and Hsieh-Wilson, L. C. 2005. Chemical approaches to deciphering the glycos-aminoglycan code. *Curr Opin Chem Biol*, 9, 609–619.

Gandhi, N. S. and Mancera, R. L. 2008. The structure of glycosaminoglycans and their inter-actions with proteins. *Chem Biol Drug Des*, 72, 455–482.

Garcia-Filipe, S., Barbier-Chassefiere, V., Alexakis, C. et al. 2007. RGTA OTR4120, a hepa-ran sulfate mimetic, is a possible long-term active agent to heal burned skin. *J Biomed Mater Res A*, 80A, 75–84.

Gauthier, S., Aisen, P., Ferris, S. et al. 2009. Effect of tramiprosate in patients with mild-to-moderate alzheimer's disease: Exploratory analyses of the MRI sub-group of the alphase study. *J Nutr Health Aging*, 13, 550–557.

Geerts, H. 2004. NC-531 (Neurochem). *Curr Opin Investig Drugs*, 5, 95–100.

Gervais, F., Morissette, C., and Kong, X. 2003. Proteoglycans and amyloidogenic proteins in peripheral amyloidosis. *Curr Med Chem Immunol Endocr Metab Agents*, 3, 361–370.

Gervais, F., Paquette, J., Morissette, C. et al. 2007. Targeting soluble Aβ peptide with Tramiprosate for the treatment of brain amyloidosis. *Neurobiol Aging*, 28, 537–547.

Gesslbauer, B. and Kungl, A. J. 2006. Glycomic approaches toward drug development: Therapeutically exploring the glycosaminoglycanome. *Curr Opin Mol Ther*, 8, 521–528.

Greenberg, S. M., Rosand, J., Schneider, A. T. et al. 2006. A phase 2 study of tramiprosate for cerebral amyloid angiopathy. *Alzheimer Dis Assoc Disord*, 20, 269–274.

Grootenhuis, P. D. J. and Van Boeckel, C. A. A. 1991. Constructing a molecular model of the interaction between antithrombin III and a potent heparin analog. *J Am Chem Soc*, 113, 2743–2747.

Gueret, P., Krezel, C., Fuseau, E. et al. 2009a. Pharmacokinetics and pharmacodynamics of EP42675 a new synthetic anticoagulant with a dual mechanism of action, In *The XXII Congress of the International Society of Thrombosis and Haemostasis* (Pharma, E., Ed.), Boston, USA.

Gueret, P., Krezel, C., Giersbergen, P. v. et al. 2009b. EP42675, a new dual action anticoagulant: Pharmacodynamic and pharmacokinetic profile, and interactions with acetylsalicylic acid, clopidogrel, and unfractionated heparin, In *The 51st Annual Meeting of the American Society of Hematology Congress* (Pharma, E., Ed.), New Orleans, Louisiana, USA.

Gueret, P., Krezel, C., Giersbergen, P. L. M. v. et al. 2010. First human study with EP217609, a new synthetic parenteral neutralizable dual action anticoagulant, In *21st International Congress on Thrombosis* (Pharma, E., Ed.), Milan, Italy.

Hammond, E., Bytheway, I., Handley, P. et al. 2009. The dual angiogenesis/heparanase inhibitor PG545 inhibits solid tumor progression in models of breast, prostate and liver cancer: A comparative assessment of once versus twice weekly administration sched-ules, In *AACR-NCI-EORTC International Conference: Molecular Targets and Cancer Therapeutics, Mol Cancer Ther* (Meeting Abstract Supplement), Boston, MA.

Harenberg, J. 2009. Development of idraparinux and idrabiotaparinux for anticoagulant ther-apy. *Thromb Haemost*, 102, 811–815.

Harenberg, J., Jörg, I., Vukojevic, Y., Mikus, G., and Weiss, C. 2008. Anticoagulant effects of Idraparinux after termination of therapy for prevention of recurrent venous thrombo-embolism: Observations from the van Gogh trials. *Eur J Clin Pharmacol*, 64, 555–563.

Harenberg, J., Vukojevic, Y., Mikus, G., Joerg, I., and Weiss, C. 2008. Long elimination half-life of idraparinux may explain major bleeding and recurrent events of patients from the van Gogh trials. *J Thromb Haemost*, 6, 890–892.

Harenberg, J., Wehling, M., Mikus, G., and Weiss, C. 2009. The anticoagulant Idraparinux: Is the extensive half life of 60 days the cause of bleeding complications. *Br J Clin Pharmacol*, 68, 21–21.

Hekman, A. 1971. Association of lactoferrin with other proteins, as demonstrated by changes in electrophoretic mobility. *Biochim Biophys Acta*, 251, 380–387.

Herbert, J. M., Herault, J. P., Bernat, A. et al. 1998. Biochemical and pharmacological properties of SANORG 34006, a potent and long-acting synthetic pentasaccharide. *Blood*, 91, 4197–4205.

Herbert, J. M., Hérault, J. P., Bernat, A. et al. 2001. SR123781A, a synthetic heparin mimetic. *Thromb Haemost*, 85, 852–860.

Herbert, J. M., Petitou, M., Lormeau, J. C. et al. 1997. SR 90107A/Org 31540, a novel anti-Factor Xa antithrombotic agent. *Cardiovasc Drug Rev*, 15, 1–26.

Hermodson, M., Schmer, G., and Kurachi, K. 1977. Isolation, crystallization, and primary amino acid sequence of human platelet factor 4. *J Biol Chem*, 252, 6276–6279.

Hirsh, J. 1991. Drug therapy: Heparin. *N Engl J Med*, 324, 1565–1574.

Hirsh, J., O'Donnell, M., and Eikelboom, J. W. 2007. Beyond unfractionated heparin and warfarin: Current and future advances. *Circulation*, 116, 552–560.

Hjelm, R. and Schedin-Weiss, S. 2007. High affinity interaction between a synthetic, highly negatively charged pentasaccharide and alpha- or beta-antithrombin is predominantly due to nonionic interactions. *Biochemistry*, 46, 3378–3384.

Hoppensteadt, D., Jeske, W., Walenga, J., and Fareed, J. 2008a. AVE5026, a new hemisynthetic ultra low molecular weight heparin (ULMWH) with enriched anti-Xa activity and enhanced antithrombotic activity for management of cancer associated thrombosis. *J Clin Oncol (Meeting Abstracts)*, 26, 14653.

Hoppensteadt, D., Jeske, W., Walenga, J., and Fareed, J. 2008b. AVE5026, a new hemisynthetic ultra low molecular weight heparin (ULMWH) with enriched anti-Xa activity and enhanced antithrombotic activity for management of cancer associated thrombosis, *J Clin Oncol* (Meeting Abstracts) 26, 14653.

Huber, R. and Carrell, R. W. 1989. Implications of the three-dimensional structure of alpha 1-antitrypsin for structure and function of serpins. *Biochemistry*, 28, 8951–8966.

Ian, D. T., John, E. T., and Richard, D. S.-G. 1995. Effect of nonanticoagulant heparin (Astenose) on restenosis after balloon angioplasty in the atherosclerotic rabbit. *J Vasc Interv Radiol*, 6, 365–378.

Iozzo, R. V. and San Antonio, J. D. 2001. Heparan sulfate proteoglycans: Heavy hitters in the angiogenesis arena. *J Clin Invest*, 108, 349–355.

Jairajpuri, M. A., Lu, A., Desai, U. et al. 2003. Antithrombin III phenylalanines 122 and 121 contribute to its high affinity for heparin and its conformational activation. *J Biol Chem*, 278, 15941–15950.

Jeske, W., Brubaker, A., Liu, D. et al. 2007. In vitro characterization of the neutralization of unfractionated heparin and low molecular weight heparin by novel salicylamide derivatives, In *American Society of Hematology*, Atlanta, GA.

Jeske, W., Brubaker, A., Liu, D. et al. 2009. Novel antagonists for low molecular weight heparin and heparin-like drugs, In *American Society of Hematology*, New Orleans, LA.

Jin, L., Abrahams, J. P., Skinner, R. et al. 1997. The anticoagulant activation of antithrombin by heparin. *Proc Natl Acad Sci USA*, 94, 14683–14688.

Johnson, D. J. and Huntington, J. A. 2003. Crystal structure of antithrombin in a heparin-bound intermediate state. *Biochemistry*, 42, 8712–8719.

Johnson, D. J. D., Langdown, J., Li, W. et al. 2006. Crystal structure of monomeric native antithrombin reveals a novel reactive center loop conformation. *J Biol Chem*, 281, 35478–35486.

Johnson, D. J. D., Li, W., Adams, T. E., and Huntington, J. A. 2006. Antithrombin-S195A factor Xa-heparin structure reveals the allosteric mechanism of antithrombin activation. *EMBO J*, 25, 2029–2037.

Joyce, J. A., Freeman, C., Meyer-Morse, N., Parish, C. R., and Hanahan, D. 2005. A functional heparan sulfate mimetic implicates both heparanase and heparan sulfate in tumor angiogenesis and invasion in a mouse model of multistage cancer. *Oncogene*, 24, 4037–4051.

Kaandorp, S., Di Nisio, M., Goddijn, M., and Middeldorp, S. 2009. Aspirin or anticoagulants for treating recurrent miscarriage in women without antiphospholipid syndrome. *Cochrane Database Syst Rev*, CD004734.

Kaandorp, S. P., Goddijn, M., van der Post, J. A. et al. 2010. Aspirin plus heparin or aspirin alone in women with recurrent miscarriage. *N Engl J Med*, 362, 1586–1596.

Karoli, T., Liu, L., Fairweather, J. K. et al. 2005. Synthesis, biological activity, and preliminary pharmacokinetic evaluation of analogues of a phosphosulfomannan angiogenesis inhibitor (PI-88). *J Med Chem*, 48, 8229–8236.

Kelton, J., Smith, J., Warkentin, T. et al. 1994. Immunoglobulin G from patients with heparin-induced thrombocytopenia binds to a complex of heparin and platelet factor 4. *Blood*, 83, 3232–3239.

Khammari Chebbi, C., Kichenin, K., Amar, N. et al. 2008. Pilot study of a new matrix therapy agent (RGTA OTR4120®) in treatment-resistant corneal ulcers and corneal dystrophy. *J Fr Ophtalmol*, 31, 465–471.

Khasraw, M., Pavlakis, N., McCowatt, S. et al. 2010. Multicentre phase I/II study of PI-88, a heparanase inhibitor in combination with docetaxel in patients with metastatic castrate-resistant prostate cancer. *Ann Oncol*, 21, 1302–1307.

Kishimoto, T. K., Qi, Y. W., Long, A. et al. 2009. M118–A rationally engineered low-molecular-weight heparin designed specifically for the treatment of acute coronary syndromes. *Thromb Haemost*, 102, 900–906.

Kisilevsky, R. 2000. The relation of proteoglycans, serum amyloid P and Apo E to amyloidosis current status, 2000. *Amyloid*, 7, 23–25.

Kisilevsky, R., Lemieux, L. J., Fraser, P. E. et al. 1995. Arresting amyloidosis in vivo using small-molecule anionic sulphonates or sulphates: Implications for Alzheimer's disease. *Nat Med*, 1, 143–148.

Klement, P. and Rak, J. 2006. Emerging anticoagulants: Mechanism of action and future potential. *Vnitr Lek*, 52, 119–122.

Kragh, M., Binderup, L., Vig, H. et al. 2005. Non-anti-coagulant heparin inhibits metastasis but not primary tumor growth. *Oncol Rep*, 14, 99.

Kreuger, J., Matsumoto, T., Vanwildemeersch, M. et al. 2002. Role of heparan sulfate domain organization in endostatin inhibition of endothelial cell function. *EMBO J*, 21, 6303–6311.

Kreuger, J., Spillmann, D., Li, J.-p., and Lindahl, U. 2006. Interactions between heparan sulfate and proteins: The concept of specificity. *J Cell Biol*, 174, 323–327.

Krishnaswamy, A., Lincoff, A. M., and Cannon, C. P. 2010. The use and limitations of unfractionated heparin. *Crit Pathw Cardiol*, 9, 35–40.

Kudchadkar, R., Gonzalez, R., and Lewis, K. D. 2008. PI-88: A novel inhibitor of angiogenesis. *Expert Opin Investig Drugs*, 17, 1769–1776.

Langdown, J., Belzar, K. J., Savory, W. J., Baglin, T. P., and Huntington, J. A. 2009. The critical role of hinge-region expulsion in the induced-fit heparin binding mechanism of antithrombin. *J Mol Biol*, 386, 1278–1289.

Lanza, T. J., Durette, P. L., Rollins, T. et al. 1992. Substituted 4,6-diaminoquinolines as inhibitors of C5a receptor binding. *J Med Chem*, 35, 252–258.

Laremore, T. N., Zhang, F., Dordick, J. S., Liu, J., and Linhardt, R. J. 2009. Recent progress and applications in glycosaminoglycan and heparin research. *Curr Opin Chem Biol*, 13, 633–640.

Lassen, M. R., Dahl, O. E., Mismetti, P., DestrÉE, D., and Turpie, A. G. G. 2009. AVE5026, a new hemisynthetic ultra-low-molecular-weight heparin for the prevention of venous thromboembolism in patients after total knee replacement surgery—TREK: A dose-ranging study. *J Thromb Haemost*, 7, 566–572.

Le Templier, G. and Rodger, M. A. 2008. Heparin-induced osteoporosis and pregnancy. *Curr Opin Pulm Med*, 14, 403–407.

Lee, D., Lee, S., Kim, S. et al. 2009. Antiangiogenic activity of orally absorbable heparin derivative in different types of cancer cells. *Pharm Res*, 26, 2667–2676.

Lee, E., Pavy, M., Young, N., Freeman, C., and Lobigs, M. 2006. Antiviral effect of the heparan sulfate mimetic, PI-88, against dengue and encephalitic flaviviruses. *Antiviral Res*, 69, 31–38.

Lefkou, E., Khamashta, M., Hampson, G., and Hunt, B. J. 2010. Review: Low-molecular-weight heparin-induced osteoporosis and osteoporotic fractures: A myth or an existing entity? *Lupus*, 19, 3–12.

Leveugle, B., Scanameo, A., Ding, W., and Fillit, H. 1994. Binding of heparan sulfate glycosaminoglycan to [beta]-amyloid peptide: Inhibition by potentially therapeutic polysulfated compounds. *NeuroReport*, 5, 1389–1392.

Li, W., Johnson, D. J. D., Esmon, C. T., and Huntington, J. A. 2004. Structure of the antithrombin-thrombin-heparin ternary complex reveals the antithrombotic mechanism of heparin. *Nat Struct Mol Biol*, 11, 857–862.

Lindahl, U. 2007. Heparan sulfate-protein interactions: A concept for drug design? *Thromb Haemost*, 98, 109–115.

Lisman, T., Bijsterveld, N. R., Adelmeijer, J. et al. 2003. Recombinant factor VIIa reverses the in vitro and ex vivo anticoagulant and profibrinolytic effects of fondaparinux. *J Thromb Haemost*, 1, 2368–2373.

Liu, C.-J., Lee, P.-H., Lin, D.-Y. et al. 2009. Heparanase inhibitor PI-88 as adjuvant therapy for hepatocellular carcinoma after curative resection: A randomized phase II trial for safety and optimal dosage. *J Hepatol*, 50, 958–968.

Liu, H., Zhang, Z., and Linhardt, R. J. 2009. Lessons learned from the contamination of heparin. *Nat Prod Rep*, 26, 313–321.

Lolkema, M. P., Lockley, M., Zhou, H. et al. 2010. M402, a novel heparan sulfate mimetic, synergizes with Gemcitabine to improve survival and reduce metastasis and epithelial-to-mesenchymal transition (EMT) in a genetically engineered mouse model for pancreatic cancer, In *American Association for Cancer Research (AACR)*, Washington, DC.

Ma, Q. and Fareed, J. 2004. Idraparinux sodium. *IDrugs*, 7, 1028–1034.

Magnani, H. N. 2010. An analysis of clinical outcomes of 91 pregnancies in 83 women treated with danaparoid (Organan®). *Thromb Res*, 125, 297–302.

Magnani, H. and Gallus, A. 2006. Heparin-induced thrombocytopenia (HIT)—A report of 1,478 clinical outcomes of patients treated with danaparoid (Organan) from 1982 to mid-2004. *Thromb Haemost*, 95, 917–1051.

Mak, A., Cheung, M. W., Cheak, A. A., and Ho, R. C. 2010. Combination of heparin and aspirin is superior to aspirin alone in enhancing live births in patients with recurrent pregnancy loss and positive anti-phospholipid antibodies: A meta-analysis of randomized controlled trials and meta-regression. *Rheumatology (Oxford)*, 49, 281–288.

Manetti, F., Corelli, F., and Botta, M. 2000. Fibroblast growth factors and their inhibitors. *Curr Pharm Design*, 6, 1897–1924.

Manenti, L., Tansinda, P., and Vaglio, A. 2008. Eprodisate in amyloid A amyloidosis: A novel therapeutic approach? *Expert Opin Pharmacother*, 9, 2175–2180.

Mangoni, M., Yue, X., Morin, C. et al. 2009. Differential effect triggered by a heparan mimetic of the RGTA family preventing oral mucositis without tumor protection. *Int J Radiat Oncol Biol Phys*, 74, 1242–1250.

Martelly, I., Singabraya, D., Vandebrouck, A. et al. 2010. Glycosaminoglycan mimetics trigger IP3-dependent intracellular calcium release in myoblasts. *Matrix Biol*, 29, 317–329.

Matzsch, T., Bergqvist, D., Hedner, U., Nilsson, B., and Ostergaard, P. 1990. Effects of low molecular weight heparin and unfragmented heparin on induction of osteoporosis in rats. *Thromb Haemost*, 63, 505–509.

McCoy, A. J., Pei, X. Y., Skinner, R., Abrahams, J. P., and Carrell, R. W. 2003. Structure of beta-antithrombin and the effect of glycosylation on antithrombin's heparin affinity and activity. *J Mol Biol*, 326, 823–833.

McKenzie, E. A. 2007. Heparanase: A target for drug discovery in cancer and inflammation. *Br J Pharmacol*, 151, 1–14.

McLaurin, J., Franklin, T., Zhang, X., Deng, J., and Fraser, P. E. 1999. Interactions of Alzheimer amyloid-β peptides with glycosaminoglycans. *Eur J Biochem*, 266, 1101–1110.

Meddahi, A., Alexakis, C., Papy, D., Caruelle, J.-P., and Barritault, D. 2002. Heparin-like polymer improved healing of gastric and colic ulceration. *J Biomed Mater Res A*, 60, 497–501.

Meddahi, A., Benoit, J., Ayoub, N., Sézeur, A., and Barritault, D. 1996. Heparin-like polymers derived from dextran enhance colonic anastomosis resistance to leakage. *J Biomed Mater Res*, 31, 293–297.

Meddahi, A., Blanquaert, F., Saffar, J. L. et al. 1994. New approaches to tissue regeneration and repair. *Pathol Res Pract*, 190, 923–928.

Meddahi, A., Brée, F., Papy-Garcia, D. et al. 2002. Pharmacological studies of $RGTA_{11}$, a heparan sulfate mimetic polymer, efficient on muscle regeneration. *J Biomed Mater Res*, 62, 525–531.

Meddahi, A., Lemdjabar, H., Caruelle, J. P., Barritault, D., and Hornebeck, W. 1995. Inhibition by dextran derivatives of FGF-2 plasmin-mediated degradation. *Biochimie*, 77, 703–706.

Meddahi, A., Lemdjabar, H., Caruelle, J.-P., Barritault, D., and Hornebeck, W. 1996. FGF protection and inhibition of human neutrophil elastase by carboxymethyl benzylamide sulfonate dextran derivatives. *Int J Biol Macromol*, 18, 141–145.

Mehta, S. R., Granger, C. B., Eikelboom, J. W. et al. 2007. Efficacy and safety of fondaparinux versus enoxaparin in patients with acute coronary syndromes undergoing percutaneous coronary intervention: Results from the OASIS-5 trial. *J Am Coll Cardiol*, 50, 1742–1751.

Mehta, S. R., Steg, P. G., Granger, C. B. et al. 2005. Randomized, blinded trial comparing fondaparinux with unfractionated heparin in patients undergoing contemporary percutaneous coronary intervention: Arixtra study in percutaneous coronary intervention: A randomized evaluation (ASPIRE) pilot trial. *Circulation*, 111, 1390–1397.

Meuleman, D. G. 1992. Orgaran (Org 10172): Its pharmacological profile in experimental models. *Pathophysiol Haemost Thromb*, 22, 58–65.

Miao, H.-Q., Liu, H., Navarro, E., Kussie, P., and Zhu, Z. 2006. Development of heparanase inhibitors for anti-Cancer therapy. *Curr Med Chem*, 13, 2101–2111.

Millward, M., Hamilton, A., Thomson, D., Gautam, A., and Wilson, E. 2007. Final results of a phase I study of daily PI-88 as a single agent and in combination with dacarbazine (D) in patients with metastatic melanoma. *J Clin Oncol (Meeting Abstracts)*, 25, 8532.

Monreal, M., Vinas, L., Monreal, L. et al. 1990. Heparin-related osteoporosis in rats. A comparative study between unfractioned heparin and a low-molecular-weight heparin. *Haemostasis*, 20, 204–207.

Morad, N., Ryser, H. J. P., and Shen, W.-C. 1984. Binding sites and endocytosis of heparin and polylysine are changed when the two molecules are given as a complex to chinese hamster ovary cells. *Biochim Biophys Acta*, 801, 117–126.

Mourey, L., Samama, J. P., Delarue, M. et al. 1990. Antithrombin III: Structural and functional aspects. *Biochimie*, 72, 599–608.

Mousa, S. A., Linhardt, R., Francis, J. L., and Amirkhosravi, A. 2006. Anti-metastatic effect of a non-anticoagulant low-molecular-weight heparin versus the standard low-molecular-weight heparin, enoxaparin. *Thromb Haemost*, 96, 816–821.

Nath, N., Lowery, C., and Niewiarowski, S. 1975. Antigenic and antiheparin properties of human platelet factor 4 (PF4). *Blood*, 45, 537–550.

Neugroschl, J. and Sano, M. 2010. Current treatment and recent clinical research in Alzheimer's disease. *Mt Sinai J Med*, 77, 3–16.

Olson, S., Swanson, R., Raub-Segall, E. et al. 2004. Accelerating ability of synthetic oligosaccharides on antithrombin inhibition of proteinases of the clotting and fibrinolytic systems. Comparison with heparin and low-molecular-weight heparin. *Thromb Haemost*, 92, 929.

Oosta, G. M., Gardner, W. T., Beeler, D. L., and Rosenberg, R. D. 1981. Multiple functional domains of the heparin molecule. *Proc Natl Acad Sci USA*, 78, 829–833.

Osborn, H. M. I., Evans, P. G., Gemmell, N., and Osborne, S. D. 2004. Carbohydrate-based therapeutics. *J Pharm Pharmacol*, 56, 691–702.

Panchal, R. G., Hermone, A. R., Nguyen, T. L. et al. 2004. Identification of small molecule inhibitors of anthrax lethal factor. *Nat Struct Mol Biol*, 11, 67–72.

Papy-Garcia, D., Barbier-Chassefiere, V., Rouet, V. et al. 2005. Nondegradative sulfation of polysaccharides. Synthesis and structure characterization of biologically active heparan sulfate mimetics. *Macromolecules*, 38, 4647–4654.

Papy-Garcia, D., Barbosa, I., Duchesnay, A. et al. 2002. Glycosaminoglycan mimetics (RGTA) modulate adult skeletal muscle satellite cell proliferation in vitro. *J Biomed Mater Res*, 62, 46–55.

Parish, C. R. 2006. The role of heparan sulphate in inflammation. *Nat Rev Immunol*, 6, 633–643.

Parish, C. R., Freeman, C., Brown, K. J., Francis, D. J., and Cowden, W. B. 1999. Identification of sulfated oligosaccharide-based inhibitors of tumor growth and metastasis using novel in vitro assays for angiogenesis and heparanase activity. *Cancer Res*, 59, 3433–3441.

Paty, I., Trellu, M., Destors, J. M. et al. 2010. Reversibility of the anti-FXa activity of idrabio-taparinux (biotinylated idraparinux) by intravenous avidin infusion. *J Thromb Haemost*, 8, 722–729.

Peterson, S., Frick, A., and Liu, J. 2009. Design of biologically active heparan sulfate and heparin using an enzyme-based approach. *Nat Prod Rep*, 26, 610–627.

Petitou, M. and Boeckel, C. A. A. v. 2004. A synthetic antithrombin III binding pentasaccharide is now a drug! What comes next? *Angew Chem Int Ed Engl*, 43, 3118–3133.

Petitou, M., Duchaussoy, P., Lederman, I. et al. 1987. Synthesis of heparin fragments: A methyl α-pentaoside with high affinity for antithrombin III. *Carbohydr Res*, 167, 67–75.

Petitou, M., Nancy-Portebois, V., Dubreucq, G. et al. 2009. From heparin to EP217609: The long way to a new pentasaccharide-based neutralisable anticoagulant with an unprecedented pharmacological profile. *Thromb Haemost*, 102, 804–810.

Portmann, A. F. and Holden, W. D. 1949. Protamine (Salmine) sulphate, heparin, and blood coagulation. *J Clin Invest*, 28, 1451–1458.

Powell, A. K., Yates, E. A., Fernig, D. G., and Turnbull, J. E. 2004. Interactions of heparin/heparan sulfate with proteins: Appraisal of structural factors and experimental approaches. *Glycobiology*, 14, 17R–30.

Prandoni, P., Tormene, D., Perlati, M., Brandolin, B., and Spiezia, L. 2008. Idraparinux: Review of its clinical efficacy and safety for prevention and treatment of thromboembolic disorders. *Expert Opin Investig Drugs*, 17, 773–777.

Rafii, M. and Aisen, P. 2009. Recent developments in Alzheimer's disease therapeutics. *BMC Med*, 7, 7.

Rajgopal, R., Bear, M., Butcher, M. K., and Shaughnessy, S. G. 2008. The effects of heparin and low molecular weight heparins on bone. *Thromb Res*, 122, 293–298.

Rao, S. V., Melloni, C., Myles-DiMauro, S. et al. 2010. Evaluation of a New Heparin Agent in Percutaneous Coronary Intervention: Results of the Phase 2 Evaluation of M118 IN pErcutaNeous Coronary intErvention (EMINENCE) Trial. *Circulation*, 121, 1713–1721.

Rek, A., Krenn, E., and Kungl, A. 2009. Therapeutically targeting protein–glycan interactions. *Br J Pharmacol*, 157, 686–694.

Revill, P., Serradell, N., and Bolos, J. 2006. Eprodisate Sodium. *Drugs Future*, 31, 576–578.

Ricci, G., Giolo, E., and Simeone, R. 2010. Heparin's 'potential to improve pregnancy rates and outcomes' is not evidence-based. *Hum Reprod Update*, 16, 225–227; author reply 227–228.

Roan, N. R., Sowinski, S., Münch, J., Kirchhoff, F., and Greene, W. C. 2010. Aminoquinoline surfen inhibits the action of SEVI (Semen-derived Enhancer of Viral Infection). *J Biol Chem*, 285, 1861–1869.

Robert, F. 2010. The potential benefits of low-molecular-weight heparins in cancer patients. *J Hematol Oncol*, 3, 3.

Roberts, A. L. K., Thomas, B. J., Wilkinson, A. S., Fletcher, J. M., and Byers, S. 2006. Inhibition of glycosaminoglycan synthesis using rhodamine B in a mouse model of mucopolysaccharidosis type IIIA. *Pediatr Res*, 60, 309–314.

Rosenberg, R. D., Shworak, N. W., Liu, J., Schwartz, J. J., and Zhang, L. 1997. Heparan sulfate proteoglycans of the cardiovascular system. Specific structures emerge but how is synthesis regulated? *J Clin Invest*, 99, 2062–2070.

Rosenthal, M. A., Rischin, D., McArthur, G. et al. 2002. Treatment with the novel anti-angiogenic agent PI-88 is associated with immune-mediated thrombocytopenia. *Ann Oncol*, 13, 770–776.

Rostand, K. S. and Esko, J. D. 1997. Microbial adherence to and invasion through proteoglycans. *Infect Immun*, 65, 1–8.

Rouet, V., Hamma-Kourbali, Y., Petit, E. et al. 2005. A synthetic glycosaminoglycan mimetic binds vascular endothelial growth factor and modulates angiogenesis. *J Biol Chem*, 280, 32792–32800.

Rouet, V., Meddahi-Pellé, A., Miao, H.-Q. et al. 2006. Heparin-like synthetic polymers, named RGTAs, mimic biological effects of heparin in vitro. *J Biomed Mater Res A*, 78A, 792–797.

Samama, M. and Gerotziafas, G. 2010. Newer anticoagulants in 2009. *J Thromb Thrombolysis*, 29, 92–104.

Santa-Maria, I., Hernandez, F., Del Rio, J., Moreno, F., and Avila, J. 2007. Tramiprosate, a drug of potential interest for the treatment of Alzheimer's disease, promotes an abnormal aggregation of tau. *Mol Neurodegener*, 2, 17.

SARL, N. L. I. 2008. Withdrawal Assessment Report for Kiacta, (CHMP) European Medicines Agency, London.

Saumier, D., Aisen, P., Gauthier, S. et al. 2009. Lessons learned in the use of volumetric MRI in therapeutic trials in Alzheimer's disease: The Alzhemedtm (Tramiprosate) experience. *J Nutr Health Aging*, 13, 370–372.

Savi, P., Herault, J. P., Duchaussoy, P. et al. 2008. Reversible biotinylated oligosaccharides: A new approach for a better management of anticoagulant therapy. *J Thromb Haemost*, 6, 1697–1706.

Schick, B. P., Maslow, D., Moshinski, A., and Antonio, J. D. S. 2004. Novel concatameric heparin-binding peptides reverse heparin and low-molecular-weight heparin anticoagulant activities in patient plasma in vitro and in rats in vivo. *Blood*, 103, 1356–1363.

Schiele, F. 2010. Fondaparinux and acute coronary syndromes: Update on the OASIS 5–6 studies. *Vasc Health Risk Manag*, 6, 179–187.

Schiele, F., Meneveau, N., Seronde, M. F. et al. 2010. Routine use of fondaparinux in acute coronary syndromes: A 2-year multicenter experience. *Am Heart J*, 159, 190–198.

Schindewolf, M., Magnani, H., and Lindhoff-Last, E. 2007. Danaparoid in pregnancy in cases of heparin intolerance-use in 59 cases. *Hamostaseologie*, 27, 89.

Schonberger, O., Horonchik, L., Gabizon, R. et al. 2003. Novel heparan mimetics potently inhibit the scrapie prion protein and its endocytosis. *Biochem Biophys Res Commun*, 312, 473–479.

Schreuder, H. A., de Boer, B., Dijkema, R. et al. 1994. The intact and cleaved human antithrombin III complex as a model for serpin-proteinase interactions. *Nat Struct Biol*, 1, 48–54.

Schuksz, M., Fuster, M. M., Brown, J. R. et al. 2008. Surfen, a small molecule antagonist of heparan sulfate. *Proc Natl Acad Sci U S A*, 105, 13075–13080.

Serina, G., Mirjoleta, J. F., Nancy-Porteboisb, V. et al. 2010. Antitumor activity of EP80061, a small-glyco drug in preclinical studies, In *AACR Meeting*, Washington DC.

Sharma, T., Mehta, P., and Gajra, A. 2010. Update on Fondaparinux: Role in Management of Thromboembolic and Acute Coronary Events. *Cardiovasc Hematol Agents Med Chem*, 8, 96–103.

Shriver, Z., Raguram, S., and Sasisekharan, R. 2004. Glycomics: A pathway to a class of new and improved therapeutics. *Nat Rev Drug Discov*, 3, 863–873.

Simoons, M. L., Bobbink, I. W. G., Boland, J. et al. 2004. A dose-finding study of fondaparinux in patients with non-ST-segment elevation acute coronary syndromes: The Pentasaccharide in Unstable Angina (PENTUA) study. *J Am Coll Cardiol*, 43, 2183–2190.

Sobel, M., Bird, K. E., Tyler-Cross, R. et al. 1996. Heparins Designed to Specifically Inhibit Platelet Interactions With von Willebrand Factor. *Circulation*, 93, 992–999.

Sternbergh, C. I., Sobel, M., and Makhoul, R. G. 1995. Heparinoids with low anticoagulant potency attenuate postischemic endothelial cell dysfunction. *J Vasc Surg*, 21, 477–483.

Stevenson, J. L., Choi, S. H., and Varki, A. 2005. Differential metastasis inhibition by clinically relevant levels of heparins—Correlation with selectin Inhibition, not antithrombotic activity. *Clin Cancer Res*, 11, 7003–7011.

Sutton, A., Friand, V., Papy-Garcia, D. et al. 2007. Glycosaminoglycans and their synthetic mimetics inhibit RANTES-induced migration and invasion of human hepatoma cells. *Mol Cancer Ther*, 6, 2948–2958.

Tan, K., Duquette, M., Liu, J.-h. et al. 2006. The structures of the Thrombospondin-1 N-terminal domain and its complex with a synthetic pentameric heparin. *Structure*, 14, 33–42.

Taniguchi, S., Suzuki, N., Masuda, M. et al. 2005. Inhibition of heparin-induced tau filament formation by phenothiazines, polyphenols, and porphyrins. *J Biol Chem*, 280, 7614–7623.

Tardieu, M., Gamby, C., Avramoglou, T., Jozefonvicz, J., and Barritault, D. 1992. Derivatized dextrans mimic heparin as stabilizers, potentiators, and protectors of acidic or basic FGF. *J Cell Physiol*, 150, 194–203.

The AMADEUS Investigators, MG, B., Bouthier J. et al. 2008. Comparison of idraparinux with vitamin K antagonists for prevention of thromboembolism in patients with atrial fibrillation: A randomised, open-label, non-inferiority trial. *The Lancet*, 371, 315–321.

The PERSIST Investigators 2002. A novel long-acting synthetic factor Xa inhibitor (idraparinux sodium) to replace warfarin for secondary prevention in deep vein thrombosis. A phase II evaluation. *Blood*, 100, 301.

Tiwari, V., O'Donnell, C., Copeland, R. J. et al. 2007. Soluble 3-*O*-sulfated heparan sulfate can trigger herpes simplex virus type 1 entry into resistant Chinese hamster ovary (CHO-K1) cells. *J Gen Virol*, 88, 1075–1079.

Tong, M., Tuk, B., Hekking, I. M. et al. 2009. Stimulated neovascularization, inflammation resolution and collagen maturation in healing rat cutaneous wounds by a heparan sulfate glycosaminoglycan mimetic, OTR4120. *Wound Repair Regen*, 17, 840–852.

Turpie, A. G. G. 2004. Fondaparinux: A Factor Xa inhibitor for antithrombotic therapy. *Expert Opin Pharmacother*, 5, 1373–1384.

Umber, F., Störring, F. K., and Föllmer, W. 1938. Erfolge mit Einem Neuartigen Depotinsulin Ohne Protaminzusatz (Surfen-Insulin). *J Mol Med*, 17, 443–446.

van Boeckel, C. A., Grootenhuis, P. D., and Haasnoot, C. A. 1991. Specificity in the recognition process between charged carbohydrates and proteins. *Trends Pharmacol Sci*, 12, 241–243.

van Horssen, J., Wesseling, P., van den Heuvel, L. P. W. J., de Waal, R. M. W., and Verbeek, M. M. 2003. Heparan sulphate proteoglycans in Alzheimer's disease and amyloid-related disorders. *Lancet Neurol*, 2, 482–492.

Vanwildemeersch, M., Olsson, A.-K., Gottfridsson, E. et al. 2006. The Anti-angiogenic His/Pro-rich Fragment of Histidine-rich Glycoprotein Binds to Endothelial Cell Heparan Sulfate in a Zn2+-dependent Manner. *J Biol Chem*, 281, 10298–10304.

Vetter, A., Perera, G., Leithner, K., Klima, G., and Bernkop-Schnürch, A. 2010. Development and in vivo bioavailability study of an oral fondaparinux delivery system. *Eur J Pharm Sci*. in Press, Corrected Proof.

Veyrat-Follet, C., Vivier, N., Trellu, M., Dubruc, C., and Sanderink, G. J. 2009. The pharmacokinetics of idraparinux, a long-acting indirect factor Xa inhibitor: Population pharmacokinetic analysis from phase III clinical trials. *J Thromb Haemost*, 7, 559–565.

Viskov, C., Just, M., Laux, V., Mourier, P., and Lorenz, M. 2009. Description of the chemical and pharmacological characteristics of a new hemisynthetic ultra-low-molecular-weight heparin, AVE5026. *J Thromb Haemost*, 7, 1143–1151.

Vlodavsky, I., Ilan, N., Naggi, A., and Casu, B. 2007. Heparanase: Structure, biological functions, and inhibition by heparin-derived mimetics of heparan sulfate. *Curr Pharm Des*, 13, 2057–2073.

Vogt, A. M., Pettersson, F., Moll, K. et al. 2006. Release of Sequestered Malaria Parasites upon Injection of a Glycosaminoglycan. *PLoS Pathog*, 2, e100.

Volovyk, Z., Monroe, D., Qi, Y., Becker, R., and Hoffman, M. 2009. A rationally designed heparin, M118, has anticoagulant activity similar to unfractionated heparin and different from Lovenox in a cell-based model of thrombin generation. *J Thromb Thrombolysis*, 28, 132–139.

Volpi, N. 2006. Therapeutic applications of glycosaminoglycans. *Curr Med Chem*, 13, 1799–1810.

Wall, D., Douglas, S., Ferro, V., Cowden, W., and Parish, C. 2001. Characterisation of the anticoagulant properties of a range of structurally diverse sulfated oligosaccharides. *Thromb Res*, 103, 325–335.

Wang, J. and Rabenstein, D. L. 2006. Interaction of heparin with two synthetic peptides that neutralize the anticoagulant activity of heparin. *Biochemistry*, 45, 15740–15747.

Warkentin, T. E., Sheppard, J.-A. I., Horsewood, P. et al. 2000. Impact of the patient population on the risk for heparin-induced thrombocytopenia. *Blood*, 96, 1703–1708.

Weitz, J., Klement, P., Liao, P. et al. 2003. GH9001, a novel anti-thrombotic agent, is more effective than low-molecular-weight heparin, fondaparinux, or hirudin in rabbit models, In *XIX International ISTH Congress, J Thromb Haemost*.

Westerduin, P., van Boeckel, C. A. A., Basten, J. E. M. et al. 1994. Feasible synthesis and biological properties of six 'non-glycosamino' glycan analogues of the antithrombin III binding heparin pentasaccharide. *Bioorg Med Chem*, 2, 1267–1280.

Wienbergen, H. and Zeymer, U. 2007. Management of acute coronary syndromes with fondaparinux. *Vasc Health Risk Manag*, 3, 321–329.

Wright, T. M. 2006. Tramiprosate. *Drugs Today*, 42, 291–298.

Xu-Song, Z. and Bing-ren, X. 2009. Discontinued drugs in 2008: Cardiovascular drugs. *Expert Opin Investig Drugs*, 18, 875–885.

Yamauchi, H., Desgranges, P., Lecerf, L. et al. 2000. New agents for the treatment of infarcted myocardium. *FASEB J*, 14, 2133–2134.

Yip, G. W., Smollich, M., and Götte, M. 2006. Therapeutic value of glycosaminoglycans in cancer. *Mol Cancer Ther*, 5, 2139–2148.

Yoshitomi, Y., Nakanishi, H., Kusano, Y. et al. 2004. Inhibition of experimental lung metastases of Lewis lung carcinoma cells by chemically modified heparin with reduced anticoagulant activity. *Cancer Lett*, 207, 165–174.

Young, E. 2008. The anti-inflammatory effects of heparin and related compounds. *Thromb Res*, 122, 743–752.

Zhou, H., Dussault, N., Cochran, E. et al. 2009. M-ONC 402, a non anticoagulant low molecular weight heparin inhibits tumor metastasis., In *100th Annual Meeting of American Association for Cancer Research (AACR)*, Denver, CO.

Zhou, H., Roy, S., Cochran, E. et al. 2010. M402, a novel heparan sulfate proteoglycan mimetic targeting tumor-host interactions, In *American Association for Cancer Research (AACR)*, Washington, DC.

Zhu, X., Hsu, B. T., and Rees, D. C. 1993. Structural studies of the binding of the anti-ulcer drug sucrose octasulfate to acidic fibroblast growth factor. *Structure*, 1, 27–34.

Zuijdendorp, H. M., Smit, X., Blok, J. H. et al. 2008. Significant reduction in neural adhesions after administration of the regenerating agent OTR4120, a synthetic glycosaminoglycan mimetic, after peripheral nerve injury in rats. *J Neurosurg*, 109, 967–973.

Index

A

A and B blood groups antigens, carbohydrate structures of, 217
ABD, *see* AT-binding domain
ABH antigens, 126
Ab initio calculations, 78
Ab initio methods, 16, 19
ABO antigens, 90
ABO blood groups, 216
ABO polymorphism, 223
AB phenotype, 216
AB$_5$ toxins, 13
N-Acetylglucosamine (*N*-GlcNAc), 33
N-Acetylglucosaminidase (OGA), 186
N-Acetyllactosamine-derived inhibitor, 120
N-Acetylmuramic acid (*N*-MurNAc), 33
Acute coronary syndrome (ACS), 268
AD, *see* Alzheimer's disease
AEAB, *see* 2-Amino-*N*-(2-aminoethyl) benzamide
Allosteric inhibition, 267, 269
Allotransplantation, 215, 216
Alzheimer's disease (AD), 276
AMADEUS phase III trial, 270
AMBER force field, 78
AMIGO-I, II and III protein, crystallography of, 17–18
Amino acids, 6
    potential, 219
2-Amino-*N*-(2-aminoethyl)benzamide (AEAB), 168
Anchored binding, of carbohydrate ligands, 8–10
Anhydrous hydrazine, 146
Animal models, 273
Anti-ABH lectins, 126
Anti-αGal antibody
    clinical relevance of, 222
    genes, 218–219
    responses, 220–222
    structure, 219–220
Anti-αGal monoclonal antibodies (mAbs), 218
Antibodies, 90–91, 166
    carbohydrate docking into, 114–115
    crystal structures of, 12
    natural, *see* Natural antibodies
    site maps, 130

Antibody–carbohydrate crystal structures, 130
Antibody–carbohydrate docking, 127
Antibody–carbohydrate recognition, *see* Carbohydrate–antibody recognition
Antibody–carbohydrate structure, 127–128
Antibody 2G12, 248
Antibody–peptide docking, 127
Anticoagulants, 261, 263
    AVE5026, 271–272
    EP42675 and EP217609, 265, 272–273
    fondaparinux, 266–269
    heparin/HS mimetics as, 264, 265
    idrabiotaparinux, 270–271
    idraparinux, 269–270
    M118, 272
    SR123781, 271
Anti-Gal antibodies, 126
Antigen-binding fragment (Fab) crystals of glycosylate, 5
Antigens, 216
    mucin, 185
    for natural antibodies, 217–218
Antithrombin (AT), 260
Apo H, *see* Apolipoprotein H
Apolipoprotein H, conformations of, 18–19, 20
Apoptosis, 29
Aquaporin-0 (AQP0), 56–57
Aquaporins, 56–57
Arixtra, 267, 268
A-specific α1,3-*N*-acetylgalactosaminyltransferase (GTA), 223
AT, *see* Antithrombin
AT-binding domain (ABD), 264
Atomic partial charges, 79, 80
ATT, *see* 2,6-Azathiothymine
AutoDock, 114, 115
    implemented scoring function, 83
    NMR and docking simulations in, 119
    use Genetic algorithms, 82
AutoDock 3.06, β-glucosides using, 119
AutoDock 3 program, 115, 116
AutoDock 4 program, 120
Automated in silico docking, ligand sampling and scoring, 112
Automated molecular docking, 116
AVE5026 anticoagulant, 271–272